JAPANESE

Radio, Record, & Tape Player

Schematic/Servicing Manual

By Homer Davidson

TAB BOOKS
Blue Ridge Summit, Pa. 17214

FIRST EDITION

FIRST PRINTING— APRIL 1973

Printed in the United States
of America

Leatherette Edition: International Standard Book No. 0-8306-3642-X

Paperbound Edition: International Standard Book No. 0-8306-2642-5

Library of Congress Card Number: 72-97211

Preface

Today, the Japanese solid-state radios, and tape and record players are commonplace in the U.S. market place. In fact, major American manufacturers are importing or selecting Japanese firms to make such units to be marketed under famous U.S. brand names. Also, many Japanese manufacturers are importing Japanese units through their own U.S.-based operations. Now, millions of these solid-state units require service.

If you are not servicing these solid-state units, now is the time to get in on the million-dollar bandwagon. It is the purpose of this book to help you to service and align the most popular Japanese solid-state radios, and tape and record players. You may also find valuable service information in TAB's book No. 532, "How To Repair Solid-State Imports."

The biggest drawback in servicing Japanese units continues to be obtaining replacement components. In many of these Japanese units the original part number must be secured because of physical size and working characteristics. Today, the situation has greatly improved. In the Appendix to this book, each manufacturer's parts location address is given, with a listing of various Japanese manufacturers and importers, and instructions etc., as to how to order the required component(s).

Of course, because of size limitations, no book can cover all units. But with the Appendix information, you should be able to obtain schematics, service information, and parts, even for those sets not included herein.

I wish to express my gratitude for the excellent cooperation extended by all the manufacturers who contributed material for this book. Special thanks to all National Service Managers and other personnel who gave permission and suggestions in the use of their test information and illustrative material. Without their assistance the preparation of this book would not have been possible.

Homer L. Davidson

Contents

Chapter 1
Servicing Multiband Radios and Amplifiers

The multiband radio receiver may cover from 2 to 8 frequency bands and consist of 5 to 23 transistors. A typical 8-band receiver covers the frequency from 540 to 1600 kHz and 1.6 to 175 MHz (Fig. 1-1). The multiband portable can be serviced in the same manner as all solid-state radios.

Most multiband radios employ conventional superheterodyne circuitry, with audio drivers and push-pull output stages (Fig. 1-2). A maze of coils and bandswitching gear can be confusing, but can be quickly located with correct service diagrams and information. In the newer models you will also notice IC circuits in the IF, FM multiplex, and audio circuits. Actually, it is a matter of isolating and troubleshooting the defective band. Of course, in many cases, the trouble may be common to all bands.

About eighty-five percent of troubles in transistorized units are caused by the transistors themselves. Thus, it would be wise for you to first

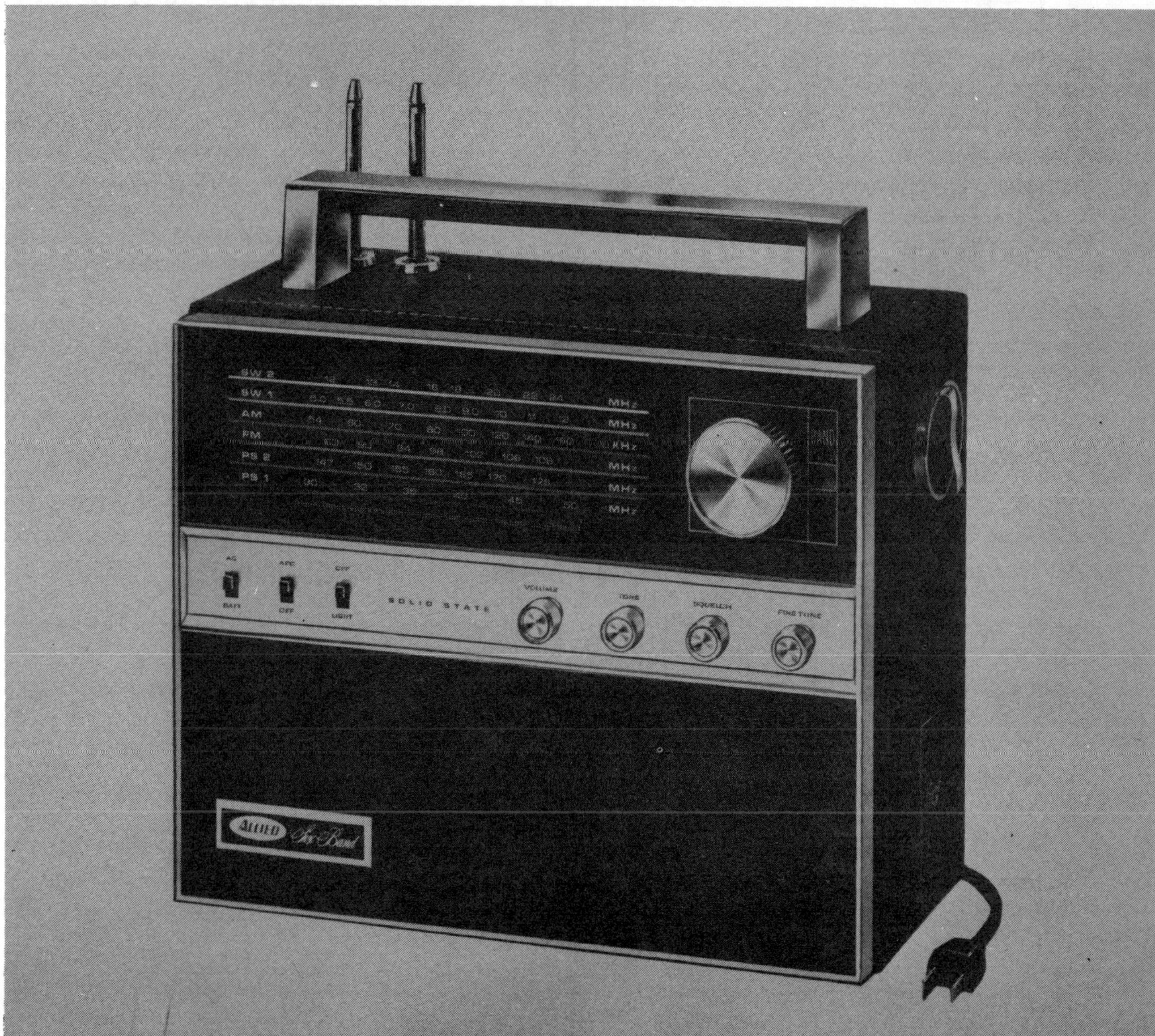

Fig. 1-1. Typical multiband radio. (Courtesy Allied Radio Shack, Tandy Corp.)

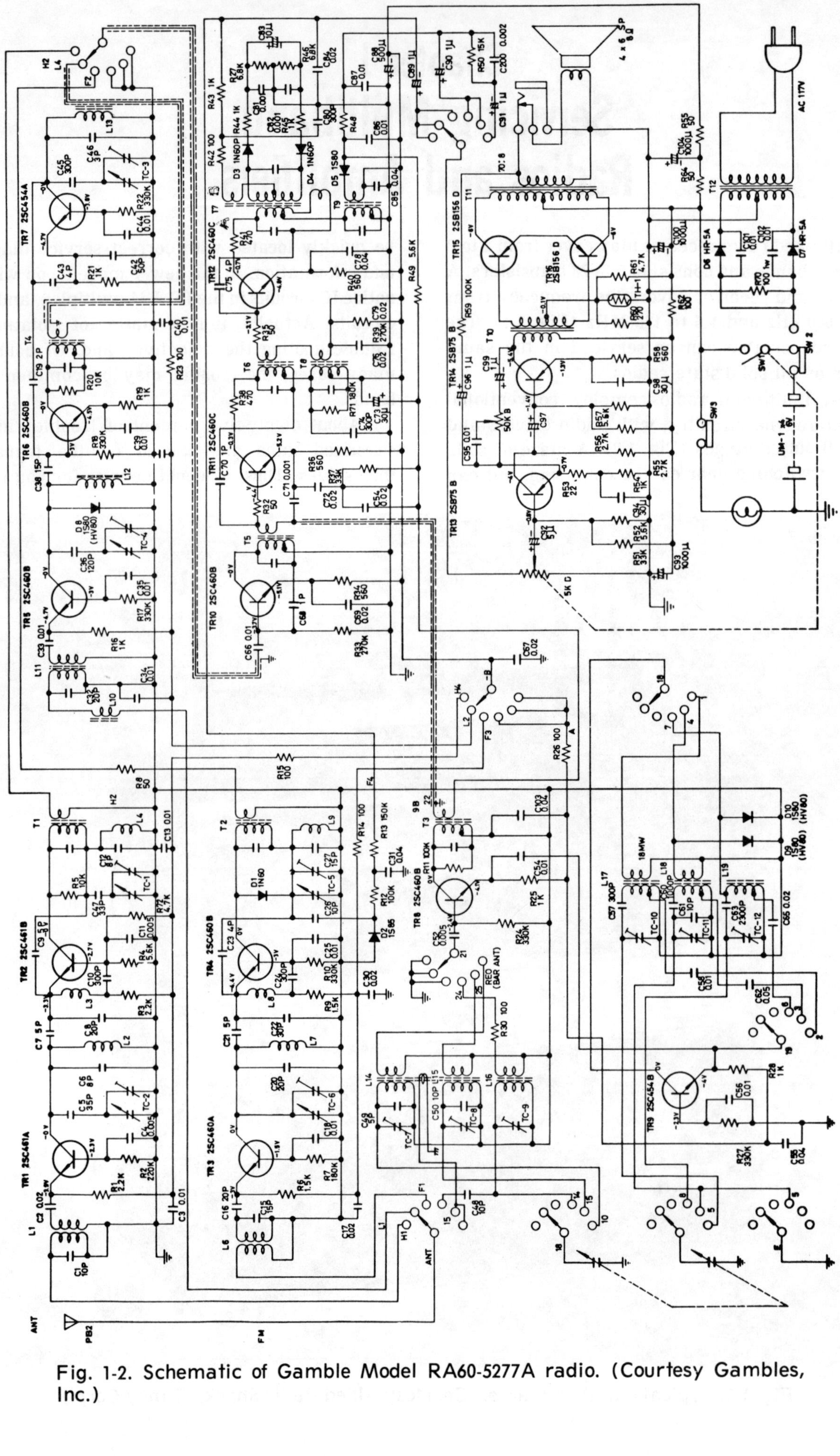

Fig. 1-2. Schematic of Gamble Model RA60-5277A radio. (Courtesy Gambles, Inc.)

check and eliminate a possibly defective transistor. Defective transistors are easily located if proper servicing methods are used. Removing and replacing the transistor can be as easy or difficult as you make it. Simply use the most logical approach as in any troubleshooting method:

1. Analyze the transistorized chassis to determine in what section or sections the trouble would occur. For instance, if a multiband radio had no FM and the AM section was good, the trouble must be in the FM section.

2. Try to locate the stage or section by eliminating stages, through use of simple deductions. When only local AM and FM stations can be heard on a multiband radio, the trouble is common to both bands. An open FM or AM RF transistor would cause weak reception on each band, but an open AM-FM IF transistor would likely be the culprit.

3. Pinpoint the trouble in one stage to a single component using signal tracing, waveforms, voltage, and transistor testing methods.

4. Be careful when removing and replacing the defective transistor. Use a low-heat suction type iron or Solder-Wik (Fig. 1-3). Be very careful not to damage the etched wiring.

REQUIRED TEST EQUIPMENT

1. VTVM or FET-VOM
2. AM signal generator

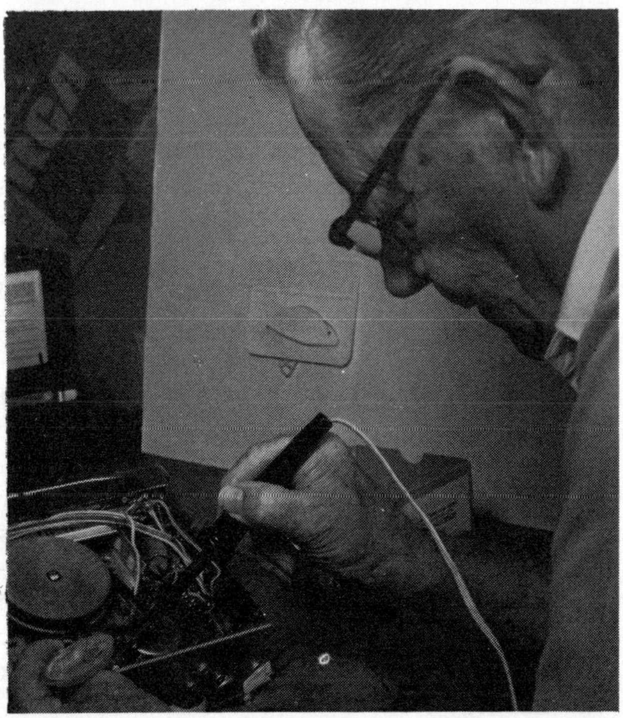

Fig. 1-3. Using Solder-Wik to remove excess solder.

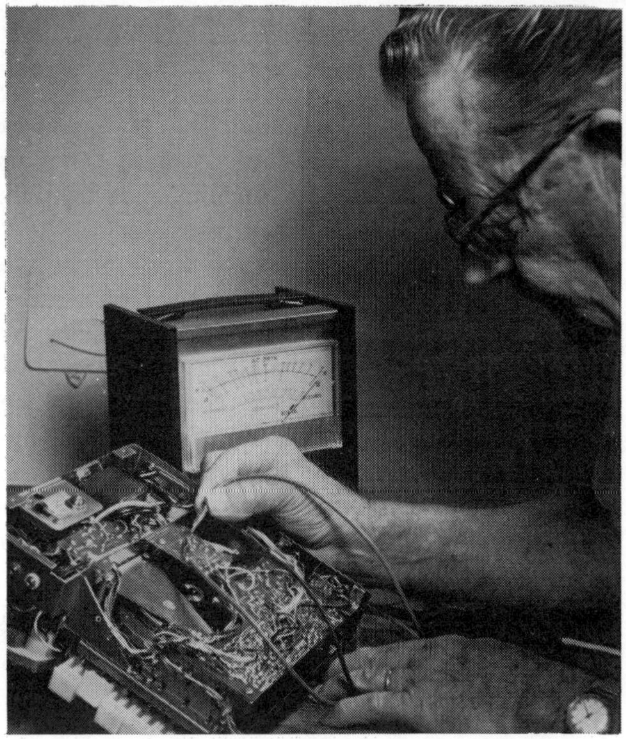

Fig. 1-4. In-circuit transistor checking.

3. AM-FM and FM MUX signal generator
4. Noise and audio signal generator
5. Oscilloscope
6. Sweep-marker generator
7. Transistor tester

Of course, all of these test instruments are not required in every service problem. But each one will help to quickly locate the defective component. Good test equipment with correct circuit knowledge can solve most solid-state problems.

VOLTAGE MEASUREMENTS

Voltages at the transistor terminals are quite low compared with those at tubes. The voltages may be either positive or negative with respect to ground potential. You can locate a defective transistor in the circuit with accurate low voltage measurements. Use either a VTVM or FET-VOM for low voltage readings.

First, check forward bias on the suspected transistor. An NPN transistor base voltage will be 0.6 to 0.8 volt more positive than the emitter terminal; the voltage is of a reverse polarity on a PNP silicon transistor. Forward bias on a germanium transistor must be 0.1 to 0.2 volt. If there is no forward bias voltage, from base to emitter terminals, the transistor or components between base and emitter terminals are defective. If in doubt, you can check the suspected transistor in the circuit with an in-circuit transistor tester (Fig. 1-4).

When the measured collector voltage is below the supply voltage, the transistor is drawing current. If the collector voltage is the same or higher than the supply voltage, the transistor is open. Be very careful not to accidentally touch or short the collector and base terminals, as this can quickly damage the suspected transistor.

In-circuit resistance measurements from the terminal of a transistor to ground are inaccurate. Check for correct bias resistor resistance by removing one lead from the circuit. Also, remove one end of a diode to make correct resistance measurements. Silicon diodes will measure approximately 10 ohms in one direction and no resistance (infinity) in reverse direction. A germanium diode will measure from 10 to 60 ohms in one direction and 100 to 150K in reverse direction. Start with the ohmmeter at the lowest scale to check for a shorted or open diode.

SIGNAL TRACING

To locate a defective stage, a noise, RF, or audio signal generator is quite helpful. For instance, in a multiband radio, a noise generator will quickly locate a defective stage from audio to RF section. When no signal is heard in a certain stage, this isolates the faulty stage and you should then proceed to determine if the transistor or components are faulty.

Signal tracing should always start at the volume control and proceed either way. You can save time by cutting the circuit in half at this point. Proceed from collector to base of each transistor until the defective stage is located. RF and IF alignment procedures also can quickly locate a defective stage. A defective IF transformer is uncovered when no change in output signal is noted during alignment procedure.

In-Circuit Tests

Transistors can be tested within or out of the circuit. Simply connect the three test leads from the in-circuit beta tester to the suspected transistor (Fig. 1-4). No reading means the transistor is open. Double check for NPN or PNP polarity of the suspected transistor. Most transistors will show some leakage on an in-circuit transistor tester, but in case of a dead short, remove the transistor and check it out of the circuit. Before installing a new transistor make an out-of-circuit test.

Circuit Isolation

Always try to isolate the trouble to a certain stage and component. For instance, if the FM section is working and the AM section is dead, the trouble most likely is a defective AM converter transistor. If the receiver is very weak on all bands, determine if the loss of signal is located in the front end or audio section. Inject an audio signal at the volume control to check if audio section is working normally. If the audio section is normal proceed with noise or modulated RF-IF signal toward the front end. When the injected signal disappears you have located the defective stage.

For weak or intermittent reception, suspect the RF stage. In most multiband radios, the FM RF and converter stages are separated from the shortwave and AM bands. A weak FM station may be caused by the RF transistor or dipole antenna circuit. When an intermittent RF transistor is suspected, apply several coats of cold-spray upon the transistor. This will cause the transistor to act up.

The usual symptom of a defective oscillator stage is a loud rushing noise with no tunable stations. You can quickly test the oscillator stage by placing another multiband radio close by and tuning through the band. You should hear a loud whistling noise when the two oscillators are close to the same frequency. When replacing a defective transistor found in the FM section, be sure to site it in the original physical position. Double-check the physical layout and locate the defective transistor upon the etched circuit board. Otherwise, you may pull out the wrong transistor and be faced with a realignment job.

MULTIBAND RECEIVER ALIGNMENT

Before attempting multiband receiver alignment, locate and mark all coils and transformers with a felt pen. Always align or check the AM band before shortwave alignment. If the AM band seems to be working normally, just touch up the IF transformers. Most multiband receivers can be aligned properly with an RF AM/FM signal generator, FM IF sweep generator (10.7 MHz), output meter, and oscilloscope.

Always keep the signal generator output (input to the unit under test) as low as possible. Reduce the input signal after each alignment. An output indicator can be a VTVM, output meter, or oscilloscope. The VTVM can be connected across the volume control or the speaker voice coil terminals. Sometimes the speaker can be disconnected and a low value resistor used as the speaker load. The output meter is plugged into the earphone jack with this load test.

A typical multiband alignment procedure is shown in Fig. 1-5.

ALIGNMENT PROCEDURE

INSTRUMENTS REQUIRED:

1. RF AM/FM signal generator
2. FM IF sweep generator (10.7Mc)
3. Oscilloscope
4. Output meter

GENERAL ALIGNMENT CONDITIONS:

1. Signal input should be kept as low as possible.
2. Standard modulation is 400c/s 30% for AM.
3. Standard modulation is 400c/s 22.5 Kc div. for FM.
4. Standard output is 50 mw.

Step	Signal gen. coupling	Signal gen. frequency	Radio dial setting	Indicator connect to	Adjust	Remarks
FM-IF alignment (switch @ position 2)						
1	Input signal directly to FM ant. and to ground	Sweep generator center freq. @ 10.7 Mc	Point of non-interference	Oscilloscope Ⓐ common to ground	T1 T2 T3 T4	"S" curve of max. symmetrical response @ equal heights.
2	,,	,,	,,	,,	T5	Symmetrical "S" curve response centered @ 10.7 Mc
3	Repeat steps (1) and (2) as necessary to obtain "S" curve linearity					
AM-IF alignment (switch @ position 3)						
4	A standard radiating loop	Signal generator @ 455 Kc	Point of non-interference	Output meter Across two speaker voice coils (4 ohms)	T6 T7 T8	Adjust for maximum and repeat as necessary
FM-RF alignment (switch @ position 2)						
5	Input signal directly to FM ant. and ground	87 Mc	87 Mc (low end)	Output meter Across two speaker voice coils (4 ohms)	L4	Adjust for maximum
6	,,	109 Mc	109 Mc (high end)	,,	C7D	,,
7	Repeat steps (5) and (6) as necessary					
8	,,	90 Mc	Equidistant between 88 and 92	,,	L1 L2	,,
9	,,	106 Mc	106 Mc	,,	C7B	,,
10	Repeat steps (8) and (9) as necessary					
BC-RF alignment (switch @ position 3)						
11	A standard radiating loop	Signal generator @ 525 Kc	525 Kc (low end)	Output meter Across the speaker voice coils (4 ohm)	L6	Maximum
12	,,	1650 Kc	1650 Kc (high end)	,,	C7H	,,
13	Repeat steps (11) and (12) as necessary					
14	,,	600 Kc	600 Kc	,,	L5	,,
15	,,	1400 Kc	1400 Kc	,,	C7F	,,
16	Repeat steps (14) and (15) as necessary					
SW-RF alignment (switch @ position 4)						
17	A standard radiating loop	Signal generator @ 3.8 Mc	3.8 Mc (low end)	Output meter Across two speaker voice coils (4 ohms)	L9	Maximum
18	,,	12.2 Mc	12.2 Mc (high end)	,,	C40B	,,
19	Repeat steps (17) and (18) as necessary					
20	,,	4.5 Mc	4.5 Mc	,,	L8	,,
21	,,	10.3 Mc	10.3 Mc (right side of 10 Mc)	,,	C40A	,,
22	Repeat steps (20) and (21) as necessary					

Fig. 1-5. Typical multiband receiver alignment chart. (Coronado RA80-5262A, Courtesy Gambles, Inc.)

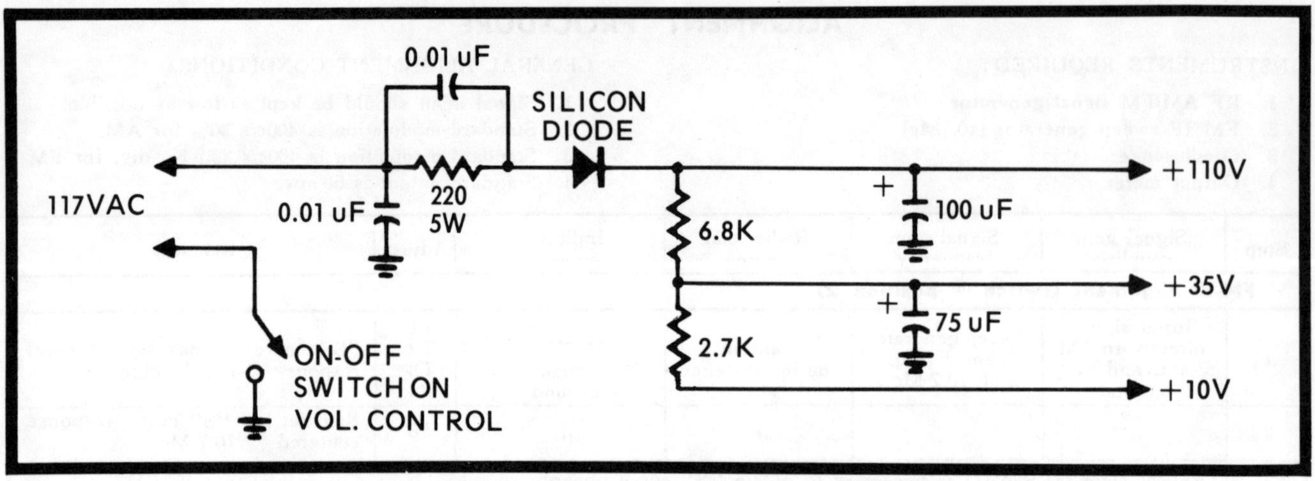

Fig. 1-6. Typical half-wave power supply for AC-DC solid-state radios.

SERVICING THE AC-DC RADIO CHASSIS

The AC-DC radio chassis is aligned and serviced in the same manner as are the multiband radios. Basically, the circuits are the same with an AC line power supply and high voltage operated output transistor. These high voltage output transistor circuits are also found in portable phonograph players.

AC Power Supply

The AC-DC solid-state radio connects directly to the power line through a fusible resistor and silicon diode. A typical half-wave rectifier circuit consists of a silicon diode and an input capacitor-resistance filter network (Fig. 1-6). The fusible resistor (220 ohms) lowers the B plus voltage to the output transistor. When there's no sound, you will generally find this transistor is open. Also, check the silicon diode for open or shorted conditions. A loud 60-Hz hum indicates a defective filter capacitor.

High Voltage Output Circuit

Generally, two or three direct-coupled transistors are found in the audio circuitry of an AC-DC radio or record player. The voltages in these circuits are higher than for most solid-state audio circuits (Fig. 1-7). Since the output transistor was designed to operate at such high voltages, replace with original type or high voltage audio replacement transistors.

Excessive distortion and weak reception can be caused by a defective high voltage output tran-

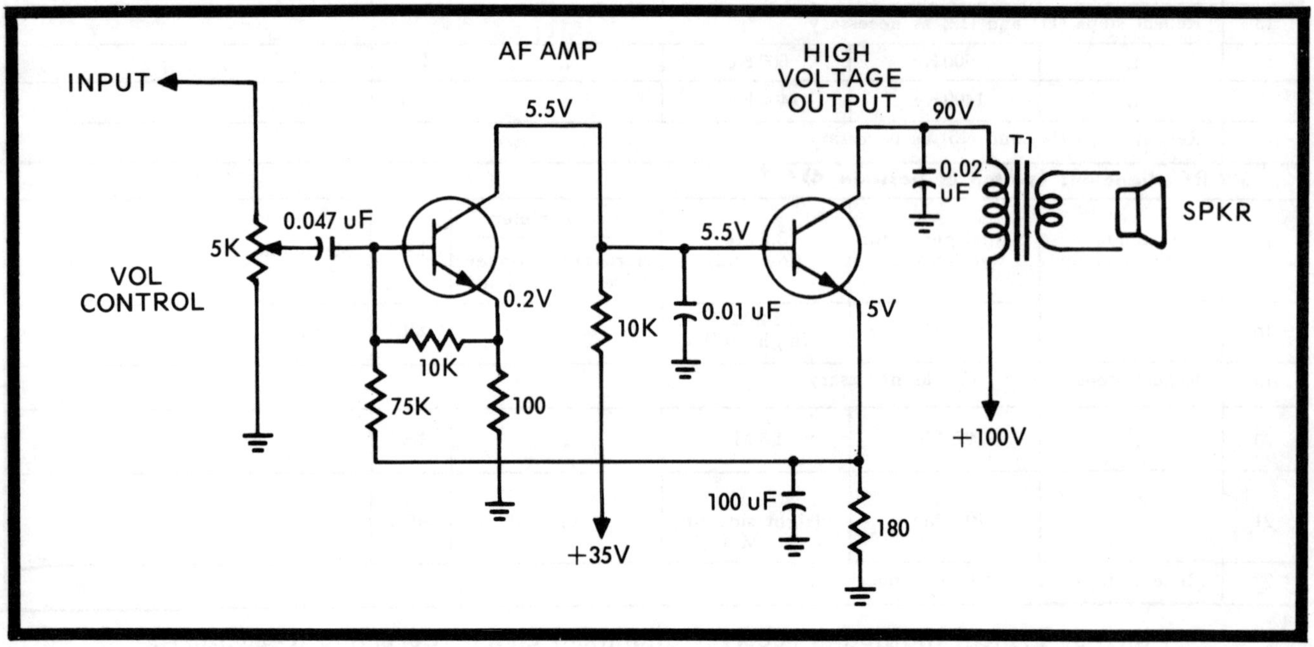

Fig. 1-7. High voltage output circuit for solid-state AC-DC radios.

Fig. 1-8. AM-FM-MUX component unit. (Courtesy Allied Radio Shack, Tandy Corp.)

sistor; this same transistor can produce low level hum. When a directly driven audio output transistor opens, unusual voltage readings are found on the AF or driver transistor. First check each transistor for open condition with an in-circuit beta transistor tester. For in-circuit leakage tests, remove the collector terminal from the circuit. Check both transistors when one transistor is found defective in a directly-driven stage.

SERVICING THE MULTIPLEX RECEIVER CHASSIS

The multiplex receiver is a deluxe AM-FM chassis with a separate multiplex and elaborate stereo sound section. You will find the multiplex chassis in large expensive consoles, table top models and separate components units (Fig. 1-8). The multiplex chassis is incorporated with stereo 8 tape, cassette and phonograph models. This chassis is diagnosed and serviced like any AM-FM section. The multiplex section is quite different and sometimes IC chips are found in some models.

First, isolate the defect to the multiplex section. For instance, if FM stereo music is only coming out of one speaker see if the trouble is in the stereo amplifier or the multiplex unit. Check the stereo amplifier section by switching to phono or stereo 8 tape position (Fig. 1-9). If both speakers are functioning in phono or tape position, the multiplex section is at fault.

Most troubles found in the multiplex section are related to defective transistors and diodes. Voltage and transistor testing will uncover most "poor separation" problems. No FM stereo light can be caused by a defective stereo pilot light or defect in multiplex circuit. After FM stereo light has been replaced, but will not glow, check transistors and diodes within the FM stereo light circuitry.

Complete FM multiplex alignment may be necessary to correct the condition where you have proper stereo tuning but poor separation. Always follow the manufacturers FM MUX alignment procedure. The FM MUX alignment procedure for Automatic Radio Model OMX-9843 is shown in Fig. 1-10.

SERVICING THE SOLID-STATE AUDIO AMPLIFIER

The solid-state audio amplifier may consist of three to sixteen transistors. A three-transistor direct-coupled phonograph amplifier is shown in Fig. 1-11. The audio section of an AM-FM MUX stereo 8 player (Fig. 1-12) has sixteen transistors. The latest audio stages consist of directly driven transistors and transformerless audio output circuits. In these circuits the speakers are coupled to the audio output transistors with electrolytic capacitors. Also, IC input circuits are found in some solid-state sound stages.

Isolation and Comparison Techniques

When trouble is encountered in one channel of a stereo amplifier simply compare the voltage and signal of the defective channel with the good channel. Use a noise or audio signal generator to

Fig. 1-9. Physical layout of AM-FM-MUX chassis with stereo 8 tape player. Notice the different sections.

find the defective stage. Then take voltage measurements around the suspected transistor. Use an in-circuit transistor tester to locate open transistors.

Weak stages are sometimes difficult to locate. Open or leaky transistors and electrolytic capacitors produce most weak conditions. For signal tracing, use an audio signal generator and start at the output transistors. Compare the signal on base and collector terminals on each transistor. When the signal disappears you have located the defective stage. Now narrow it down by first checking forward bias voltage, then test the transistor for open conditions. If the transistor is good and there's no or low forward bias voltage, suspect a leakage between components in base and emitter circuits. Don't overlook those bypass and coupling electrolytic capacitors.

Distortion problems in audio circuits are caused by leaky transistors, burned or improper bias resistors, and leaky electrolytic coupling capacitors. Use voltage measurements and the audio signal generator to locate distortion conditions. An audio signal generator and scope is a must in locating difficult distortion problems (Fig. 1-13).

Direct-Coupled Stages

A direct-coupled solid-state circuit consists of transistors in cascade. The collector of an AF amplifier is tied directly to the base of another audio transistor, or the emitter of an AF amplifier or driver transistor is connected directly to the base of a succeeding transistor. These directly driven stages are found in the portable solid-state phonographs and high powered amplifiers (Fig. 1-14).

Voltage measurements in direct-coupled circuits can sometimes become quite confusing. Fig. 1-15 is a schematic of a direct-coupled circuit with a high voltage output transistor. This particular portable phono came in with a dead right and normal left channel. Notice the supply voltage on the output transistor is 130 volts.

When making voltage measurements compare them with the same stage in the other channel. Notice the collector voltage of the output transistor is the same as supply voltage. This indicates no voltage drop or current flow in the output transformer. We could say the output transistor is open. Also, notice how far off the voltages are in the defective AF amplifier stages. Could it be that Q4 or Q5 was leaky as the voltages on all terminals of each transistor are quite close?

Before jumping off at the wrong end, check each transistor for open conditions. Use an in-circuit beta transistor tester to check each transistor for open conditions. Leakage readings within the circuit are inaccurate, so forget this test. Start with the output transistors since they run quite warm and produce many problems. In this case, the high voltage output transistor was open. Direct-coupled circuits can be quickly serviced by using

Circuit Alignment	Equipment Connection	Step	Gen. Freq.	Dial Setting	Adjustments
67Kc SCA FILTER	Connect Gen. to C-58 , and VTVM to Base of Q-8	1	67Kc (50–100mv)	—	Adjust L-7 (Black) for minimum indicated.
CHAN-NEL SEPA-RATION	FM STEREO GEN. Composite out connect to Ext. Mod. of FM Signal Generator FM SIGNAL GEN. Connect to FM (Ant. Terminal) VTVM connect to voice coil	2	98Mc	98Mc	Adjust T-11, T-12 and VR-52 for minimum output from left (right) channel when right (left) channel is modulated. Channel separation should not be less than 25 db.

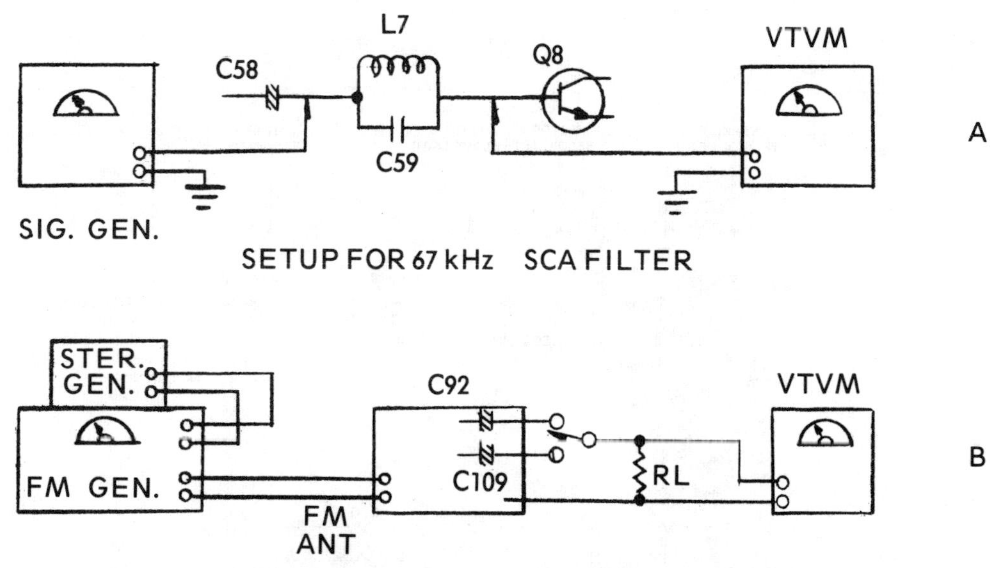

Fig. 1-10. Alignment procedure and test setups for FM MUX circuits of Model OMX-9843. (Courtesy Automatic Radio.)

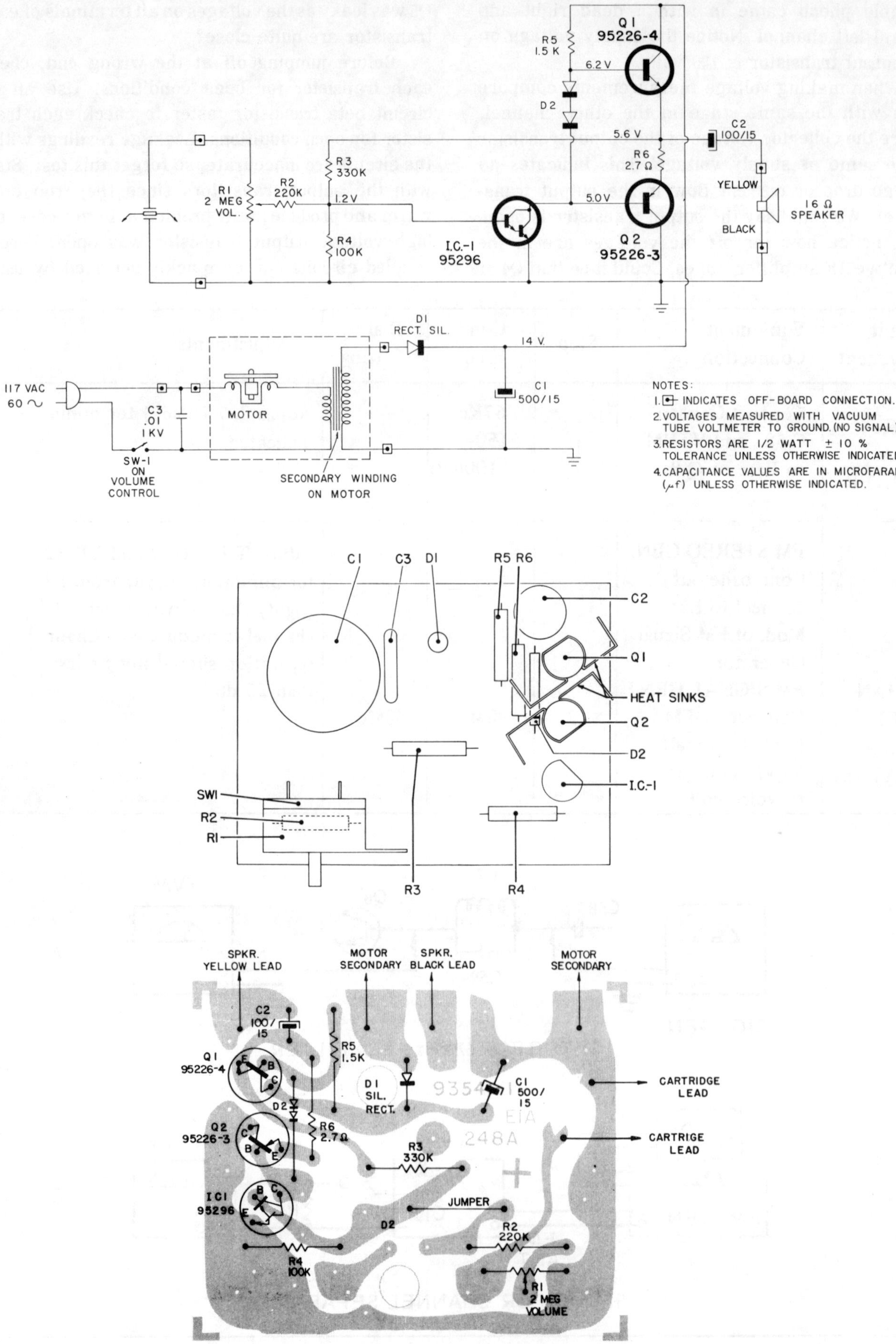

Fig. 1-11. Simple transistor phonograph audio amplifier with power supply. (Courtesy Gambles, Inc.)

voltage comparison measurements and in-circuit beta transistor tests.

Leaky transistors in directly driven circuits produce weak and distorted reception. Voltage measurement and out of circuit transistor tests will locate leaky transistors. Remove the collector terminal lead for in-circuit transistor leakage tests. Always test other directly connected transistors when one is found to be shorted or leaky.

For instance, in the particular Japanese amplifier shown in Fig. 1-16, the left channel was weak and distorted. Q1 had a leakage of 500 microamps and was replaced. This solved the weak and distorted problem, but within a half-hour the new Q1 ran quite hot. The same type transistor (Q1) in the Right channel ran real cool. Another transistor was installed in the left channel amp and the results were the same.

Voltage measurements were quite close on all terminals of Q1. Only a difference of 0.2 volt was noticed on the emitter terminal of Q1 with respect to the base. Resistance measurements were accurate in the Q1 circuitry. Q2 was removed and found to have a 750 microamp leakage. Although weak volume and distortion were cured by replacing Q1, its life could easily be shortened by leaky transistor Q2, so it too was replaced.

In summary then, in stereo amplifiers, comparison voltages, in-circuit open, and out-of-circuit leakage transistor tests will help to solve the most difficult sound problems.

TRANSISTOR REPLACEMENT METHODS

Before removing the defective transistor mark all E, B and C terminals with a felt pen. Be careful,

Fig. 1-12. AM-FM-MUX-stereo 8 tape player, locating audio amplifier and power output transistors.

Fig. 1-13. Use an audio signal generator and a scope to pinpoint faulty part or parts that cause distortion.

as all Japanese transistors are not mounted in order. Check the correct terminals with an in-circuit beta transistor tester, voltage measurements, or corresponding components tied to each terminal.

Use a low wattage pencil soldering iron to remove those tiny transistors. You can eliminate excessive heat damage to the transistor and prevent circuit wiring from "popping off" of the board by using heat sinks. Don't use a large iron tip, it will cause excessive solder to run between etched wiring, resulting in transistor or circuit damage.

Before replacing the defective transistor, test the new one in a beta transistor tester. When replacing those high powered output transistors, place silicone grease upon the transistor and the mounting surface. Check to see if the clean piece of insulation is used and in place. Be sure the mounting surface is free of dirt and small beads of solder. If not, the insulation may be punctured and ground out the power transistor. Double check that the collector terminal (outside metal case) is not grounded to the frame or heat sink. Always replace those small heat sinks, mounted on top of the TO-5 case, when replacing the defective transistor.

ORDERING PARTS

When ordering parts for Japanese units be sure to include the following in your order:

1. Part number and description.
2. Model number found on the back, side or bottom of unit.
3. If part number differs from that shown on parts list, give part number of component. Appendix 1 lists the major brands and parts depart-

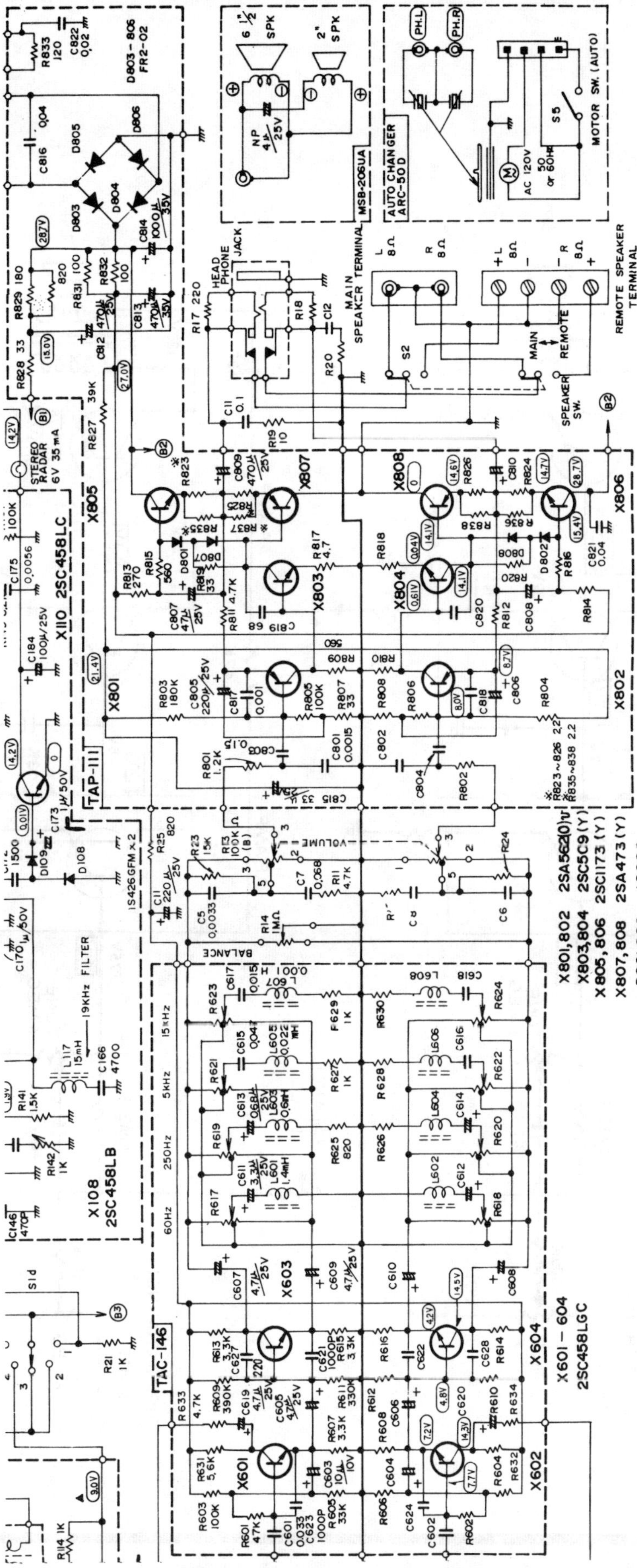

Fig. 1-14. Direct-coupled transistor stages in large audio amplifiers. (Model 4431U, Courtesy JVC America, Inc.)

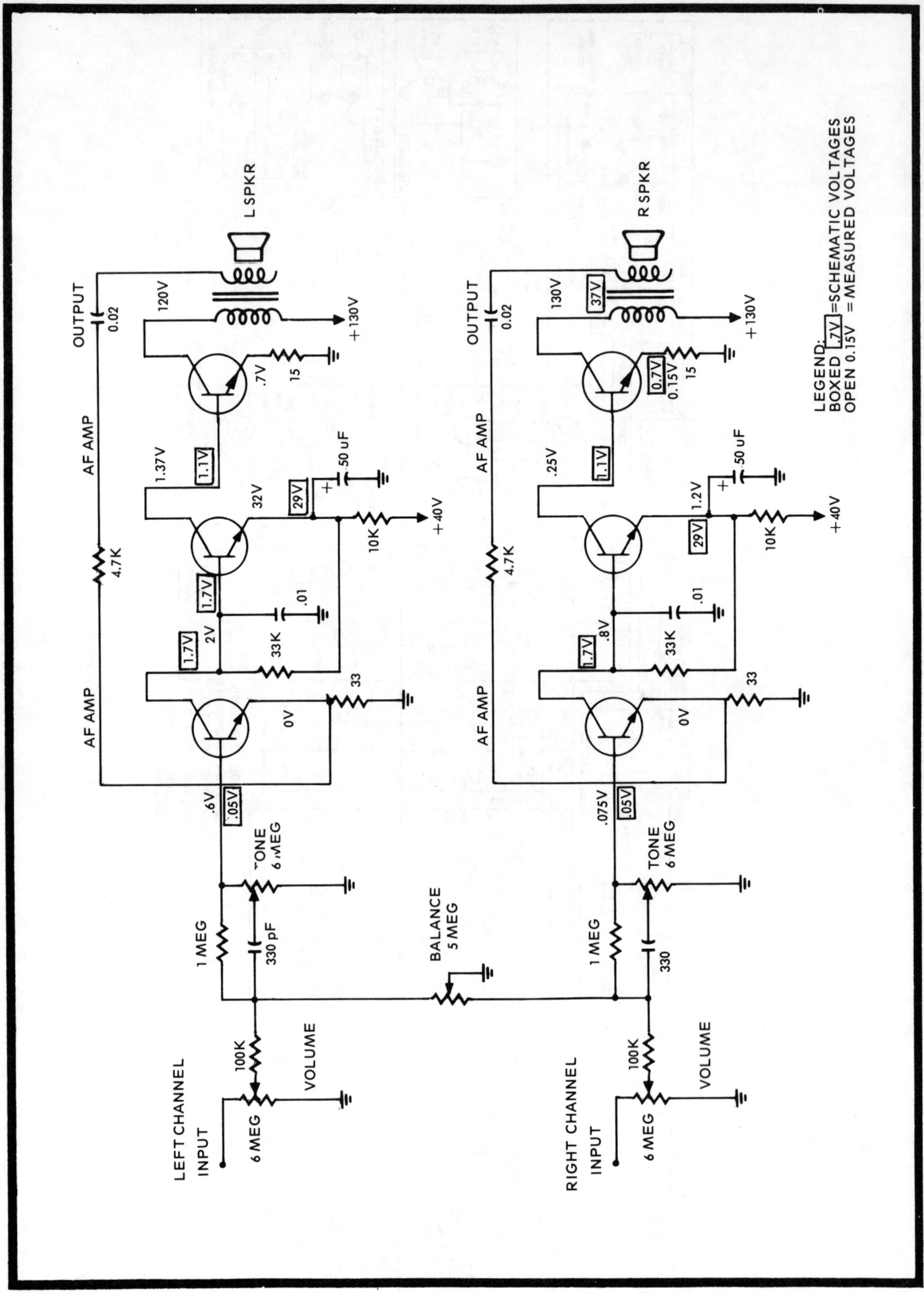

Fig. 1-15. Voltage measurements in a direct-coupled stereo amplifier.

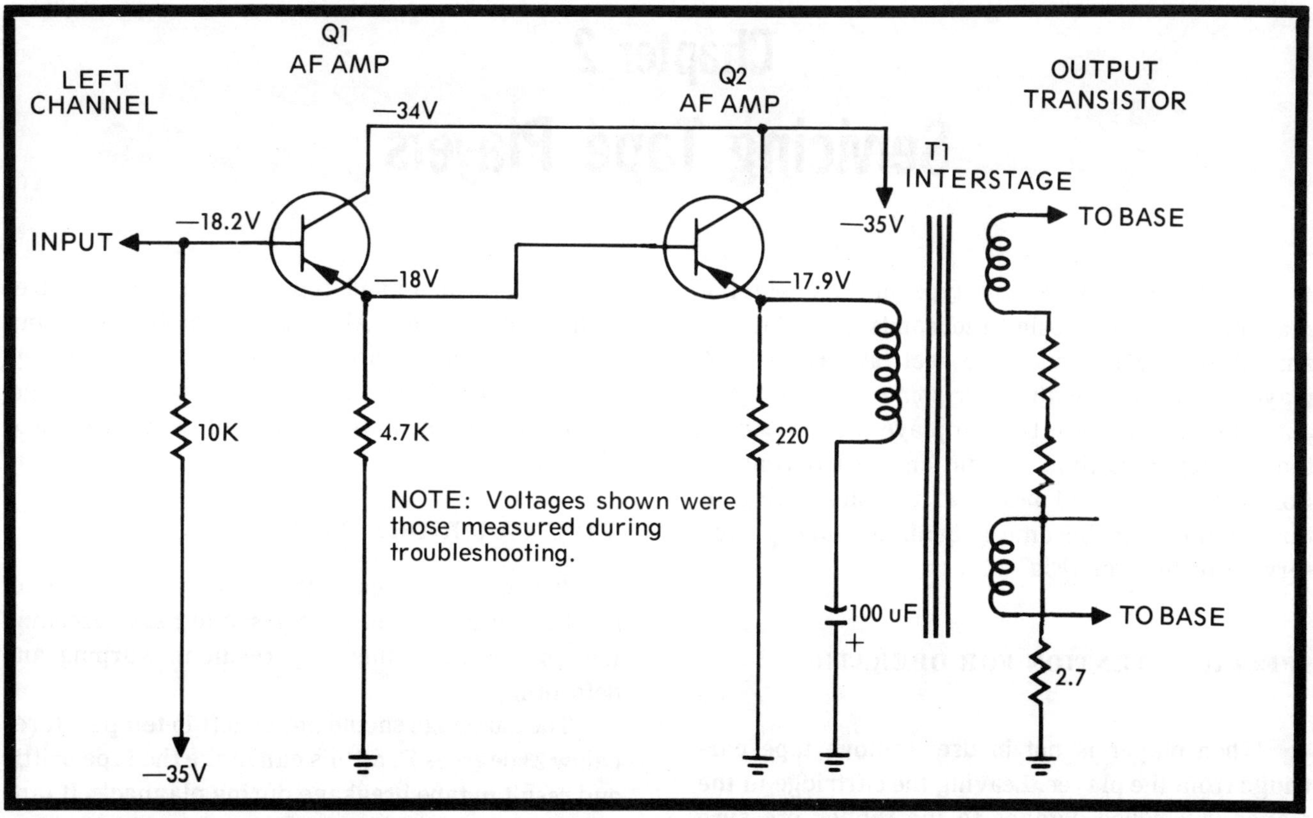

Fig. 1-16. Left channel in a stereo amplifier. Refer to text for troubleshooting discussion.

ment addresses. Appendix 2 lists units by brand name, company, and address.

Some firms will honor business checks or bank draft checks. Generally, when checks are accepted, the shipment is not made until sufficient time has elapsed for the check to clear the local bank, unless you have an open account with that firm. Other firms will send a letter indicating parts cost plus shipping and handling charges and will request a check in advance. Some firms will ship parts on Honor Billing if the list price of the component is under $5.00. Many firms will send the parts C.O.D. and others will not.

Most parts are shipped within 5 or 6 days and sent parcel post or United Parcel Service (UPS). Large orders, of course, would be shipped via truck freight. Orders shipped from across the country should be requested to be sent via Air Freight.

Perhaps the best way to order is to call the firm, requesting correct shipping orders, parts availability, and how payment is made.

Chapter 2
Servicing Tape Players

If you do not service any type of tape playing machines, you are losing thousands of dollars in annual service revenue. The reel tape recorder / players have been with us for some time; now the auto stereo and cassette tape players are flooding the market. It is time you got on the bandwagon. Follow the hints and practical troubleshooting in this chapter and become a qualified tape player service technician (Fig. 2-1).

SPECIAL ATTENTION FOR OPERATION

When player is not in use, remove tape cartridge from the player. Leaving the cartridge in the player will cause damage to the rubber pressure roller and may cause irregular rotation, resulting in wow, flutter, and poor sound.

Use the recommended fuse for the tape unit. A higher rated fuse may cause severe damage to the player.

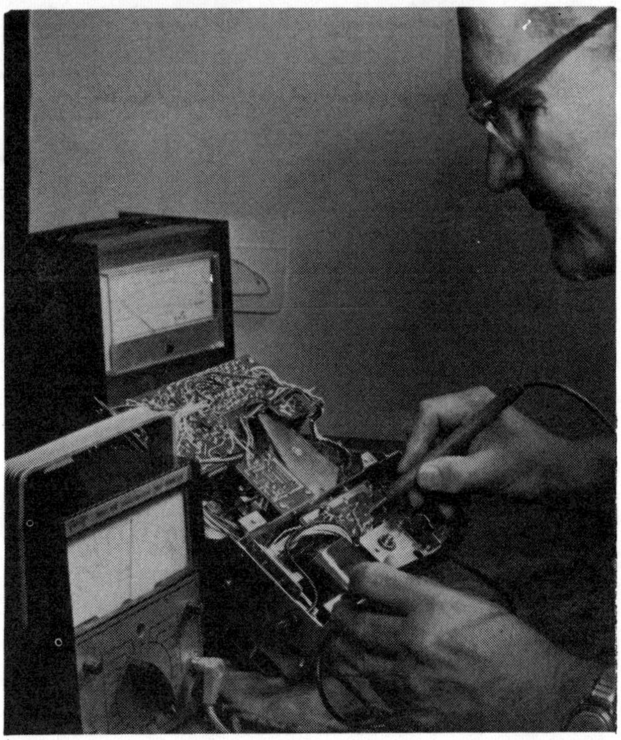

Fig. 2-1. Using FET VOM and transistor beta tester to troubleshoot solid-state tape unit.

Clean the head and capstan with a soft lint-free cloth and alcohol. Also, clean rubber pressure roller, flywheel, neoprene belt and motor pulley. Also, clean heads with tape head cleaning cartridge.

Keep foreign material out of tape head door opening.

TAPE CARTRIDGE CARE

The cartridge case is made of high impact plastic and should not be exposed to heat exceeding 130 degrees F, as this may result in warping and deforming.

The cartridge should not be left in temperatures below 23 degrees F, as this can make the tape brittle and result in tape breakage during playback. If tape has been exposed to a cold atmosphere, it should be allowed to warm up at normal temperature (65-70 degrees F) for 30 minutes before being played.

Never place tape in the vicinity of a strong magnetic field (such as electric motors, loudspeakers, TV sets, radios or other electrical appliances), as this may result in erasure or distortion of the recording on the tape.

Tape cartridges should be placed in plastic bags or stored in places that are not exposed to excessive heat (do not leave in sun), cold or humidity. Proper care will ensure longer tape life and more satisfactory performance.

Before servicing the tape player unit, always check for the possibility of a defective prerecorded tape cartridge.

TAPE HEAD CLEANING

Magnetic tapes and tape players require proper maintenance if maximum results are to be expected. Due to the constant friction between the tape, the tape head, and the drive assembly, a residue accumulation results. This accumulation of oxide will cause poor playback. In many cases, excessive packed oxide upon the tape head will cause a loss of sound and excessive distortion. The tape head should be cleaned periodically and very thoroughly when serviced.

Commercial tape head cleaning fluids are available nationwide, or a cotton swab lightly

Fig. 2-2. Clean the drive belt with alcohol to prevent slippage and wow.

moistened with alcohol can be used to clean the tape head and the drive capstan (located to the right of the tape head). Do not allow the alcohol to run down the capstan into the drive bearings. To clean the tape head periodically, use a commercially available head cleaning tape cartridge. Generally, complete head assembly cleaning can be completed through the cartridge door opening.

When excessive packed oxide is found upon the tape head, remove the unit covers and do a complete clean-up job. Use a wedge type plastic rod to remove packed oxide from the tape head. Keep screwdriver blades away from the tape head to prevent scratching the surface and magnetizing the tape head. Clean up oxide dust beneath tape head and capstan drive bearings.

CAPSTAN DRIVE CLEANING

The capstan can be thoroughly cleaned by using the rubber tipped end of a pencil to activate the motor switch and simultaneously pressing the alcohol moistened cotton swab **gently** against the revolving capstan.

Do not use the same cotton swab to clean both the capstan and the tape head, as oils picked up off the capstan may be transferred to the sensitive tape head and cause damage to the protective coating.

Allow several minutes for the alcohol to dry completely before inserting a tape cartridge.

When excessive oxide dust is found around the capstan drive, remove flywheel and capstan drive from base unit. Clean and wash out bearings and drive assembly with alcohol or tape head cleaner. Also, clean the motor pulley and flywheel assembly. Wipe off all motor drives and tape pinch rollers. Remove drive belt and clean off with alcohol (Fig. 2-2).

Before replacing the flywheel, lubricate the bearing with a spot of phono grease or light machine oil. Go easy on the lubrication; excessive oil can cause tape slippage and wow conditions. Lubricate the tape head and guide assemblies with regular tape head solution.

TAPE HEAD ADJUSTMENT

After the tape head has been properly cleaned use a test tape and check head alignment (Fig. 2-3). Crosstalk and excessive distortion may result from poor head adjustments. Crosstalk is caused by simultaneous reproduction of two audio sources. Excessive noise or distortion may be caused by a magnetized tape head.

Head Adjustment Equipment Required

1. VTVM or VOM
2. Load resistor (8 ohms, 10W)

Fig. 2-3. Check for tape head adjustment screw locations. In this unit, the head height adjustment is at the rear of the tape head, and the azimuth adjustment is to one side.

3. DC power source or auto battery (12 volts)

4. Head alignment test tape cartridge with 1000 Hz, -10 dB recorded signals on all channels

5. Azimuth adjustment tape cartridge with 9000 Hz, -10 dB recorded signal.

Head Adjustment Sequence

1. When the head is in the highest position, the pole piece and the tape guide should be in parallel at the top edge. Adjust with height adjusting screw (number 1 channel).

2. Set for maximum output with the 8 kHz portion of the test tape, using azimuth adjusting screw.

3. Set for maximum output with the 1 kHz portion of the tape, using head height adjusting screw.

4. Repeat steps 2 and 3.

5. Check uniform output on all four programs (channels), using the 1 kHz portion of the tape.

The adjustment must be performed several times over all tracks in order to obtain perfect alignment.

Adjustment of Head Height

To adjust head height, rotate the flat machine screw located in the rear of the playback stereo head either clockwise or counterclockwise. Generally, clockwise rotation of the screw lifts the head with respect to the tape, while counterclockwise rotation lowers it.

Adjustment of Head Azimuth

The head azimuth adjusting screw is located to the right of the head height adjustment screw. As in the case of head adjustment, clockwise rotation of the screw inclines the top of the head to the right, while counterclockwise rotation tilts it to the left.

Touch up or repeat both adjustments.

Head Azimuth and Height Adjustment Using RCA Test Cartridge 321

1. Connect the VTVM (or multimeter, 10V AC range) to individual right and left output circuits of the stereo player.

2. Set the balance control to mid-position.

3. Set to the channel (on the head) by depressing the channel select button (Channels 2-6).

4. Turn the height adjustment screw back and forth to locate the position at which the meter connected to the left speaker shows minimum deflection.

5. After height adjustment has been completed, turn the azimuth adjusting screw to locate the position at which the meter connected to the right speaker shows maximum deflection.

6. Repeat these adjustments (steps 4 and 5) alternately since the height may change by turning the azimuth adjusting screw.

7. After all adjustments have been completed, securely lock the azimuth and height adjusting screws.

NOTE: A 1000 Hz signal is recorded on each side above and below track 2 (left side of the RCA test cartridge 321). Track 6 (right side) has an 8000 Hz signal with slightly wider width than standard, and is used for azimuth adjustment.

Head Adjustment Without Test Tape

If a test tape is not available, a tape with vocal selections may be used. In this case, head height adjustment is satisfactory when the pauses between the words are free from sound coming from other tracks. Recheck by turning balance control to far right and then check left channel. Be sure and check the first (channel 1) and last (channel 4) position of the tape head assembly for crosstalk.

For proper azimuth adjustment, if a test tape is not available, the background "hiss" on any cartridge may be used. It may be adjusted for maximum output with use of the azimuth adjusting screw. A balance meter can also be used as an output indicator for head adjustments.

CHECK FOR IMPROPER ADJUSTMENTS

When the adjustment screws have been turned way out of line and tape head doesn't rotate over all four channels, begin at Channel 1 position (when tape head is at highest point). Generally, at this position, excessive crosstalk is noted and only one speaker may be functioning. Select a known tape cartridge with music on position 1. Adjust height screw for music coming from both channels. Rotate balance control to far extreme right and left channels. You can use a balance meter with this test. Now trigger the channel selector (manually) to position 4. Recheck known music on last track of the tape cartridge. Readjust height screw for maximum output on both channels. Then check each position for crosstalk and correct tape head balance.

HEAD DEMAGNETIZING

Do not use magnetized screwdrivers and other tools near the head as these will magnetize the tape head. After a long period of time, the head may also acquire some residual magnetism. The magnetized head will show a definite loss of high frequency response and an increase in noise level. There are several commercially available head demagnetizers. Several types come packaged in a tape cartridge assembly and can be inserted into play position to demagnetize the tape head. A pencil type demagnetizing tool can be applied through the tape head door and is also available in demagnetizing cassette tape heads.

EXCESSIVE WOW OR FLUTTER CONDITIONS

Proper cleaning and lubrication may cure wow or flutter conditions. In many cases, oil or grease may be found upon capstan drive, flywheel, or drive belt. The drive belt may be loose or stretched. A flat or dry pinch roller can cause wow conditions. Make sure the capstan flywheel bearing is not worn or has become dry (improper lubrication). Don't overlook a defective tape cartridge; try a new one. A defective motor can cause wow or slow speed conditions.

SPEED PROBLEMS

Slow or fast speed problems are generally the biggest complaint found in servicing auto stereo or cassette tape players. Before attempting to check for slow or fast speeds, make certain that the supply voltage is around 12V. Check the slow speed problem with a new tape cartridge.

First determine if the slow or high-speed problem is related to mechanical or electronic speed control circuits. Some models have a transistorized electronic speed circuit and when it is defective it produces fast speed problems. A slow or irregular speed problem is generally caused by dry or dirty motor and flywheel bearings.

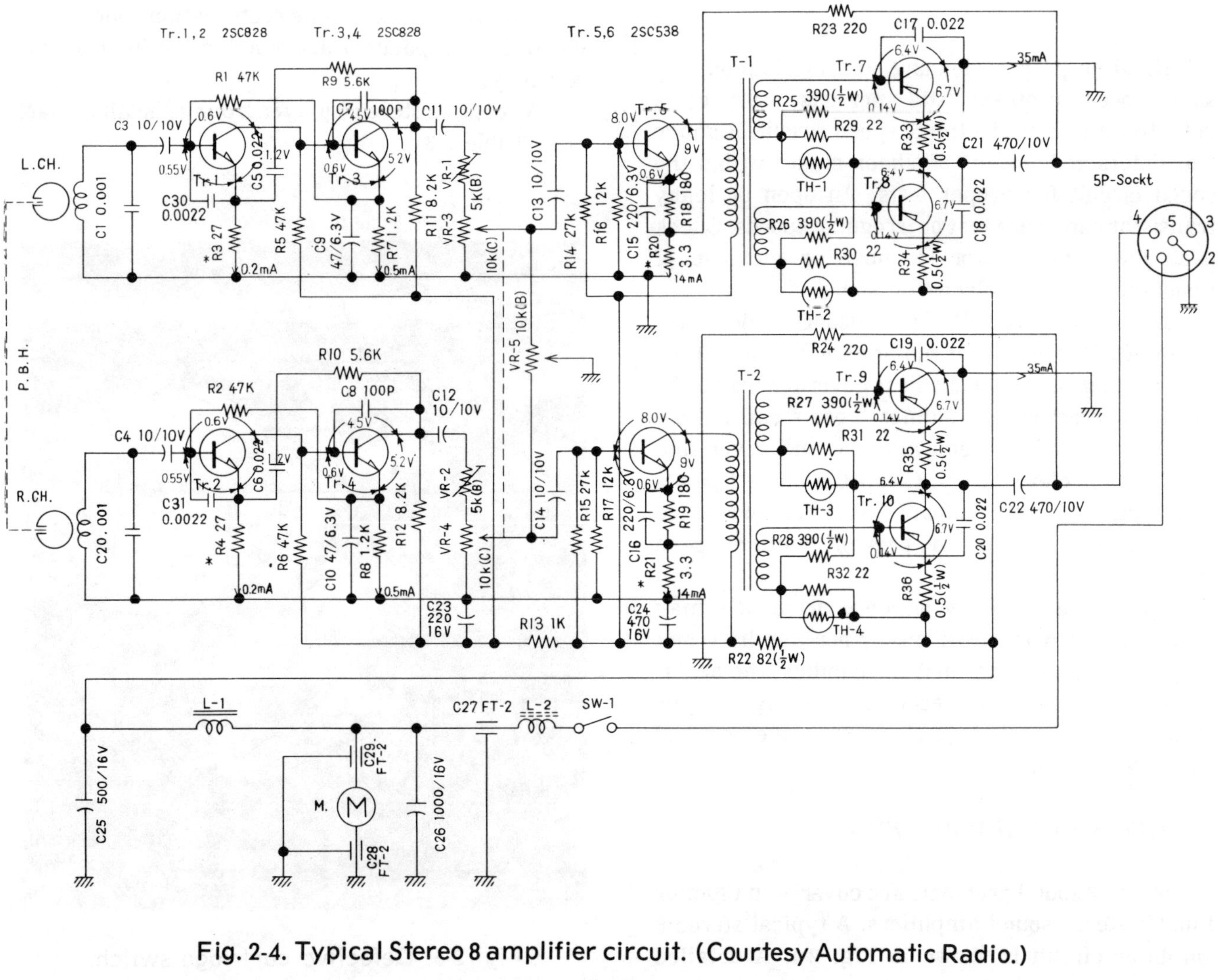

Fig. 2-4. Typical Stereo 8 amplifier circuit. (Courtesy Automatic Radio.)

Fig. 2-5. Using pencil type audio signal tracer to locate defective sound circuit in Stereo 8 tape player.

If, after proper cleaning and lubrication, the slow speed problem still persists, suspect a defective motor or electronic speed circuit. Remove transistors and one end of silicon diodes within the speed circuit for proper tests. An open or leaky transistor can cause fast or irregular speeds. Check each low ohm resistor for open or increase of resistance in these electronic speed circuits.

If the speed is still slow, suspect a defective motor. Tap the end of the motor assembly and if the speed changes, the mechanical governor inside the motor is defective. Sometimes these small motors can be repaired by removing the end cover and checking for clean points. Clean the burned points with a piece of postcard paper. It is best to replace the defective motor to prevent future speed problems.

Don't forget a dirty or loose drive belt may cause slow or wow conditions. When the drive belt keeps coming off a beveled drive pulley, the belt is too loose and should be replaced (use original type by part number). Always clean the drive belt and pulley with alcohol.

STEREO 8 SOUND PROBLEMS

Stereo 8 sound problems are covered in Chapter 1 under stereo sound amplifiers. A typical stereo 8 amplifier circuit is shown in Fig. 2-4. A detailed stereo 8 troubleshooting chart is in Table 2-1. Check Table 2-2 for sound problems. Fig. 2-5 illustrates signal tracing.

Fig. 2-6 locates a defective cartridge switch.

PORTABLE AC-DC CARTRIDGE TAPE PLAYERS

The portable stereo 8 player can be operated from the AC outlet, flashlight batteries, auto or boat battery. A typical unit is shown in Fig. 2-8. The battery player can be operated from flashlight batteries or the auto cigarette lighter. Very small portable cartridge players are battery operated and have monophonic sound.

Tape head cleaning and adjustments are the same as for auto stereo 8 players. For critical azimuth and height adjustment locations, follow the manufacturer's literature. Stereo 8 sound problems are the same in all stereo units.

A typical AC power supply circuit found in most portable tape players is shown in Fig. 2-9. The AC voltage is stepped down with a power transformer with full wave silicon diode rectification. One or two electrolytic capacitors are found in the filter output circuitry.

A portable stereo 8 player troubleshooting chart is in Table 2-3.

Fig. 2-6. Defective cartridge switch.

Table 2-1. Troubleshooting Chart for Tape Players

Symptom	Check
Fails to play No pilot light - No tape motion - No amplifier hiss.	1. Check fuse. 2. Poor power ground. 3. Defective cartridge switch. (off and on) (Fig. 2-6). 4. Burned connection wires—also, burned PC board wiring.
Pilot lamp on - No tape motion	1. Drive belt too loose or off. 2. Setscrew loose on motor pulley. 3. Motor frozen or defective. 4. Capstan drive frozen. 5. Broken motor power connection on PC board.
Excessive wow or flutter	1. Improper lubrication. 2. Defective tape or cartridge. 3. Defective or oily capstan. 4. Loose or defective drive belt. 5. Defective motor. 6. Pinch roller defective or frozen. 7. Flat spot on pinch roller. 8. Spring or rubber assembly frozen. 9. Dry or defective capstan bearing. 10. Check electronic speed circuitry.
Excessive noise	1. Noisy motor bearings. 2. Capstan frozen or seized up. 3. Dry capstan bearing. 4. Defective cartridge. 5. Ignition noise. 6. Defective motor. 7. Check to see if noise is in the amplifier or in both channels. 8. Check playback head and demagnetize.
No or low output (any or all tracks)	1. Speaker or defective wiring. 2. Defective tape. 3. Defective amplifier components. 4. Defective tape head or out of adjustment. 5. Excessive oxide deposits on tape head. 6. Low battery voltage (battery operated units). 7. Use audio signal generator to locate weak stages.
Excessive crosstalk	1. Head height not adjusted. 2. Defective tape. 3. Defective cam assembly. 4. Foreign material jammed in cartridge opening.
Low treble	1. Oxide deposits on tape head. 2. Head surface worn. 3. Head out of alignment. 4. Amplifier component defect.
Will not change channels (automatically)	1. Shorted or open solenoid circuit. 2. Shorted or open selector switch. 3. Defective pawl. 4. Defective cam mechanism. 5. Open or shorted solenoid coil. 6. Head index cam frozen.
Will not change channels (manually)	1. Defective manual switch. 2. Dirty switch contacts. 3. Broken wires to manual switch. 4. Indicator arm binding.
Blows fuses when changing channels	1. Check diode across solenoid winding.
Excessive distortion	1. Defective speaker. 2. Defective tape. 3. Defective amplifier component. 4. Low DC supply voltage. 5. Excessive DC supply voltage. 6. Defective speaker connection.
Unbalanced audio output	1. Check balance control. 2. Tape head. 3. Check for defective speaker. 4. Incorrect height adjustment. 5. Check for weak stage in audio circuit.
Tape runs too slow	1. Flywheel binding. 2. Belt slipping. 3. Worn or dry capstan bearing. 4. Oil on belt or capstan drive. 5. Defective motor. 6. Motor pulley loose.

Table 2-2. Troubleshooting Sound Problems in Stereo 8 Players

Symptom	Check
Caused in power supply circuit	1. Faulty switch contact (cartridge). 2. Faulty output connection (cable plug). 3. Open line choke. 4. Open voltage isolation resistor to driver amplifier. 5. Leaky filter capacitors.
Sound problems in output circuits	1. Open speaker voice coil. 2. Open speaker electrolytic coupling capacitor. 3. Open or leaky output transistors. 4. Open or burned emitter bias resistor. 5. Open base driving transformer.
Sound problems in driver stage circuitry	1. Open or leaky driver transistor. 2. Open or shorted base coupling capacitor. 3. Open or erratic volume control.
Sound problems in preamp stages (Fig. 2-8)	1. Open or leaky transistors. 2. Open balance control. 3. Open or shorted electrolytic coupling capacitors. 4. Open or shorted tape head lead and circuitry.
Poor quality or noisy sound	1. Check oxide deposits on tape head.
Poor sound volume in output circuitry	1. Open or leaky driver and output transistors. 2. Open base driver resistors. 3. Open output speaker coupling capacitors.
Poor volume in preamp stages	1. Open or leaky AF transistors. 2. Defective voltage dropping resistor. 3. Reduced sensitivity or defective tape head. 4. Dried up emitter bias electrolytic capacitor.
Poor highs in preamp stages	1. Incorrect head azimuth (head angle). 2. Defective tape head. 3. Open capacitor across tape head.
Poor lows in feedback circuitry	1. Shorted feedback capacitor input and driver circuits. 2. Open feedback capacitor. 3. Open feedback resistors.
Distortion in output and driver stages (Fig. 2-7)	1. Defective output transistors. 2. Shorted or leaky speaker electrolytic coupling capacitor. 3. Shorted or leaky base capacitors. 4. Open or burned base bias resistor. 5. Open or shorted driver transformer winding. 6. Burned or open emitter bias resistor. 7. Open bias resistor in base circuit of driver transistor.
Distortion in AF and preamp stages	1. Open or burned collector resistors. 2. Leaky transistors. 3. Open or leaky electrolytic base coupling capacitors.
Abnormal oscillations produced in power supply	1. Filter capacitors. 2. Shorted or value change of isolation resistor between power and driver stages.
Abnormal oscillations in output stages	1. Open resistor across primary winding of driver transistor. 2. Shorted or change of value of feedback resistor. 3. Defective output transistor.
Abnormal oscillations in driver stages	1. Open base bypass capacitor.
Abnormal oscillations in AF stages	1. Faulty head lead wires. 2. Defective balance and volume controls.

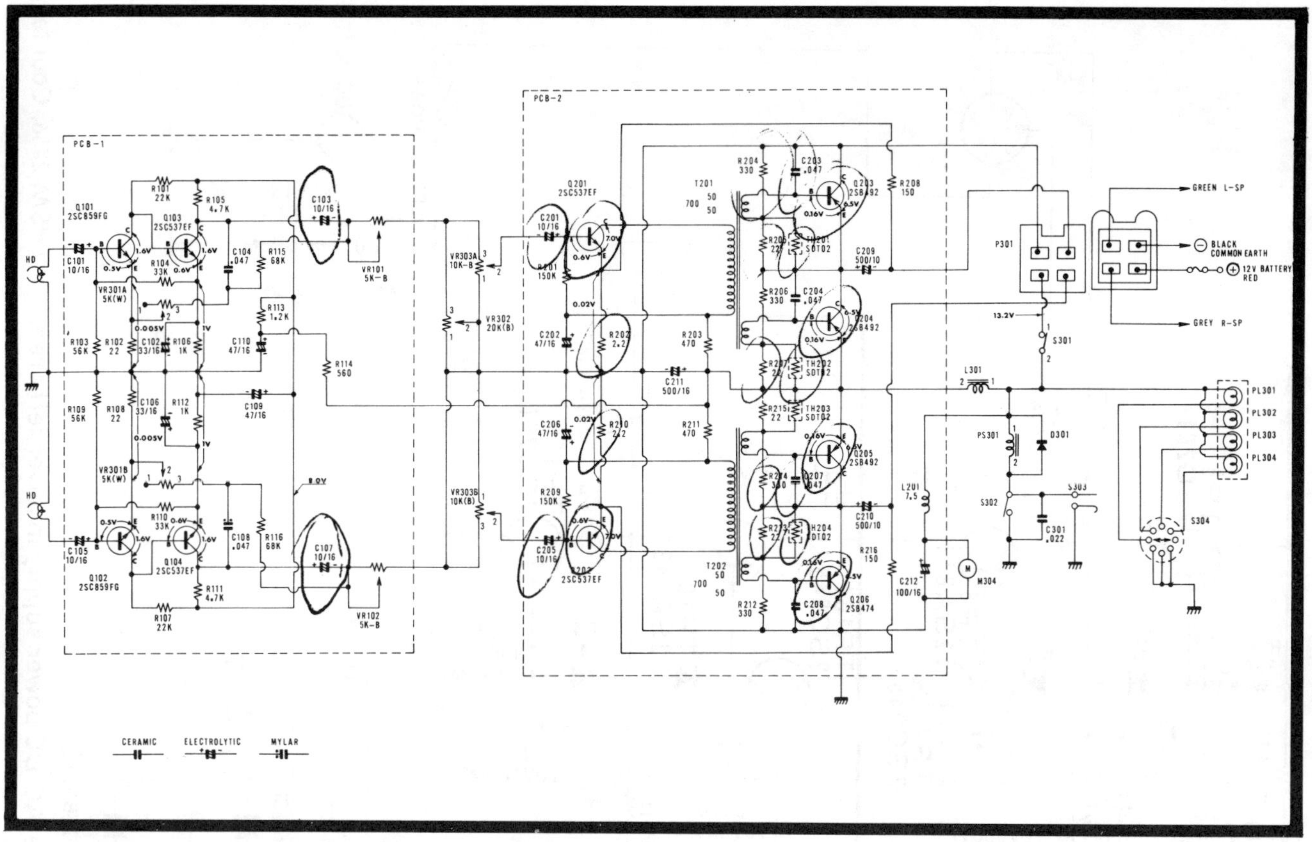

Fig. 2-7. Circled parts produce distortion. (Model SSS-2594, Courtesy Automatic Radio.)

Fig. 2-8. This typical Stereo 8 AC-DC portable tape cartridge player has a simple motor-belt-capstan drive assembly.

CASSETTE TAPE PLAYER REPAIR

The cassette tape recorder/player can be operated from batteries, AC-DC powerpack and car or boat 12 volt battery system. An AM-FM radio can be found in the cassette unit. The portable cassette player has monophonic sound while the home and auto players employ stereo operation. Fig. 2-11 shows an example unit.

Tape head cleaning and demagnetizing operations are the same as for other types of stereo 8 players. Head adjustments, bias, pinch-roller pressure, brake and miscellaneous adjustments should be made per the manufacturer's service literature.

The following adjustment procedure in Table 2-4 is taken from instructions for the Masterwork Model M401 cassette player (Courtesy CBS Records).

A typical cassette player schematic diagram is shown in Fig. 2-12. Fig. 2-13 is a typical block diagram of a cassette 4-track stereo amplifier. A troubleshooting guide is shown in Table 2-5.

REEL TYPE PLAYER MAINTENANCE

1. Cleaning

During normal operation, oxide from the tape can build up on the heads, tape guides, pinch roller capstan, and the pressure pads as the tape passes

(Text continues on p. 38.)

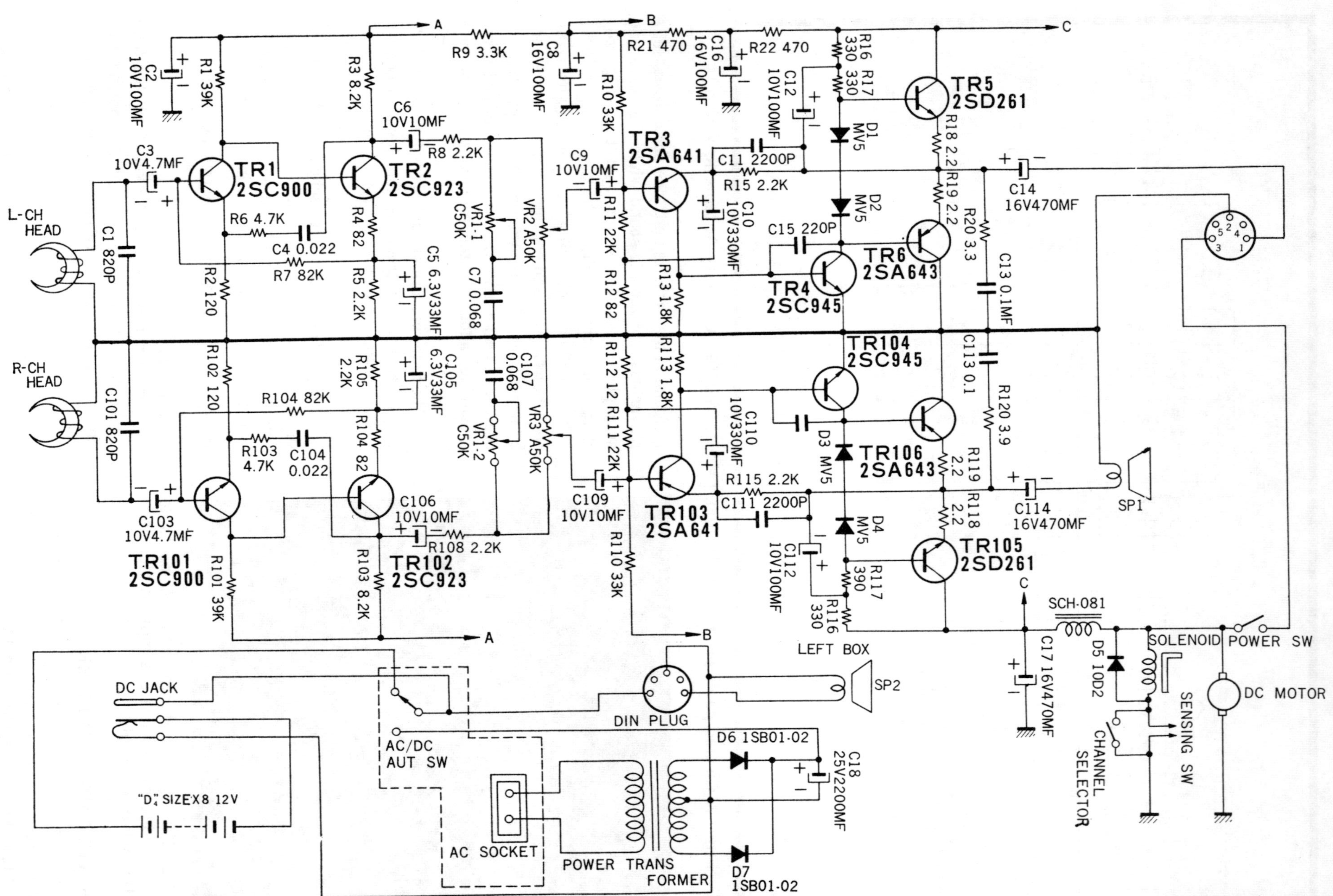

Fig. 2-9. Schematic diagram of portable Stereo 8 AC-DC players. Note AC-DC power supply in lower left area. (PSM-2456, Courtesy Automatic Radio.)

Table 2-3. Portable AC-DC Stereo 8 Player Troubleshooting Chart

Symptom	Defect
No pilot light No tape motion or amplifier hiss	1. Poor power connection. 2. Defective on-off switch (Fig. 2-10). 3. Switch not operating.
Pilot light on- No tape motion	1. Drive belt too loose. 2. Motor pulley loose. 3. Defective or frozen motor. 4. Frozen or seized up capstan assembly.
Excessive flutter or wow conditions	1. Lubrication needed. 2. Defective tape or cartridge. 3. Defective or oily capstan. 4. Defective or loose drive belt. 5. Defective motor.
Excessive noise	1. Defective motor and bearings. 2. Seized up or frozen capstan. 3. Very dry capstan. 4. Defective cartridge.
No or low output sound	1. Defective speaker or wiring. 2. Defective tape. 3. Defective amplifier components. 4. Defective head or out of adjustment. 5. Oxide deposit on heads.
Excessive crosstalk	1. Head height not adjusted. 2. Defective tape.
Low treble	1. Oxide deposit on tape head. 2. Head surface worn. 3. Head out of alignment.
Will not change channels	1. Shorted or open solenoid circuit. 2. Shorted or open selector switch.
Motor runs slow	1. Flywheel shaft binding. 2. Check battery voltage (on DC operation only).
Will not play on AC	1. Check AC cord. 2. Check for shorted or leaky silicon diodes. 3. Check AC-DC switch.
Motor runs backwards	1. Check batteries for correct polarity (in battery operations).

Fig. 2-10. Defective on-off switch in portable AC-DC Stereo 8 cartridge player.

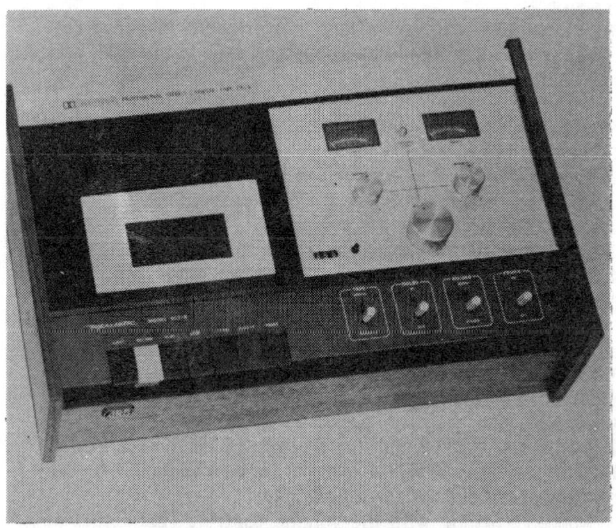

Fig. 2-11. Professional cassette recorder-player. This modern set uses the Dolby system. (Realistic Model SCT-6, Courtesy Allied Radio.)

Mechanical Adjustment

Head Alignment
1. Set function control to play position.
2. Connect an AC VTVM to the output terminals, right and left, with a 3K resistor as a dummy load.
3. Play back 6.3 kHz azimuth test tape and adjust record / playback head spring (54) for 35 mV of maximum output reading on AC VTVM. These procedures shall be applied on each of right and left channels twice.
4. The difference between outputs of each channel should not exceed 5 mV—the maximum output should be 35 mV.

Azimuth voltage should be more than 35 mV.

Miscellaneous
1. The operating force of all function levers other than play button (152) should be less than 1.8 kilograms just before locked. Play lever operating force should be less than 2.0 kilograms just before locked.
2. In play mode, take-up torque should be between 30-35 gm / cm.
3. In fast-forward and rewind modes, take-up torque should be more than 80 gm—cm.
4. Pinch roller (51) pressure against tape should be 500 gm. plus or minus 50 gm.
5. Push button over-stroke should be within 1 mm.
6. Thrust rattling for flywheel (85) should be within 0.2 to 0.5 mm.
7. In stop mode, fast-forward / rewind pulley assembly (43) should keep 2 mm distance from flywheel (85).
8. In stop mode, clutch (19) should be more than 0.5 mm distance from flywheel (85).
9. In pause mode, there should be a 0.5 mm space between capstan and pinch roller (51).
10. Actuation force ejecting the cassette should be less than 1.5 kgrs. just before the cassette holder (93) ejects.
11. Actuation force locking pause button should be less than 1.0 kgrs. just before locked, and disengaging force should be less than 1.5 kgrs.

How to Adjust Recording Bias

1. Set the unit in record mode.
2. Connect a 100 ohm resistor in parallel with the "RED" marked ground side of record / playback head (53) and connect an AC VTVM and an oscilloscope at both ends of the resistor.
3. Adjust bias trap coil (L2) for maximum bias current. These procedures should be applied on each of the right and left channels.

How to Measure Erasing Current
1. Set the unit in record mode.
2. Unsolder the low potential (ground) wire connections of erase head (55) and connect a 1 ohm resistor as a dummy load leading to the oscillator circuit.
3. Connect an AC VTVM to both ends of the 1 ohm resistor.
4. If the unit is operating normally, the reading on the VTVM should be more than 100 mV.

How to Measure Overall Gain
1. Connect an attenuator to the microphone input terminals (120) from an audio generator (oscillator) and set the audio generator with a 0 dBm signal.
2. Adjust the audio signal for -74 dBm (0.155 mV) at 1 kHz, then the audio signal -74 dBm is to be connected to the microphone input lug terminals (120).
3. Make recording with the volume controls (122) (164) at maximum.
4. Connect an AC VTVM to the output lug terminals. Then play back the recorded materials. The reading on VTVM should be 0.5 V plus or minus 0.1 V.

How to Measure Frequency Response
1. Set the unit in record mode with the volume controls (122) (164) at maximum.
2. Connect an audio generator to an attenuator leading to the microphone input lug terminals (120) and connect an oscilloscope to the attenuator.
3. Set the audio signal from the audio generator at 0 dB and adjust attenuator so that a -90 dBm signal is connected to the microphone input lug terminals (120).
4. Make recording at 100 Hz, 10 kHz and 1 kHz frequencies. For fully checking, further recordings of 200 Hz, 400 Hz, 600 Hz, 2 kHz, 4 kHz and 6 kHz is recommended.
5. Connect an AC VTVM to the output lug terminals. Then play back the recorded materials. The signal on VTVM should be 0 dB at 1 kHz difference between each frequency.

How to Measure Signal to Noise Ratio
1. Set the unit in record mode with the volume controls (122) (164) at maximum.
2. Connect an audio generator to the microphone input lug terminals (120) through an attenuator.
3. Set the audio generator with a -64 dBm (0.5 mV) at 1 kHz reading on the attenuator, then make recording.
4. When recording, no signal should be applied by the audio generator.
5. Connect an AC VTVM to the output lug terminals, then play back recorded materials. When playing back, read the output signal on VTVM on decibel scales for the difference between -64 dBm at 1 kHz and no signal.

Auxiliary Recording
1. Connect an audio generator to the auxiliary input lug terminals (AUX) and make recording with a -8 dBm signal (300 mV) at 1 kHz.
2. Connect an AC VTVM to the output lug terminals, then play back the recorded materials. When playing back, the reading on the VTVM should be approximately 0.5 V.

How to Measure Hum Level
1. Insert a cassette into the unit and set the unit in pause mode on play position.
2. Connect an AC VTVM to output lug terminals.
3. If the preamplifier sensitivity is normal, the reading on the VTVM should be less than 5 mV.

How to Measure Channel Separation
1. Set the unit in record mode with the volume controls (122) (164) at maximum.
2. Make recording with a -74 dBm at 1 kHz input to the left channel. This is also to record with no signal on the right channel simultaneously.
3. During recording, change the -74 dBm at 1 kHz signal to the right channel. This is also to change so there is no signal to the left channel.
4. Connect an AC VTVM to the output lug terminals.
5. When playing back the recorded materials, the reading on the VTVM for the signal crosstalk into the no signal recorded channel should be more than 0.28 dB.
6. The checking should be done both from left to right and from right to left.

How to Measure Crosstalk
1. Set the unit in record mode with the volume controls (122) (164) at maximum.
2. Make recording with a -74 dBm at 1 kHz signal on right channel.
3. Connect an AC VTVM to the output lug terminals.
4. When playing back the recorded materials, the reading on the VTVM should be referenced to the 0 dB basis.
5. Turn the cassette over and play the second channel. The reading on the VTVM should be more than 40 dB with 0 dB basis described in step 4.

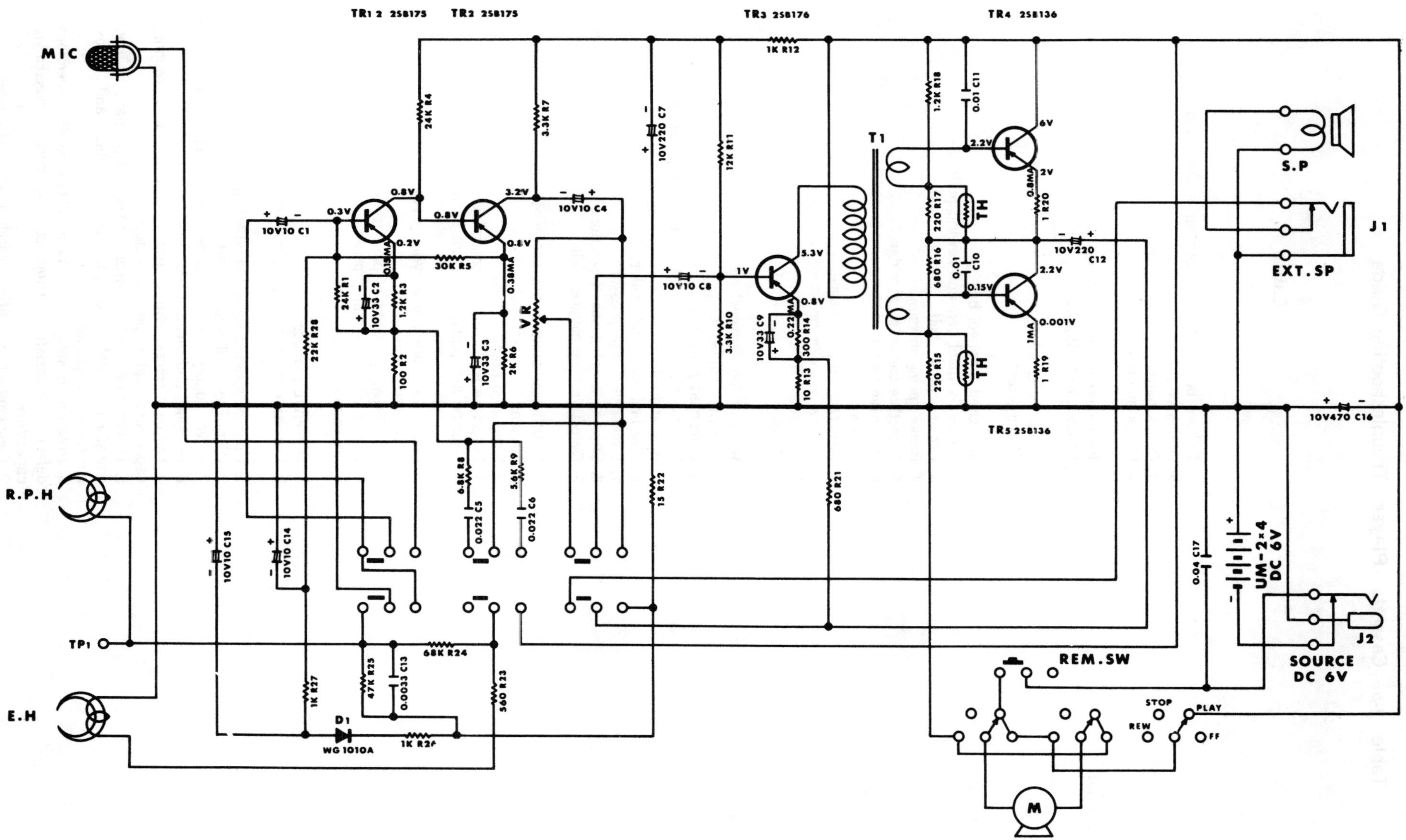

Fig. 2-12. Schematic of typical cassette recorder-player. (Courtesy Midland Int'l Corp.)

Table 2-5. Cassette Player Troubleshooting Guide

Symptom	Cause
Inoperative	1. Rubber belt of flywheel broken or stretched. 2. Defective motor. 3. Check battery. 4. Check on-off switch.
Takeup hub does not wind tape	1. Idler clutch pulley slipping. 2. Pinch roller slipping.
Pause does not function	1. Pinch roller sticks. 2. Clutch sticks.
No fast-forward or rewind	1. Stretched flywheel belt. 2. Oil on flywheel belt and rewind pulley. 3. Broken rubber belt. 4. Brake spring missing. 5. Defective cassette cartridge. 6. Defective motor. 7. Broken pushbutton. 8. Incorrect idler pressure. 9. Excessive roller pressure. 10. Incorrect roller pressure. 11. Faulty brake operation.
Excessive wow or flutter	1. Worn or dry flywheel assembly. 2. Stretched rubber belt. 3. Deformed pinch roller. 4. Insufficient or excessive pinch roller pressure. 5. Belt and idler slipping. 6. Oily belt and pinch roller. 7. Excessive brake pressure. 8. Capstan bearing out of position. 9. Defective tape counter belt.
Tape speed erratic during operation	1. Insufficient pinch roller pressure. 2. Slipping flywheel rubber belt. 3. Excessive takeup torque. 4. Defective motor on motor pulley. 5. Capstan and pinch roller not in line.
Slow tape speed	1. Check lubrication of moving parts. 2. Defective motor. 3. Insufficient pressure of pinch roller. 4. Oily pinch roller and capstan. 5. Loose motor belt.
Faulty eject operation	1. Defective plunger. 2. Defective microswitch mechanism.
No sound output	1. Open or shorted record-playback head. 2. Dirty record-playback head. 3. Power supply circuits: check microswitch, power supply connection, and bleeder resistors. 4. Output circuits: check open voice coil, output coupling capacitors, and output transistors. 5. Driver stages: check open or shorted primary of input transformer, defective transistor, coupling and bypass electrolytic capacitors. 6. Preamp circuits: check for open tape head connections, defective transistors, coupling and bypass electrolytic capacitors. 7. Poor contact on muting switch or short-circuited.

Table 2-5. Cont'd.

Sympton	Cause
Distorted sound	1. Dirty record-playback head. 2. Improper bias current. 3. Incorrect bias frequency. 4. Output stages: check output transistors, bias resistors, leaky coupling capacitors, and defective thermistor. 5. Driver stage: check transistors, secondary winding of input transformer, and burned bias resistors. 6. Preamp circuit: Check defective transistors, are burned or changed value of bias resistors. 7. Worn record-playback head. 8. Defective microphone.
Low volume	1. Dirty record-playback head. 2. Improper bias current. 3. Incorrect bias frequency. 4. Defective tape head. 5. Driver and predriver stages: transistors, gain adjustment control, leaky or dried out coupling capacitors, and bias resistors. 6. Improper head azimuth. 7. Improper record level.
Excessive noise	1. Noisy transistors in input and output circuits. 2. Defective decoupling capacitor. 3. Poor ground lead wiring. 4. Open electrolytic coupling capacitor from tape head to input preamp transistor. 5. Defective base bias resistor in preamp transistor circuit. 6. Defect in motor, or noisy bearings.
Excessive hum	1. Defective record-playback head. 2. Defective silicon diode rectifiers and filter capacitors in power supply circuitry. 3. Open or poor mike cable connection when recording. 4. Poor connections on record-playback head. 5. Defective power supply.
Oscillations	1. Defective head lead connection. 2. Defective tape head. 3. Defective base to collector bypass capacitors in output transistor stages. 4. Poor ground lead wiring. 5. Defective output transistor.
Won't record	1. Defective record-playback changing switch. 2. Open or shorted record-playback head. 3. Broken leads to tape head. 4. Defective slide switch. 5. Defective components in oscillator circuit. 6. Defective microphone. 7. Broken mike cable. 8. Defective microphone input jack. 9. Dirty record-playback tape head.
Crosstalk	1. Defective tape head. 2. Improper azimuth and height adjustment. 3. Defective tape.

Table 2-6. Typical Reel Type Player Troubleshooting Chart

Symptom	Cause
Take up reel revolves rapidly when play-record button is pressed	1. Insufficient tension on pinch roller. 2. Accumulation of oxide on capstan and pressure roller causing tape to climb. 3. Excessive takeup torque.
Supply reel does not revolve in rewind position	1. Loose reel shaft screw. 2. Rewind belt weak or broken. 3. Oil or grease on rewind belt. 4. Brake not releasing. 5. Fast forward idler not releasing. 6. Takeup idler not releasing. 7. Loose pulley on motor shaft. 8. Takeup turntable bearing gummy.
Takeup does not revolve in forward position	1. Loose reel shaft screw. 2. Fast forward idle binding on shaft. 3. Rewind (L.H.) turntable bearing gummy.
Reels do not stop when button is pressed	1. Brake lever out of adjustment. 2. Brake tension spring broken or missing. 3. Worn brake string. 4. Brake slide binding on baseplate.
Takeup reel does not revolve in play/record position	1. Pause brake not releasing. 2. Brake not releasing when play-record button is pressed. 3. Clutch belt slipping. 4. Clutch belt weak or broken. 5. Clutch belt off flywheel or clutch pulley. 6. Motor torque too low. 7. Takeup idler binding on shaft.
Unit does not turn on when "Off-On" switch is turned to "On" position	1. Make sure power cord is properly connected to AC outlet. 2. Make sure switch cord is properly connected to connector. 3. Motor shaft finding on bearing. 4. Inoperative power switch.
Wow or flutter (record or playback)	1. Oil or grease on the flywheel drive idler. 2. Brakes dragging. 3. Flat or bumpy pressure roller. 4. Build-up of oxide on capstan and / or pressure roller. 5. Flywheel is eccentric. 6. Erratic takeup torque. 7. Pause arm binding. 8. Dirty pinch roller. 9. Worn capstan drive belt. 10. Dirty tape pad. 11. Defective motor.
Record button interlock inoperative	1. Play function lever binding. 2. Interlock lever binding. 3. Worn or bent latch bar.
Pause brake inoperative	1. Broken pause lever.
Counter does not operate	1. Broken or missing belt. 2. Inoperative counter assembly.

Table 2-6. Con't.

Symptom	Cause
Erratic takeup torque	1. Foreign material on belt clutch. 2. Foreign matter on clutch disc. 3. Bent or twisted takeup idler lever. 4. Clutch throwout bearing binding. 5. Binding turntable.
Tape climbs up or down as it passes between the capstan and pressure roller	1. Capstan and/or pressure roller covered with oxide. 2. Pinch roller shaft not parallel to capstan. 3. Excessive take-up torque.
Weak or distorted sound in playback	1. Dirty heads. 2. Pressure pad not holding tape properly against the pole piece of the head. 3. Worn, open or shorted record-play head.
Weak or distorted sound in record	1. Check the amplifier. 2. Improper record bias setting.
No erase	1. Open lead to erase head (may have fallen off). 2. Open or shorted erase head.
Slow tape speed	1. Capstan drive belt not running freely. 2. Worn or stretched capstan drive belt. 3. Dirt or oil on capstan drive belt or drive surfaces. 4. Insufficient pinch roller presence. 5. Defective motor.
Poor erase or no erase	1. Defective erase head. 2. Open or shorted leads to erase head. 3. Improper erase head current. 4. Erase head out of line. 5. Insufficient tape head pressure.
Playback normal but will not record	1. Defective record-playback switch. 2. Defective microphone or radio jack. 3. Broken leads to microphone jack. 4. Defective microphone. 5. Defective amplifier circuit.
Poor sound quality in playback position	1. Dirty record-playback head. 2. Magnetized record-playback head. 3. Insufficient tape head pad pressure. 4. Defective tape. 5. Defective microphone. 6. Defective sound circuits.
Recorded tape does not playback properly on another recorder	1. Improper record-playback head position or adjustment.
Operates on battery but not on AC	1. Defective AC cord. 2. Defective AC socket. 3. Defective power transformer. 4. Shorted or open silicon rectifiers. 5. Defective filter in power supply circuit.
Does not operate on battery	1. Weak batteries. 2. Defective AC socket or switch mechanism.
Does not opera' , on remote	1. Defective remote switch on microphone. 2. Defective remote jack. 3. Broken lead to remote switch.

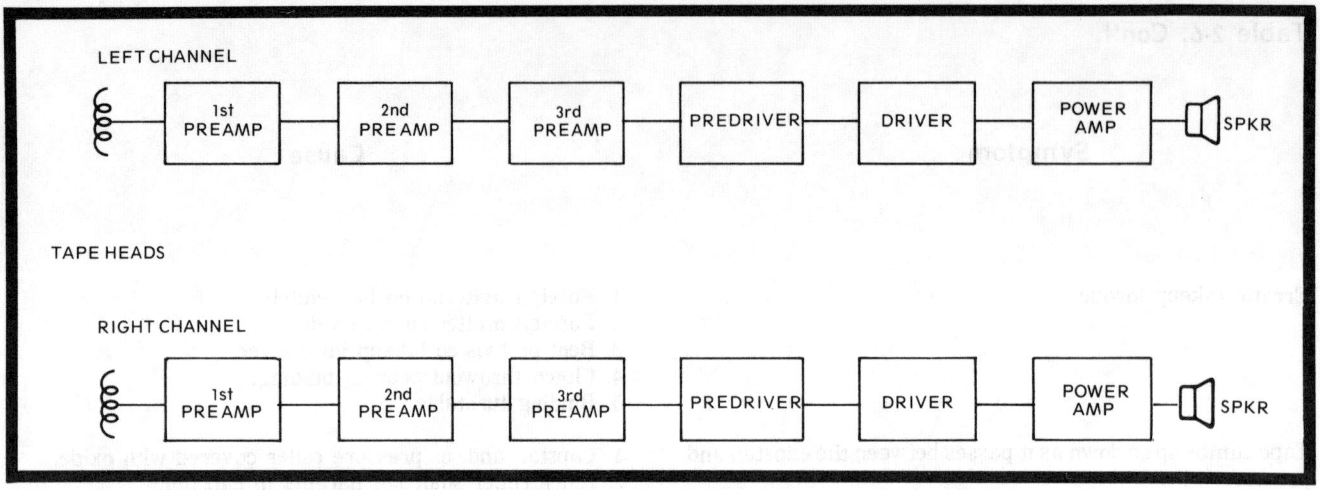

Fig. 2-13. Block diagram for cassette 4-track stereo player.

over them. This oxide accumulation should be cleaned from these parts periodically, using a soft lint-free cloth and alcohol.

All drive surfaces must be kept free of grease and oil. Use only a soft clean cloth to clean the motor pulley, drive idler, flywheel, takeup idler, fast forward idler, rewind belt, takeup belt clutch pulley, and turntable drum. Detergent and water may be used if necessary.

2. Lubrication

The mechanism is properly lubricated as it leaves the factory and additional lubrication should not be necessary for a long period of time.

Should the mechanism be disassembled or the need for lubrication become apparent, lubrication should be applied as noted below.

The capstan and flywheel bearings, the turntable bearings, the fast forward idler bearing, the rewind pulley bearing, the drive pulley bearing, and the pinch roller bearing should be lubricated with a "Molykote" or equivalent. Lubricant should also be applied between the brake slide and the guide, to all points of contact on the key button latching bar and the record lock latching bar.

3. Head Demagnetization

Head magnetization is evidenced by a high residual noise (a hissing and popping sound). Use a standard demagnetizing tool.

4. Head Azimuth

Alignment of the record/playback head for azimuth is dependent on the adjustment of screws which hold the head to the base plate. That is, adjusting the screw by rotating it in a clockwise (down) direction or in a counterclockwise (up) direction.

A standard test tape having 6 to 15 kHz tone on one track may be used for azimuth alignment.

Adjust the head as follows:

a. Remove the head cover.

b. Place a standard test tape on the instrument.

c. Connect an AC voltmeter across the speaker output.

d. Push the play key 7½".

e. Adjust the screws as previously described, for maximum meter indication.

Fig. 2-14 shows a typical 3-speed reel tape deck. Table 2-6 is a troubleshooting chart.

Fig. 2-14. Stereo 3-speed reel-to-reel tape deck. (Realistic Model 999B, Courtesy Allied Radio Shack.)

Chapter 3

Schematics, Alignment Procedures, and Parts Lists for Representative Multiband Radios

7-Band AM-FM-SW AC/DC—Craig Corp. 1306

1306 ALIGNMENT INSTRUCTIONS

Equipment required:
1 Signal Generator...300 KHz to 120 MHz - AM and FM modulation
2 Sweep Generator....10.7 MHz - at least 500 KHz sweep width
3 AC Voltmeter
4 Oscilloscope
5 IRE Loop
6 Dummy Antenna (Fig. A)
7 Detector Probe (Fig. B)

SET VOLUME CONTROL TO MAXIMUM

Function	Band Switch	Signal Input	Frequency	Dial Setting	Output	Adjustment	
FM IF	FM	Connect Sweep Generator between antenna and ground through dummy antenna. Couple marker gen. loosely to sweep. Signal level low	10.7 MHz	Lowest end	Connect scope through Det. probe between TP2 and ground	T1, T2, T3, T4, T2a, T2b	Maximum response (Fig. C)
FM Detector					Connect scope between TP3 and ground	T5	Response curve (Fig. D)
FM Osc.		Signal Generator FM modulation. (22.5KHz deviation 400Hz) Connect between antenna and ground thru. dummy antenna. Signal level low	90 MHz	90 MHz	Connect AC voltmeter across voice coil of speaker	Ctb	
			106 MHz	106 MHz		Lb	
FM RF			98 MHz	98 MHz		Cta	Maximum
AM IF	AM	Signal Generator (30% AM modulation 400 Hz) connected to IRE loop coupled loosely to antenna coil Signal level low	455 KHz	Lowest end		T6, T7, T8	Output
AM Osc.			600 KHz	600 KHz		T13	
			1400 KHz	1400 KHz		Ct14	
AM RF			600 KHz	600 KHz		L11	
			1400 KHz	1400 KHz		Ct7	
SW1 Osc.	SW1	Same	1.7 MHz	1.7 MHz	Same	T12	Same
			5.0 MHz	5.0 MHz		Ct13	
SW1 RF			1.7 MHz	1.7 MHz		L9	
			5.0 MHz	5.0 MHz		Ct6	
SW2 Osc.	SW2		6.0 MHz	6.0 MHz		T11	
			7.5 MHz	7.5 MHz		Ct12	
SW2 RF			6.0 MHz	6.0 MHz		L7	
			7.5 MHz	7.5 MHz		Ct5, Ct16	
SW3 Osc.	SW3		9.5 MHz	9.5 MHz		Ct10, Ct17	
			16 MHz	16 MHz		Ct11	
SW3 RF			9.5 MHz	9.5 MHz		Ct3, Ct4	
			16 MHz	16 MHz		Ct15	
SW4 Osc.	SW4		18 MHz	18 MHz		T10	
			23 MHz	23 MHz		Ct9	
SW4 RF			18 MHz	18 MHz		L5	
			23 MHz	23 MHz		Ct2	
SW5 Osc.	SW5		23 MHz	23 MHz		T9	
			28 MHz	28 MHz		Ct8	
SW5 RF			23 MHz	23 MHz		L4	
			28 MHz	28 MHz		Ct1	

KEEP LEADS SHORT

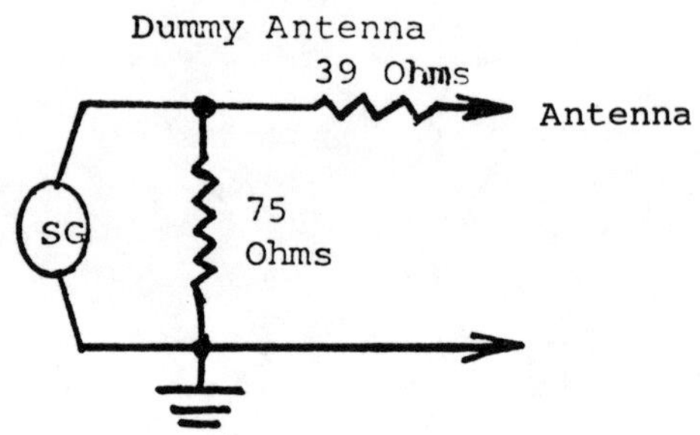

Fig. A

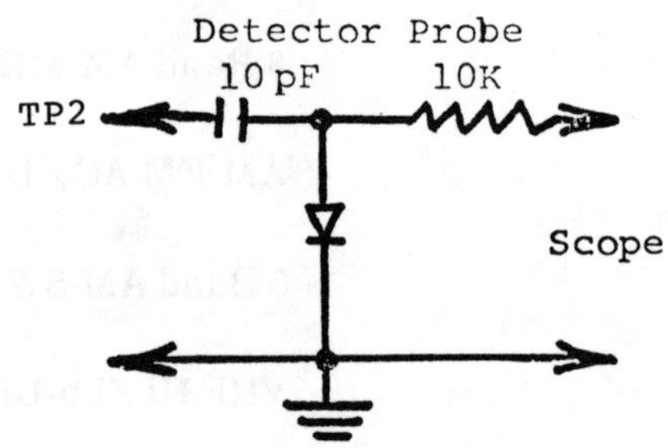

Fig. B

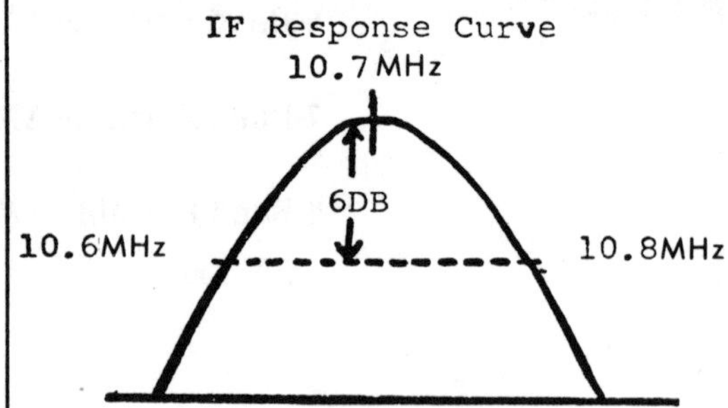

Fig. C

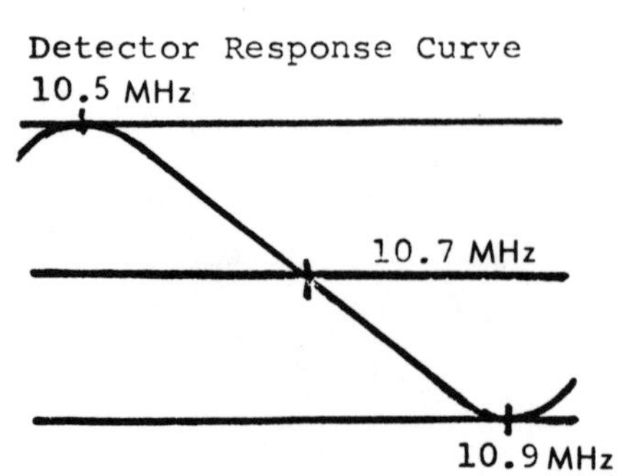

Fig. D

NOTE:
1. Repeat OSC and RF adjustments in each band until no further improvement is noted.
2. Align SW2 before SW3, as any coil adjustment on SW2 will effect the adjustments on SW3.
3. Adjust R9 (50K) semi-fixed resistor for 1.5 ma collector current, Tr2. Use an ammeter with no less than 1 **meg** series resistor.
4. Never short the base of Tr2 to ground (B+) or the base of Tr11 to common (B–), to prevent damage to these transistors. Extreme care should be taken during service or alignment of this receiver to insure these shorts not be made inadvertently by a screw driver or other test equipment.
5. R9 and R70 are both 50K ohm variable resistors. Neither of these should be set below 10K ohms whenever receiver is on, but if repair must be made to either, preset to mid range before turning receiver on.
6. Note that FM oscillator adjustments are made contrary to normal procedure. The high end is adjusted with the coil and the low end is adjusted with the trimmer capacitor, since the tuning variables are the coils.
7. The AM IF response curve is double peaked. If alignment is made with a sweep generator and oscilloscope adjust the IF transformers for maximum response and minimum valley. The 4.55 kHz marker does not necessarily fall directly in the center of the response curve. A ceramic filter is used which determines the two peaks of this response curve.

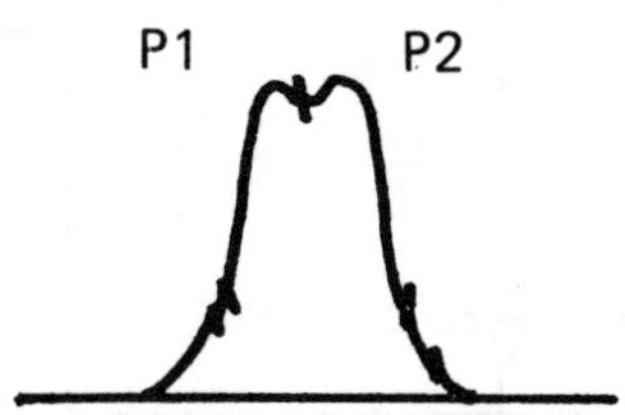

RESPONSE CURVE OF AM IF STAGE

CRAIG 1306 PARTS/PRICE LIST. SUBJECT TO CHANGE W/O NOTICE.
ITEMS W/O A CRAIG KEY NO. MUST BE ORDERED BY COMPLETE DESCRIPTION.

Ref. No.	Craig Key No.	Description	List Price
COILS & TRANSFORMERS			
	0013060090	Choke Coil	$ .30
	0013060091	Antenna Coil, SW5	.50
	0013060092	Antenna Coil, SW2, SW3, SW4	1.80
	0013060093	Antenna Coil, MW SW1	1.80
	0013060094	IF Transformer, FM	.60
	0013060095	IF Transformer, FM	.80
	0013060096	IF Transformer, FM	.80
	0013060097	IF Transformer, AM	.80
	0013060098	IF Transformer, AM	.80
	0013060099	IF Transformer, AM	.80
	0013060100	Oscillator Coil, SW5	.60
	0013060101	Oscillator Coil, SW4	$ .60
	0013060102	Oscillator Coil, SW3/SW2	.60
	0013060103	Oscillator Coil, SW1	.60
	0013060104	Oscillator Coil, MW	.60
	0013060105	Power Transformer	3.40
CONTROLS			
	0013060106	FM Converter	$18.30
	0013060107	Var. Cap. AM Tune	4.50

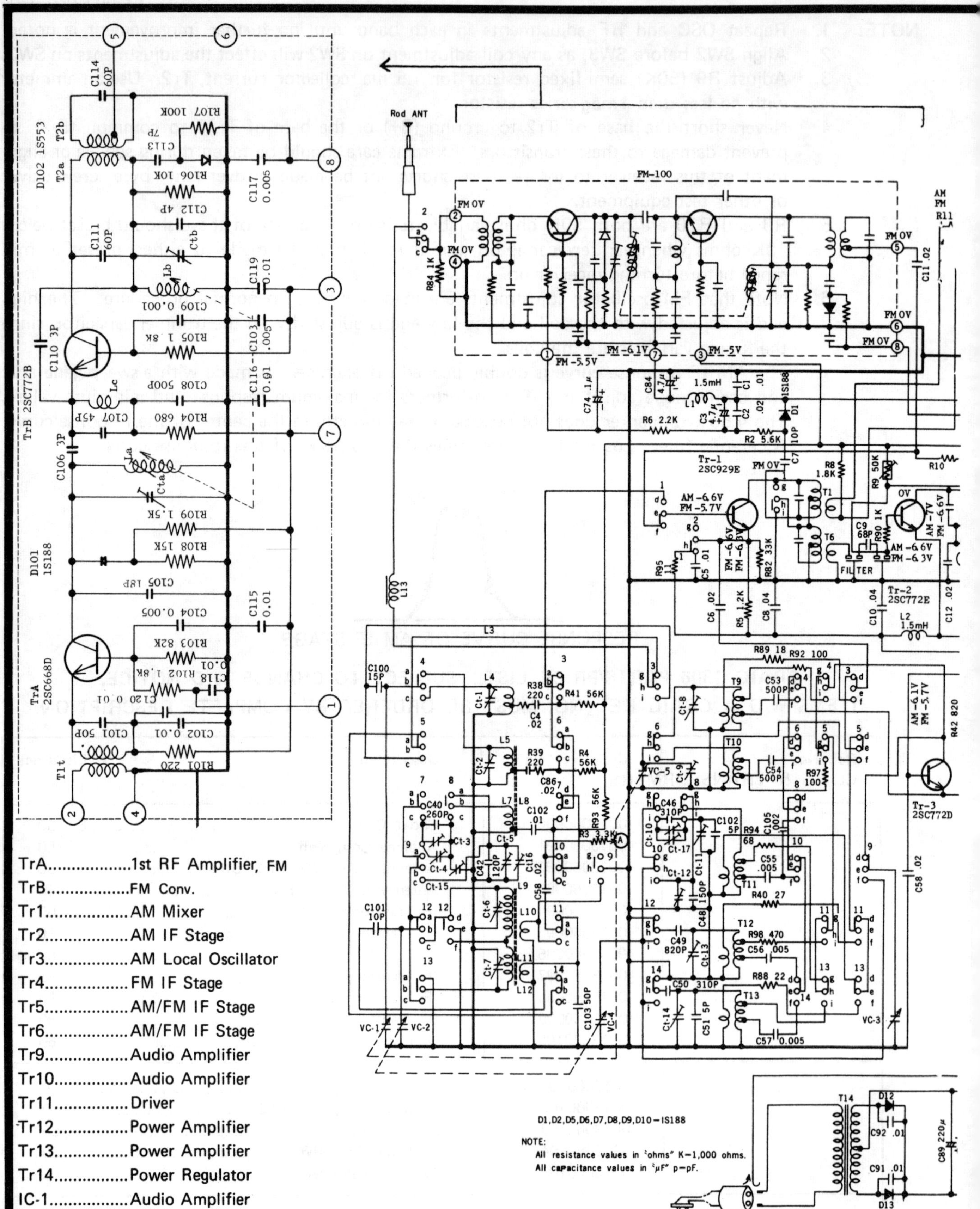

TrA..................1st RF Amplifier, FM
TrB..................FM Conv.
Tr1..................AM Mixer
Tr2..................AM IF Stage
Tr3..................AM Local Oscillator
Tr4..................FM IF Stage
Tr5..................AM/FM IF Stage
Tr6..................AM/FM IF Stage
Tr9..................Audio Amplifier
Tr10................Audio Amplifier
Tr11................Driver
Tr12................Power Amplifier
Tr13................Power Amplifier
Tr14................Power Regulator
IC-1.................Audio Amplifier

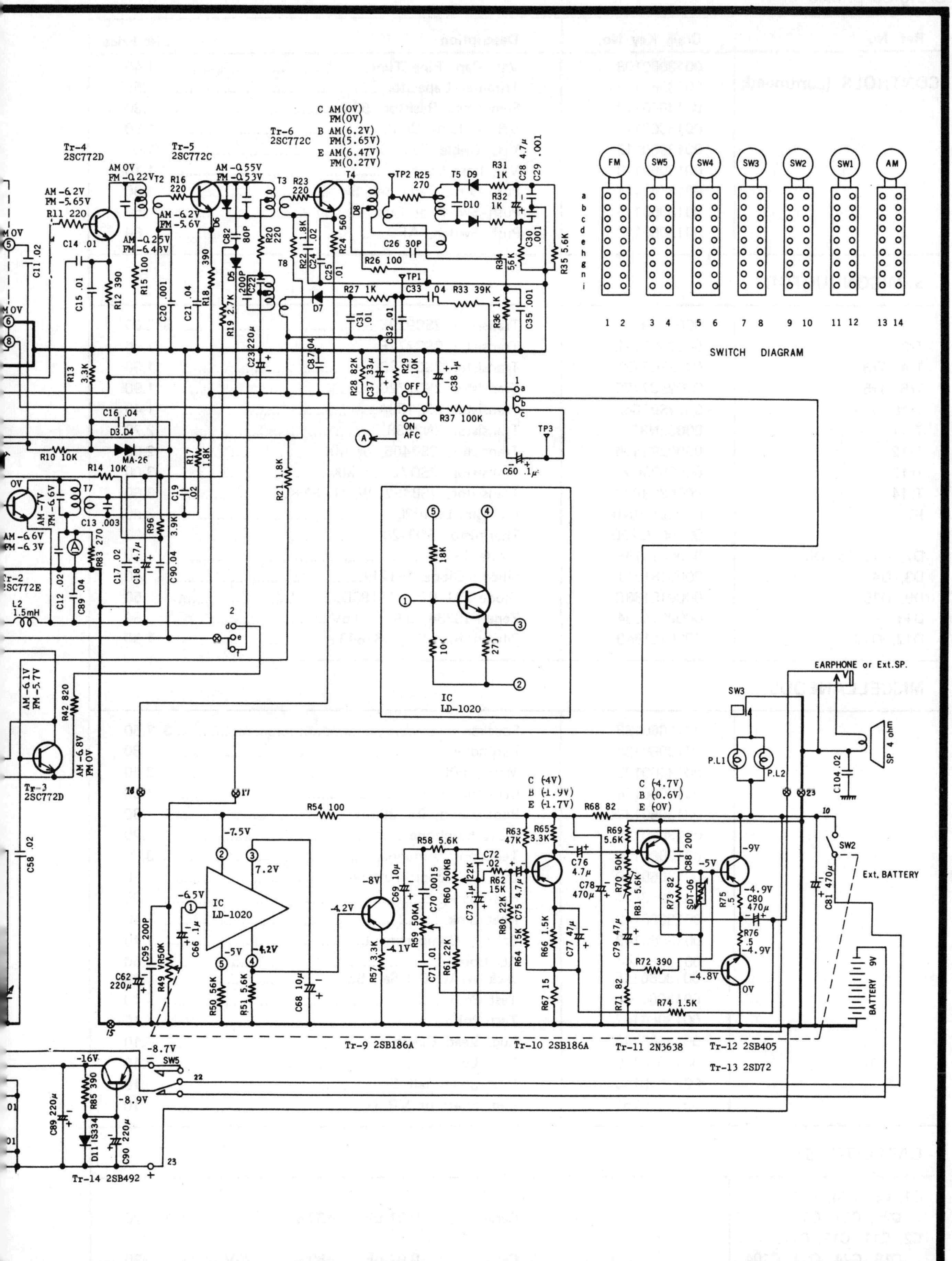

Ref. No.	Craig Key No.	Description	List Price
CONTROLS (continued)	0013060108	Var. Cap. Fine Tune	1.40
	0013060109	Trimmer Capacitor	.50
	0013060110	Semi-fixed Resistor, 50K	.30
	0013060111	VR, Volume Cont.	1.50
	0013060112	VR, Treble Cont.	1.00
	0013060113	VR, Bass Cont.	1.10
	0013060114	Push But., Band Sel. Sw.	6.80
	0013060115	Push Sw. Pilot Lite	.50
	0013060116	Push Switch, AFC	.80
SEMICONDUCTORS			
Tr1	0002SC929E	Transistor 2SC929E	$ 2.00
Tr2	0002SC772E	Transistor 2SC772E	1.90
Tr4, Tr3	0002SC772D	Transistor 2SC772D	1.90
Tr6, Tr5	0002SC772C	Transistor 2SC772C	1.90
Tr10, Tr9	0002SB186A	Transistor 2SB186A	1.40
Tr11	00002N3638	Transistor 2N3638	2.70
Tr12	00002SB405	Transistor, 2SB405, or MK	2.60
Tr13	000002SD72	Transistor, 2SD72, or MK	3.00
Tr14	00002SB492	Transistor, 2SB492, W. H. SNK.	3.60
IC	0000LD1020	Int. Cir. LD1020	3.60
	00000SDT20	Thermistor SDT-20	.30
D7, D1, D5, D6	000001S188	Diode 1S188	.50
D3, D4	00001S1211	Silicon Diode 1S1211	.40
D9, D10	00001S188D	Diode FM Disc. 1S188D	.50
D11	000001S334	Zener 1S334, 8.9 – 9.6V	.50
D12, D13	00001S1849	Silicon Rectifier, 1S1849	1.90
MISCELLANEOUS			
	0013060133	Speaker	$ 7.50
	0013060134	Earphone	.80
	0013060135	Meter, 600uA	3.60
	0013060136	Ceramic Filter	1.80
	0013060137	Pilot Lamp, 9V 30mA	.30
	0013060138	Jack, Earphone	.20
	0013060139	Telescopic Antenna	3.00
	0013060140	Lug, Telescopic Antenna	.10
	0013060141	Heat Sink,	.20
	0013060142	Heat Sink	.40
	0013060143	Battery Compartment	1.40
	0013060144	AC Cord	1.50
	0013060145	Jack W/AC DC Sel. Sw.	.60
	0013060146	Test Point	.10
	0013060147	Test Point	.10
	0013060148	Lug, Lead Clamp	.10
	0013060149	Lug, Lead Clamp	.10
	0013060150	Lug, Slide Rail Gr.	.10
	0013060151	Lug, Shaft of VR Gr.	.10
CAPACITORS			
C1, C5, C14, C15 C25, C31, C32		Ceramic 0.01 uF +80% 25V	$ .20
C2, C11, C12, C17, C19, C24, C58, C104		Ceramic 0.02 uF +80% 25V	.20

Ref. No.	Craig Key No.	Description				List Price
C3, C18, C28, C75, C76, C84		Electrol.	4.7 uF		6.3V..........	.30
C4, C6, C72, C85, C86		Mylar	0.02 uF	±20%	50V..........	.20
C7, C101		Ceramic	10 pF	±10%	25V..........	.20
C8, C16, C33		Mylar	0.04 uF	±20%	50V..........	.20
C9		Ceramic	68 pF	±5%	25V..........	.20
C10, C21, C87 C89, C90		Ceramic	0.04 uF	+80%	25V..........	.20
C13		Mylar	0.003 uF	±20%	50V..........	.20
C20, C29, C30, C35		Mylar	0.001 uF	±20%	50V..........	.20
C22, C88, C95		Ceramic	200 pF	±10%	25V..........	.20
C23, C62, C90		Electrol.	220 uF		10V..........	.20
C26		Ceramic	30 pF	±10%	25V..........	.20
C37		Electrol.	33 uF		6.3V..........	.30
C38		Electrol.	1 uF		10V..........	.30
C40		Styrol	260 pF	±5%	50V..........	.20
C42		Ceramic	120 pF	±5%	25V..........	.20
C46		Styrol	310 pF	±5%	50V..........	.20
C48		Styrol	150 pF	±5%	50V..........	.20
C49		Styrol	820 pF	±10%	50V..........	.20
C50		Styrol	310 pF	±10%	50V..........	.20
C51, C106		Ceramic	5 pF	±10%	25V N3300....	.20
C53, C64		Ceramic	500 pF	±20%	25V..........	.20
C55, C56, C57		Mylar	0.005 uF	±20%	50V..........	.20
C60, C66, C73, C74		Electrol.	0.1 uF		10V..........	.30
C68		Electrol.	10 uF		10V..........	.30
C69		Electrol.	10 uF		6.3V..........	.30
C70		Mylar	0.0015 uF	±20%	50V..........	.20
C71		Mylar	0.01 uF	±20%	50V..........	.20
C77, C79		Electrol.	47 uF		10V..........	.30
C78, C80, C81		Electrol.	470 uF		10V..........	.30
C82		Ceramic	80 pF	±5%	25V..........	.20
C89		Electrol.	220 uF		25V..........	.20
C91, C92		Ceramic	0.01 uF	+80%	50V..........	.20
C100		Ceramic	15 pF	±10%	25V..........	.20
C103		Ceramic	50 pF	±20%	25V..........	.20
C105		Mylar	0.002 uF	±20%	25V..........	.20
C107		Ceramic	100 pF	±20%	25V..........	.20
RESISTORS, OHMS, ±10%, ¼W						
R2, R34, R35, R51 R58, R69, R81		5.6K..			$	.20
R3, R13, R57, R65		3.3K..				.20
R4, R41, R50, R93		56K..				.20
R5		1.2K;..				.20
R6		2.2K..				.20
R8, R17, R21, R22		1.8K..				.20
R10, R14, R29		10K..				.20
R11, R16, R20 R23, R38, R39		220..				.20
R12, R18, R72, R85		390..				.20
R15, R26, R54, R92, R97		100..				.20
R19		2.7K..				.20
R24		560..				.20
R25, R83		270..				.20
R27, R31, R32, R36, R84, R90, R95		1K..				.20

Ref. No.	Craig Key No.	Description	List Price
FM CONVERTER (FM–100)			
T1t	0013060258	Antenna Coil	$ .50
La, Lb	0013060259	Var, Induct, Coil Ass'y.	1.80
		Bobbin	
	0013060260	Ferrite Slug (Core)	.50
		Stopper	
		Flat Copper Wire	
Lc	0013060261	Choke Coil, 0.6uH, 16T	.20
T2	0013060262	IF Transformer	.60
Cta, Ctb	0013060263	Cylinder Trimmer	.20
TrA	0002SC668D	Transistor, RF 2SC668D	2.40
TrB	0002SC772B	Transistor, Conv. 2SC772B	1.90
D101	0001S188PM	Diode 1S188FM	.50
D102	000001S553	Diode, Var. Cap. 1S553	2.40
	0013060269	Shield Case	.30
	0013060270	Bottom Casing	.20
	0013060271	Metal Brkt., Guide	.20
	0013060272	Guide Shaft	.10
	0013060273	Guide	.10
	0013060274	Tension Spring	.10
	0013060275	Screw, spec. head 3 x 6mm	.10
R101		220 ohms ±5% ¼W	.20
R102, R105		1.8K ohms ±10% ¼W	.20
R103		82K ohms ±5% ¼W	.20
R104		680 ohms ±5% ¼W	.20
R106		10K ohms ±10% ¼W	.20
R107		100K ohms ±10% ¼W	.20
R108		1.5K ohms ±10% ¼W	.20
R109		1.5K ohms ±10% ¼W	.20
C101		Ceramic 50pF ±10% 50V	.20
C102, C115, C116, C118, C119, C120		Ceramic 0.01 uF +80% 50V	.20
C103, C104, C117		Ceramic 0.005 uF +80% 50V	.20
C105		Ceramic 18 pF ±10% N220 50V	.20
C106, C110		Ceramic 3 pF ±0.25pF NPO 50V	.20
C107		Ceramic 45 pF ±10% N220 50V	.20
C108		Ceramic 500 pF ±20% 50V	.20
R28		82K	.20
R33		39K	.20
R37		100K	.20
R40		27	.20
R42		820	.20
R61, R80		22K	.20
R62, R64		15K	.20
R63		47K	.20
R66, R74		1.5K	.20
R67		15	.20
R68, R71, R73		82	.20
R75, R76		0.5 Wirewound	.30
R82		33K	.20
R88		22	.20
R89		18	.20
R91		330K	.20
R94		68	.20
R96		3.9K	.20
R98		470	.20

Craig Corp. 1306

Ref. No.	Craig Key No.	Description					List Price
C109		Ceramic	0.001 uF	±20%		50V..	.20
C111, C114		Ceramic	60 pF	±10%	N220	50V..	.20
C112		Ceramic	4 pF	±0.5pF	N330	50V..	.20
C113		Ceramic	7 pF	±10%		50V..	.20

ALIGNMENT INSTRUCTIONS

Step	Generator Frequency	Dial Setting	Indicator	Adjust	Remarks
AM Section					
1	455KHz	Tuning Gang Fully Closed	Output Meter Across Voice Coil	T3, T8, T9	Adjust For Maximum Output
2	535KHz			L17	
3	1650KHz			TC-10	
4	600KHz	Tuning Gang To Signal		L14	Adjust Coil On Loopstick Ant. For Maximum
5	1400KHz			TC-7	Adjust For Maximum Output
M.B. Section					
1	1.45MHz	Tuning Gang Fully Closed	Output Meter Across Voice Coil	L18	Adjust For Maximum
2	4.5MHz	Tuning Gang Fully Open		TC-11	
3	1.7MHz	Tune To Signal		L15	Adjust Coil On Loopstick For Maximum
4	3.6 MHz			TC-8	Adjust For Maximum
S.W. Section					
1	3.8MHz	Tuning Gang Fully Closed	Output Meter Across Voice Coil	L19	Adjust For Maximum
2	12.2MHz	Tuning Gang Fully Open		TC-12	
3	4.8MHz	Tune To Signal		L16	
4	11.2MHz			TC-9	

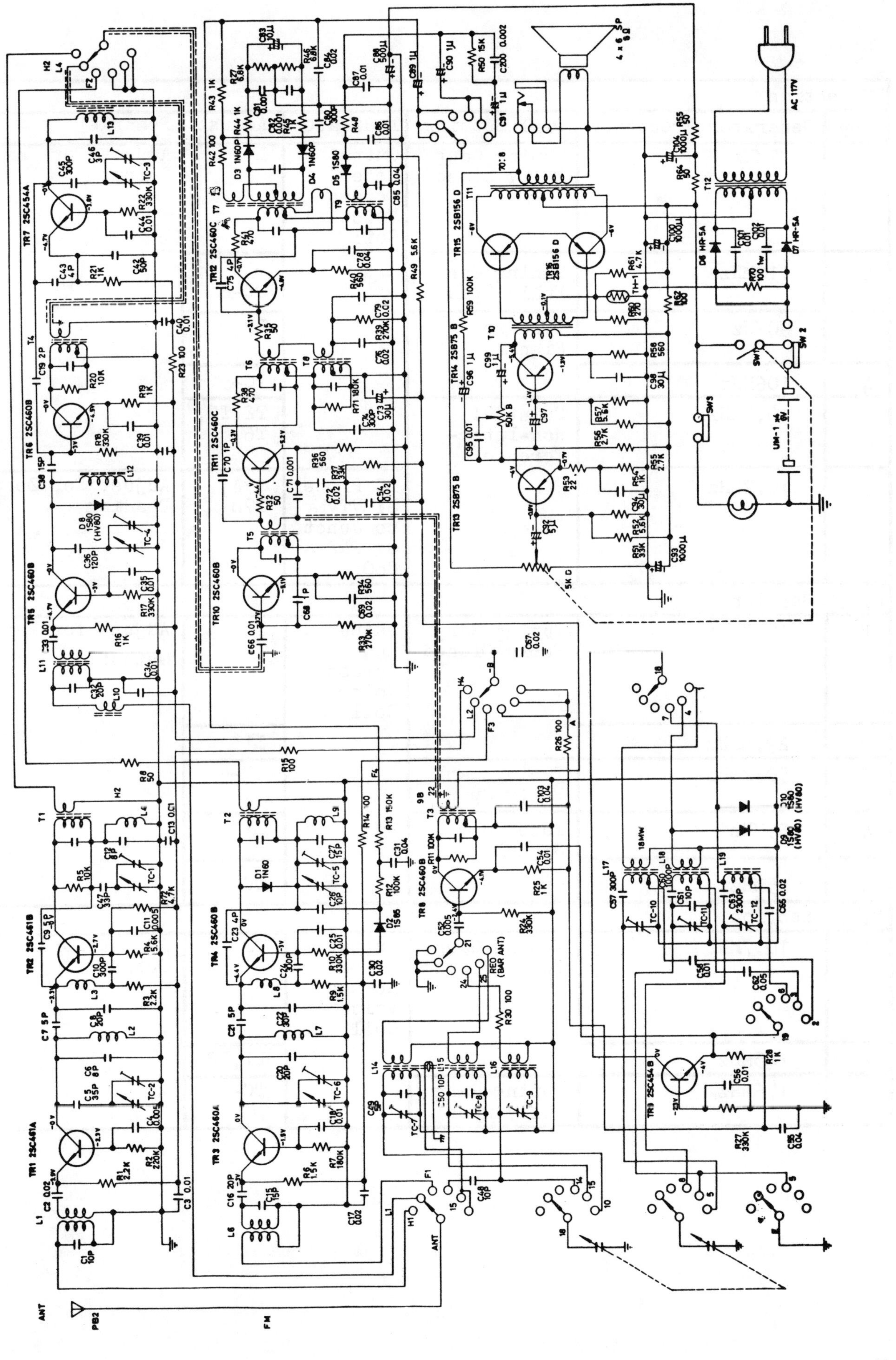

Step	Generator Frequency	Dial Setting	Indicator	Adjust	Remarks
FM Section					
1	87MHz	Tuning Gang Fully Closed	Output Meter Across Voice Coil	L9	Adjust For Maximum Output
2	110MHz	Tuning Gang Fully Open		TC-6	
3	90MHz	Tune To Signal		L7	
4	106MHz			TC-5	
5	10.7MHz	Point Of Non-Interference		T2, T5, T6	
6	10.7MHz		DC Probe of VTVM to Junct. of R42 & C80	T7A, T7B	Adjust For Zero Reading
P.B.1 Section					
1	10.7MHz	Tuning Gang Fully Closed	Output Meter Across Voice Coil	T4	Adjust for Maximum Output
2	29.5MHz			L13	
3	50.2MHz	Tuning Gang Fully Open		TC-4	
4	35MHz	Tune To Signal		L12	
5	45MHz			TC-3	
P.B.2 Section					
1	10.7MHz	Tuning Gang Fully Closed	Output Meter Across Voice Coil	T2	Adjust For Maximum Output
2	145MHz			L1, L2	
3	176MHz	Tuning Gang Fully Open		TC-2	

REF. NO.	DESCRIPTION	PART NO.
TRANSISTORS:		
TR1, 2	2SC 461	09-302005
TR3, 4, 5, 6, 8, 10, 11, 12	2SC 460	09-302004
TR7, 9	2SC 454	09-302017
TR13, 14	2SB 75	09-301005
TR15, 16	2SB 156	09-301056
DIODES:		
D1	IN60	09-306019
D2	IS85	09-306021
D3, 4	IN60P	09-306057
D5, 8, 9, 10	IS80	09-306009
D6, 7	HR-5A	09-306125
THERMISTOR:		
TH-1	D-2B	09-307028

RESISTORS:

All resistors not shown on this parts list are carbon, 1/8 watt. See schematic for specific values.

R70	Carbon film, 150 ohm/1-watt	77-804151

CAPACITORS:

All ceramic disc capacitors not shown on this parts list are 50V. See schematic for specific values.

MYLAR CAPACITORS:		
C25, 54, 58	.01MF/50V	78-651103
C62	.005MF/50V	78-651502
C65	.002MF/50V	78-651202
POLYSTYRENE CAPACITORS:		
C60	1,000PF/50V	78-351102
C63	2,300PF/50V	78-351232
ELECTROLYTIC CAPACITORS:		
C73, 94, 98	30MF/10V	77-336306
C83	10MF/10V	77-336106
C88	500MF/10V	77-336507

REF. NO.	DESCRIPTION	PART NO.
	COILS & TRANSFORMERS:	
L1	Coil, PB 2 Antenna	10-176067
L2	Coil, PB2 RF	10-176068
L3, 8	Coil, PB2 IF Trap or FM IF Trap	10-176069
L4	Coil, PB2 oscillator	10-170159
L6	Coil, FM antenna	10-176070
L7	Coil, FM RF	10-176071
L9	Coil, FM oscillator	10-170160
L10	Coil, choke	10-178013
L11	Coil, PB1 antenna	10-176072
L12	Coil, PB1 RF	10-176073
L13	Coil, PB1 oscillator	10-170161
L14, 15, 16	Antenna, loopstick	10-050078
L17	Coil, AM oscillator	10-170162
L18	Coil, MB oscillator	10-170163
L19	Coil, SW oscillator	10-170164
T1	IFT, PB2 1st	10-090244
T2	IFT, FM 1st	10-090245
T3	IFT, AM 1st	10-090246
T4	IFT, PB1 1st	10-090247
T5	IFT, FM/PB1/PB2 2nd	10-090248
T6	IFT, FM/PB1/PB2 3rd	10-090249
T7A	Discriminator, primary (pink slug)	10-093157
T7B	Discriminator, secondary (blue slug)	10-093158
T8	IFT, AM 2nd	10-090250
T9	IFT, AM 3rd	10-090251
T10	Transformer, audio input	10-096068
T11	Transformer audio output	10-096069
T12	Transformer, AC power w/cord	10-098016
	MISCELLANEOUS:	
	Antenna, telescope (FM/SW) 2 used	10-040044
	Switch, band selector	10-180021
VR1	Control, volume w/switch	10-160068
VR2	Control, tone	10-163012
	Control, fine tuning	10-166001
	Tube, battery box	10-031004
	Speaker (4" x 6") (8ohm/2W)	10-060039
	Switch, AC/DC	10-183034
	Lamp, dial scale	10-201007
	Switch, dial scale illumination	10-186005
TC1, 2, 3, 4	Capacitor, tuning (PB1-PB2)	10-120065
TC5, 6, 7, 10	Capacitor, tuning (FM-AM-MB-SW)	10-120064
TC8, 9	Trimmer, MB & SW antenna	10-123016
TC11, 12	Trimmer, MB & SW oscillator	10-123017

ELECTROLYTIC CAPACITORS:

C89, 90, 91, 96, 99	1MF/6V	77-335105
C92	5MF/6V	77-335505
C93, 100, 104	1000MF/10V	10-135001
C97	5MF/10V	77-336505

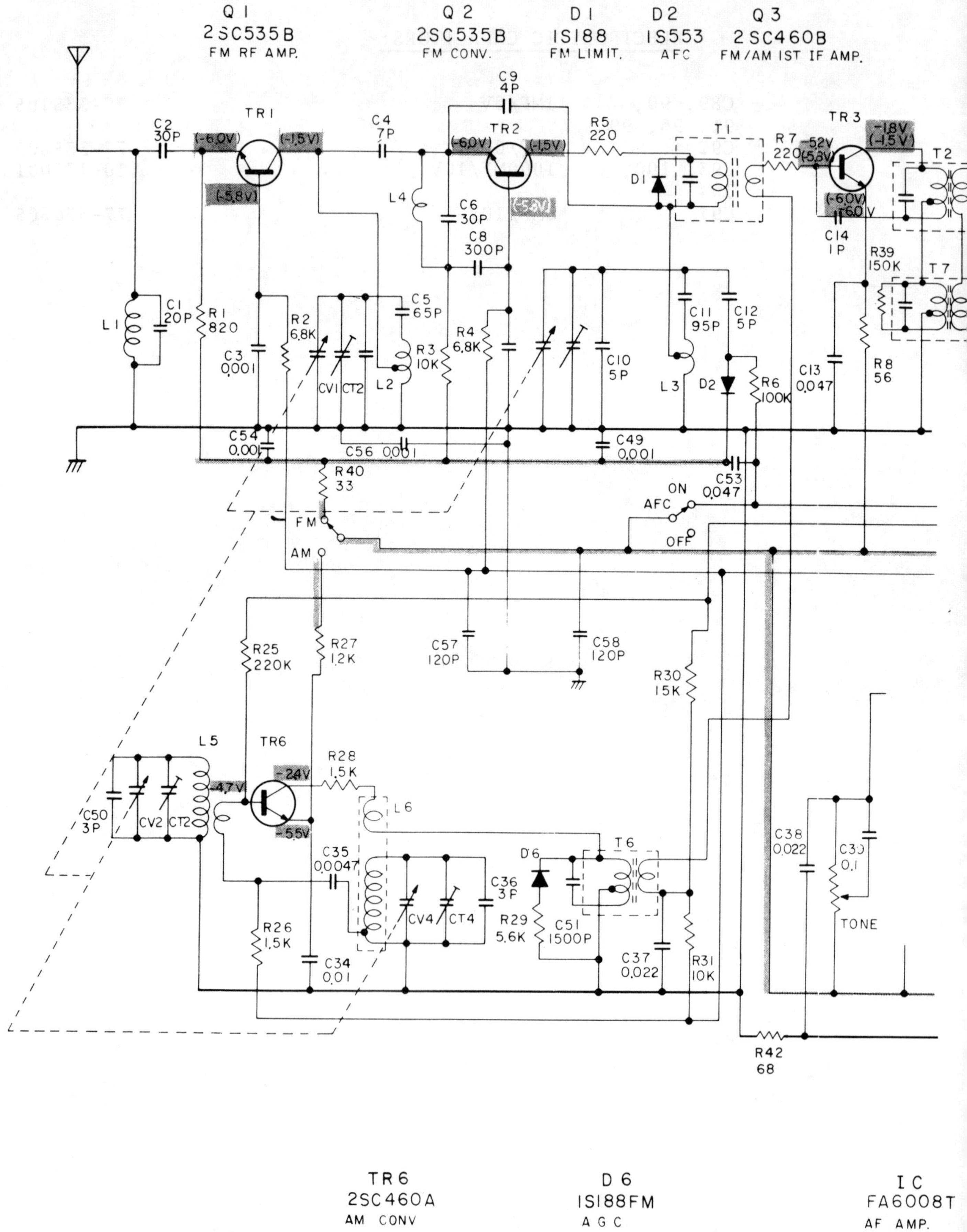
Q1
2SC535B
FM RF AMP.
Q2
2SC535B
FM CONV.
D1
IS188
FM LIMIT.
D2
IS553
AFC
Q3
2SC460B
FM/AM IST IF AMP.
TR1
TR2
TR3
C2 30P
C4 7P
C9 4P
R5 220
R7 220
T1
T2
T7
C6 30P
C8 300P
C14 1P
R39 150K
L1
C1 20P
R1 820
C3 0.001
R2 6.8K
CV1
CT2
C5 65P
L2
R3 10K
R4 6.8K
L4
C10 5P
C11 95P
C12 5P
L3
D2
R6 100K
C13 0.047
R8 56
C54 0.001
C56 0.001
C49 0.001
C53 0.047
R40 33
FM
AM
AFC
ON
OFF
C57 120P
C58 120P
R25 220K
R27 1.2K
R30 15K
L5
TR6
R28 1.5K
L6
C50 3P
CV2
CT2
C35 0.0047
CV4
CT4
C36 3P
D6
T6
R29 5.6K
C51 1500P
C37 0.022
R31 10K
R26 1.5K
C34 0.01
C38 0.022
C30 0.1
TONE
R42 68
TR6
2SC460A
AM CONV
D6
IS188FM
AGC
IC
FA6008T
AF AMP.

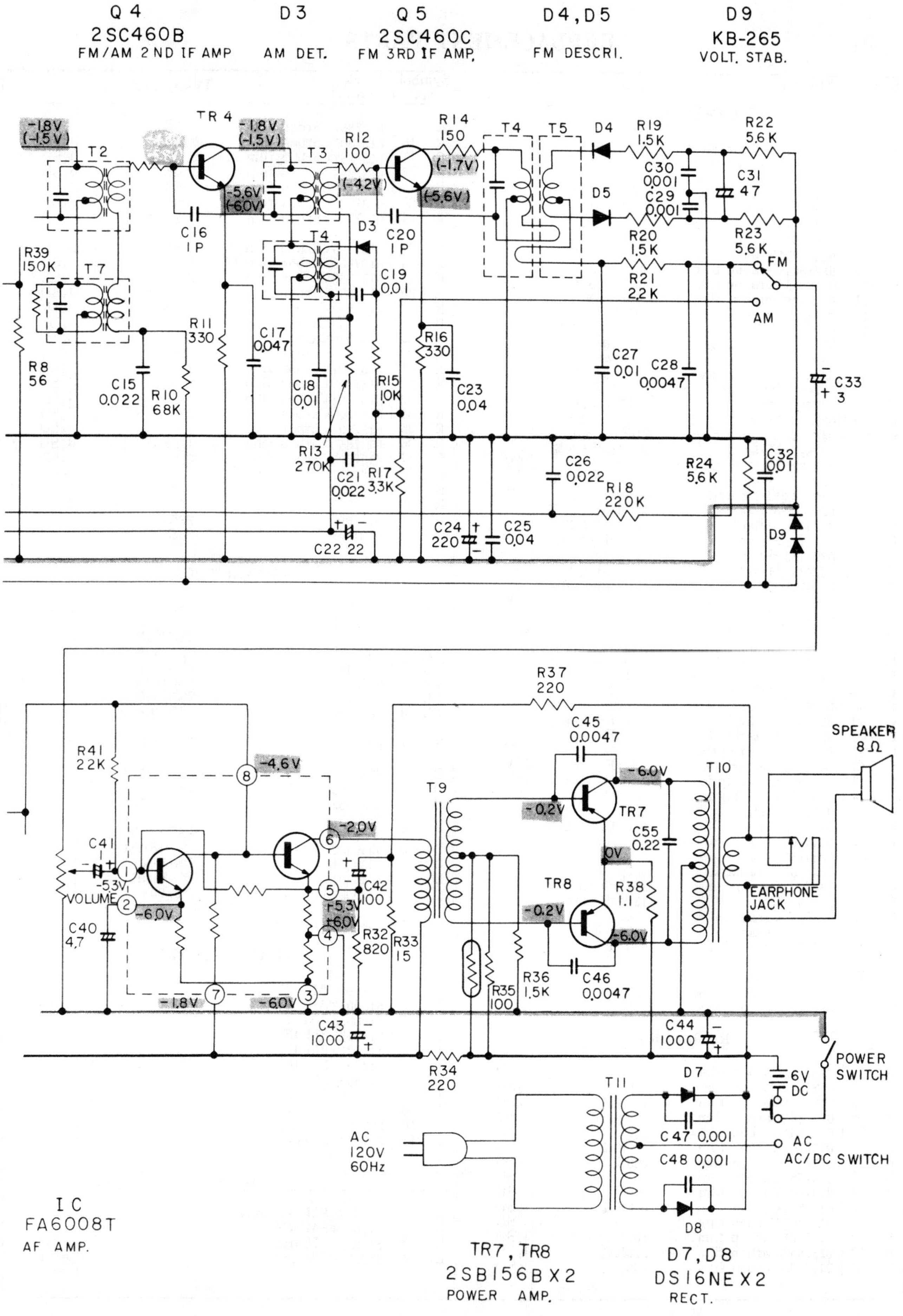
Q 4
2SC460B
FM/AM 2ND IF AMP
D 3
AM DET.
Q 5
2SC460C
FM 3RD IF AMP.
D 4, D 5
FM DESCRI.
D 9
KB-265
VOLT. STAB.
-1.8V
(-1.5V)
TR 4
-1.8V
(-1.5V)
-5.6V
(-6.0V)
(-4.2V)
(-1.7V)
(-5.6V)
T2
T3
T4
T5
T7
R14
150
R12
100
D4
D5
R19
1.5K
R22
5.6K
C30
0.001
C29
0.001
C31
47
R20
1.5K
R23
5.6K
R21
2.2K
FM
AM
C16
1P
C20
1P
C19
0.01
R39
150K
R8
56
R11
330
C17
0.047
C15
0.022
R10
68K
C18
0.01
R15
10K
R16
330
C23
0.04
C27
0.01
C28
0.0047
C33
3
R13
270K
C21
0.022
R17
3.3K
C26
0.022
R18
220K
R24
5.6K
C32
0.01
C24
220
C25
0.04
C22 22
D9
R37
220
C45
0.0047
SPEAKER
8 Ω
R41
22K
-4.6V
-6.0V
T10
T9
-0.2V
TR7
-2.0V
C41
C55
0.22
0V
-5.3V
VOLUME
-6.0V
C42
100
TR8
R38
1.1
+5.3V
+6.0V
EARPHONE
JACK
C40
4.7
R32
820
R33
15
-0.2V
-6.0V
R35
100
R36
1.5K
C46
0.0047
-1.8V
-6.0V
C43
1000
C44
1000
R34
220
D7
POWER
SWITCH
6V
DC
T11
AC
120V
60Hz
C47 0.001
C48 0.001
AC
AC/DC SWITCH
D8
I C
FA6008T
AF AMP.
TR7, TR8
2SB156B X 2
POWER AMP.
D7, D8
DS16NE X 2
RECT.

REPLACEMENT PARTS

Symbol No.	Stock No.	Description		
		CAPACITORS:		
C 1	0248667	Ceramic disc	20pF±5pF	
C 2	0248671	Ceramic disc	30pF±5%	
C 3	0244161	Ceramic disc	0.001μF	
C 4	0248647	Ceramic disc	7pF±0.5pF	
C 5	0249527	Ceramic disc	65pF±5%	
C 6	0248671	Ceramic disc	30pF±5%	
C 7	0244161	Ceramic disc	0.001μF	
C 8	0248695	Ceramic disc	300pF±5%	
C 9	0248644	Ceramic disc	4pF±0.5pF	
C 10	0248645	Ceramic disc	5pF±0.5pF	
C 11	0249528	Ceramic disc	95pF±5%	
C 12	0248645	Ceramic disc	5pF±0.5pF	
C 13	0275115	Mylar	0.047μF±20%	
C 14	0248641	Ceramic disc	1pF±0.5pF	
C 15	0275113	Mylar	0.022μF±20%	
C 16	0248641	Ceramic disc	1pF±0.5pF	
C 17	0275115	Mylar	0.047μF±20%	
C 18	0275111	Mylar	0.01μF±20%	
C 19	0275111	Mylar	0.01μF±20%	
C 20	0448641	Ceramic disc	1pF±0.5pF	
C 21	0275113	Mylar	0.022μF±20%	
C 22	0252322	Electrolytic	22μF	10V
C 23	0275115	Mylar	0.047μF±20%	
C 24	0252332	Electrolytic	220μF	10WV
C 25	0275115	Mylar	0.047μF±20%	
C 26	0275113	Mylar	0.022μF±20%	
C 27	0275111	Mylar	0.01μF±20%	
C 28	0274115	Mylar	0.0047μF±20%	
C 29	0274111	Mylar	0.001μF±20%	
C 30	0274111	Mylar	0.001μF±20%	
C 31	0252615	Electrolytic	47μF	25V
C 32	0275114	Mylar	0.01μF±20%	
C 33	0252613	Electrolytic	3.3μF	25V
C 34	0275111	Mylar	0.01μF±20%	
C 35	0274115	Mylar	0.0047μF±20%	
C 36	0248643	Ceramic disc	3pF±0.5pF	
C 37	0275113	Mylar	0.022μF±20%	
C 38	0275113	Mylar	0.022μF±20%	
C 39	0276111	Mylar	0.1μF±20%	
C 40	0252615	Electrolytic	4.7μF	
C 41	0252611	Electrolytic	1μF	25V
C 42	0252331	Electorlytic	100μF	10V
C 43	0256003	Electorlytic	1000μF	10V
C 44	0256003	Electorlytic	1000μF	10V
C 45	0274115	Mylar	0.0047μF±20%	
C 46	0274115	Ceramic disc	0.0047μF±20%	
C 47	0244161	Ceramic disc	0.001μF	
C 48	0244161	Ceramic pisc	0.001μF	
C 49	0244161	Ceramic disc	0.001μF	
C 50	0246423	Ceramic disc	3pF±0.5pF	
C 51	0221339	Styrol	1500pF±5%	
C 52	0246430	Ceramic disc	10pF±0.5pF	
C 53	0244175	Ceramic disc	0.047μF	
C 54	0244161	Ceramic disc	0.01μF	
C 55	0276013	Mylar	0.22μF	
C 56	0274111	Mylar	0.001μF	
C 57	0247896	Ceramic disc	120pF	
C 58	0247896	Ceramic disc	120pF	
		RESISTORS:		
R 1	0137812	Carbon film	820Ω	¼SD
R 2	0137861	Carbon film	6.8KΩ	¼SD
R 3	0137851	Carbon film	1.0KΩ	¼SD
R 4	0137861	Carbon film	6.8KΩ	¼SD
R 5	0137805	Carbon film	220Ω	¼SD
R 6	0137951	Carbon film	100KΩ	¼SD
R 7	0137805	Carbon film	220Ω	¼DS
R 8	0137768	Carbon film	56Ω	¼SD
R 9	0137805	Carbon film	220Ω	¼SD
R 10	0137861	Carbon film	6.8KΩ	¼SD
R 11	0137807	Carbon film	330Ω	¼SD
R 12	0137801	Carbon film	100Ω	¼SD
R 13	0137956	Carbon film	270KΩ	¼SD
R 14	0137803	Carbon film	150Ω	¼SD
R 15	0137851	Carbon film	1.0KΩ	¼SD
R 16	0137807	Carbon film	330Ω	¼SD
R 17	0137857	Carbon film	3.3KΩ	¼SD
R 18	0137955	Carbon film	220KΩ	¼SD
R 19	0137853	Carbon film	1.5KΩ	¼SD
R 20	0137853	Carbon film	220KΩ	¼SD
R 21	0137855	Carbon film	2.2KΩ	¼SD
R 23	0137860	Carbon film	5.6KΩ	¼SD
R 24	0137860	Carbon film	5.6KΩ	¼SD
R 25	0137955	Carbon film	220KΩ	¼SD
R 26	0137853	Carbon film	1.5KΩ	¼SD
R 27	0137852	Carbon film	1.2KΩ	¼SD
R 28	0137853	Carbon film	1.5KΩ	¼DS
R 29	0137860	Carbon film	5.6KΩ	¼SD
R 30	0137903	Carbon film	15KΩ	¼SD
R 31	0137901	Carbon film	10KΩ	¼DS
R 32	0137812	Carbon film	820Ω	¼SD
R 33	0137761	Carbon film	15Ω	¼SD
R 34	0137805	Carbon film	220Ω	¼SD
R 35	0137801	Carbon film	100Ω	¼SD
R 36	0137853	Carbon film	1.5KΩ	¼SD
R 37	0137805	Carbon film	220Ω	¼SD
R 38	0183801	Carbon film	1.1Ω	½GF
R 39	0137953		150KΩ	¼SD
R 40	0137765		33Ω	¼SD
R 41	0137905		22KΩ	¼SD
R 42	0137769		68Ω	¼SD
		TRANSISTORS:		
TR 1	0573510		2SC535B	
TR 2	0573510		2SC535B	
TR 3	0573486		2SC460B	
TR 4	0573486		2SC460B	
TR 5	0573487		2SC460C	
TR 6	0573485		2SC460A	
TR 7	0573002		2SB156B	
TR 8	0673002		2SB156B	
IC 1	5353031		FA-6008T	
D 1	5330332		1S-188AM	
D 2	5340082		1S-553	
D 3	5330332		1S-188AM	
D 4	5330331		1S-188FM	
D 5	5330331		1S-188FM	
D 6	5330332		1S-188AM	
D 7	5330301		DS-16NE	
D 8	5330301		DS-16NE	
D 9	5340051	Varistor	KB-265	
TH 1	5340061	Thermister		
		TRANSFORMERS:		
T 1	5130034	AM IF		
T 2	5130043	AM IF		
T 3	5130033	AM IF		
T 4	5140031	FM IF		
T 5	5140019	FM IF		
T 6	5148051	FM IF		
T 7	5148052	FM IF		
T 8	0324003	IF trap		
T 9	5230061	In put		
T 10	5250071	Out put		
T 11	5211023	Power		
		COILS:		
L 1	5126233	FM Antenna		
L 2	5126234	FM RF		
L 3	5126235	FM osc		
L 4	5120275	AM osc		

TO ALIGN THE SET (Refer to Fig. A

Power Source :	DC 6V
Output Measuring :	Speaker Terminal (Impedance=8Ω)
Tone Control :	"HIGH" position
Squelch Control :	"0" position
AFC Switch :	"OFF" position

AM IF& RF Alignment

Volume Control :	Maximum
Input Signal :	Modulation 400Hz modulated to 30%

Step	Frequency Band	Input Signal		Place to be aligned	Set the V. Capacitor to
		Frequency	Given to		
1	AM	455kHz	Loop Antenna	L32, 33, 34	Minimum
2		Repeat the Step 1, and adjust for no further improvement.			
3	AM	520kHz	Loop Antenna	L21	Maximum
4		1650kHz		C74	Minimum
5		Repeat the Step 3 & 4, and adjust for maximum output.			
6		600kHz	Loop Antenna	L19	600kHz Signal
7		1400kHz		C68	1400kHz Signal
8		Repeat the Step 6 & 7, and adjust for no further improvement.			
9	SW	5.9MHz	Loop Antenna	L21	Maximum
10		18.6MHz		C77	Minimum
11		Repeat the Step 9 & 10, and adjust for maximum output.			
12		6MHz	Loop Antenna	L20	6MHz Signal
13		18MHz		C67	18MHz Signal
14		Repeat the Step 12 & 13, and adjust for no further improvement.			

Note : Keep the output level always on about 50mW (0.6V/8Ω) when aligning.

PARTS ARRANGEMENT FOR ALIGNMENT

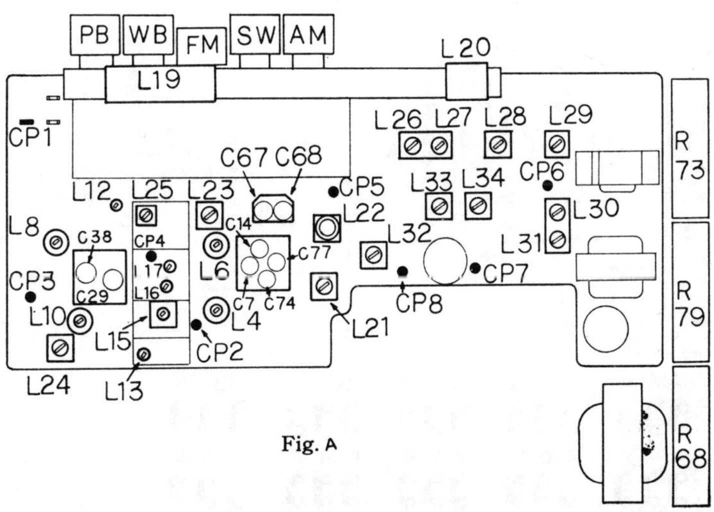

Fig. A

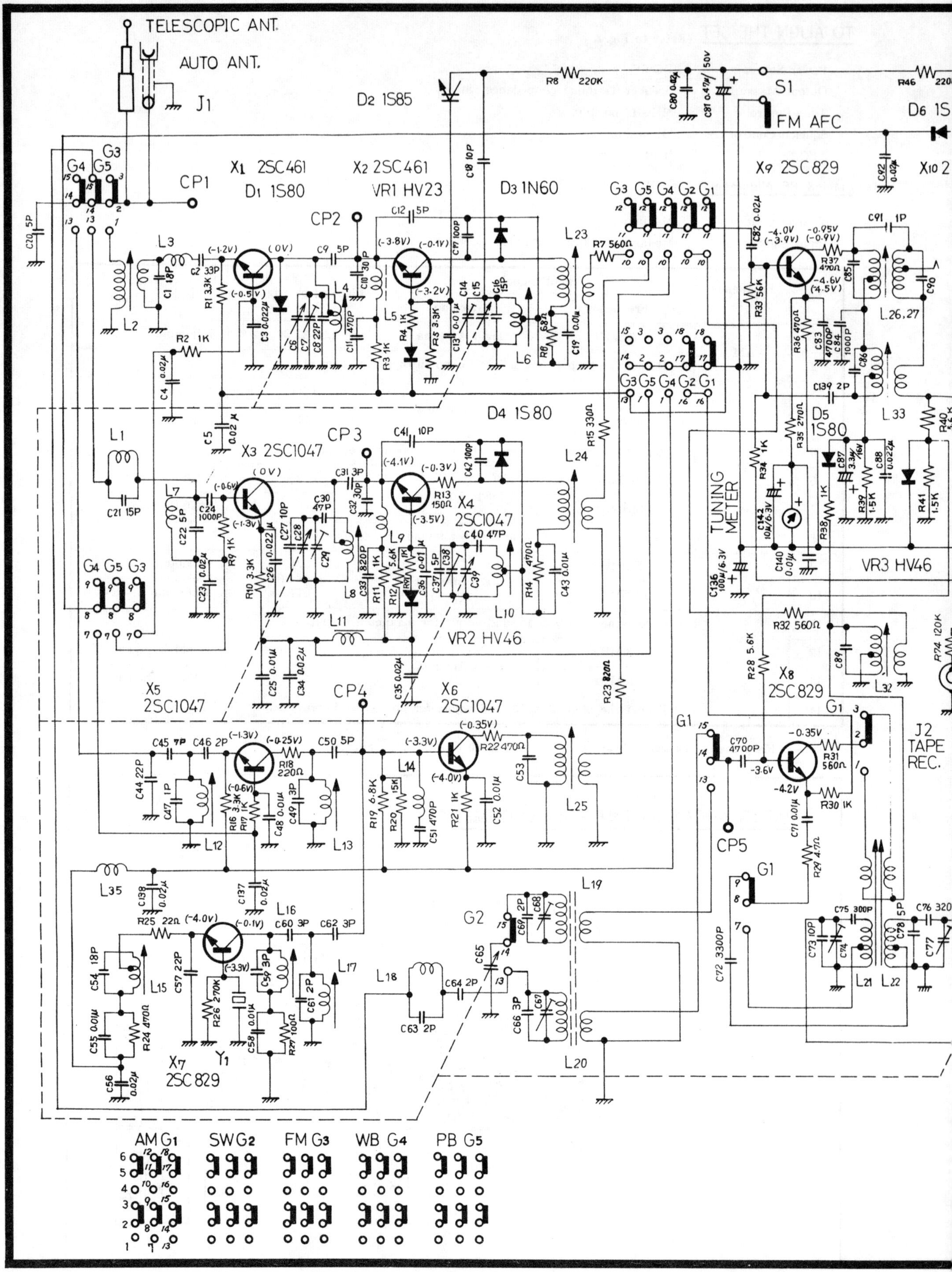
TELESCOPIC ANT.
AUTO ANT.
J1
D2 1S85
R8 220K
S1
FM AFC
R46 220K
D6 1S
CP1
X1 2SC461
D1 1S80
X2 2SC461
VR1 HV23
D3 1N60
X9 2SC829
X10 2
CP2
L3
L2
L4
L5
L6
L23
L26,27
L1
CP3
X3 2SC1047
X4 2SC1047
D4 1S80
L24
D5 1S80
TUNING METER
L33
VR3 HV46
L7
L8
L9
L10
L11
VR2 HV46
R32 560Ω
L32
X5 2SC1047
CP4
X6 2SC1047
X8 2SC829
J2 TAPE REC.
L12
L13
L14
L25
CP5
L35
L16
L15
L17
L18
L19
L20
L21
L22
X7 2SC829
Y1
AM G1
SW G2
FM G3
WB G4
PB G5

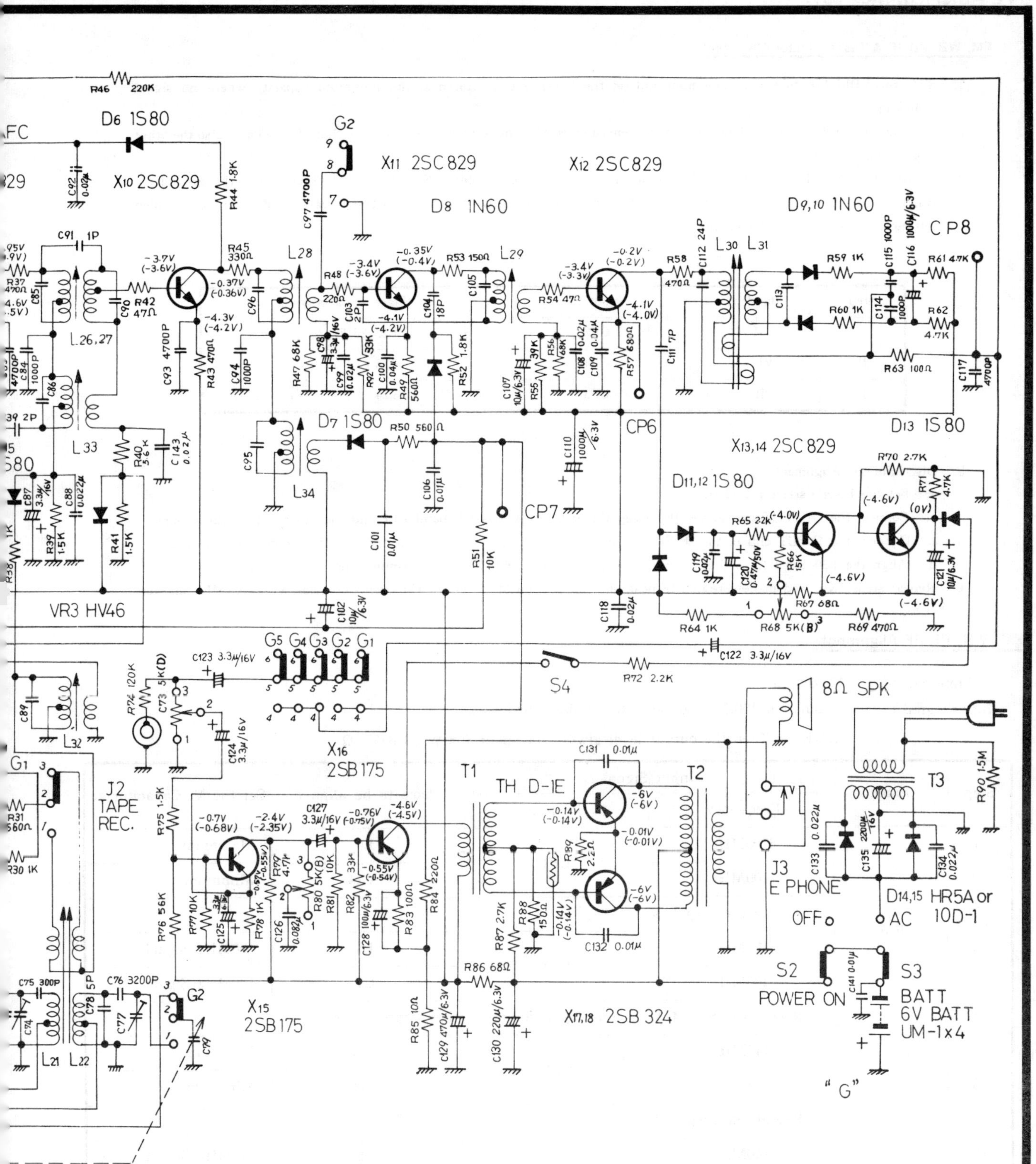

NOTE :

1. Band switches (G1~5) show "OFF" position.
2. Voltages are measured with V. T. V. M. from "G", battery should hold within ±20% with new batteries, volume control at minimum & no signal.
3. Voltage values in parenthesis are taken with G3 in FM position.

FM, WB, PB IF & Discriminator Alignment

1. Slide the VOLUME control to minimum and set the Variable Capacitor near the minimum capacity where no signal comes in.
2. Connect the OUTPUT clip of the IF Sweep Generator to the check point (refer to the chart below) and also the other clip to the ground.
3. Connect the INPUT clip of the IF Sweep Generator to the check point CP6 and also the other clip to the ground.
4. Align as following the chart below so that the maximum sensitivity and symmetrical wave mode will be obtained setting the marker 10.7MHz on the peak.

Step	Band	Input to	Place to be aligned
1	FM	CP2	L23, 26, 27, 28, 29
2	WB	CP4	L25
3	PB	CP3	L24

5. Discriminator Alignment
   a. Set the BAND selector to FM.
   b. Connect the OUTPUT clip of the IF Sweep Generator to the check point CP2 and the other clip to the ground.
   c. Connect the INPUT clip of the IF Sweep Generator to the check point CP8 and the other clip to the ground.
   d. Align the L30 and L31 so that the response of "S" curve will become maximum and symmetry.
6. Repeat steps 4&5, and make sure that IF peak and center point of "S" curve will exactly coincide with 10.7MHz marker.

FM, WB, PB RF Alignment

Volume control : Maximum

Input signal : Modulation 400Hz modulated to 22.5kHz deviation.

NOTE : Keep the output level always on about 50mW (0.63V/8Ω)

Step	Band	Input Signal		Place to be aligned	Set the V. Capacitor to
		Frequency	Given to		
1	FM	87.5MHz	CP1	L6	Maximum
2		109MHz		C14	Minimum
3		Repeat the Step 1 & 2.			
4		90MHz	CP1	L4	90MHz Signal
5		106MHz		C7	106MHz Signal
6		Repeat the Step 4 & 5, and adjust for no further improvement.			
7	PB	145MHz	CP1	L10	Maximum
8		178MHz		C38	Minimum
9		Repeat the Step 7 & 8.			
10		146MHz	CP1	L8	146MHz Signal
11		176MHz		C29	176MHz Signal
12		Repeat the Step 10 & 11, and adjust for no further improvement.			
13	WB	162.55MHz	CP1	L15, 16, 17	—
14		Repeat the Step 13, and adjust for maximum output.			
15		162.55MHz	CP1	L12, 13	—
16		Repeat the Step 15, and adjust for no further improvement.			—

Replacement Parts

Parts No.	Parts Name	Description
Coils & Transformers		
V03047-9	Ant. Coil (FM)	L1
03161-16	〃	L2
V03047-16	〃	L3
V03041-23	RF Coil (FM)	L4
03226-1	Trap Coil	L5, 9, 11, 14
V03041-16	Osc Coil (FM)	L6
V03047-15	Ant Coil	L7, 18 (PB & SW)
V03041-24	RF Coil (PB)	L8
V03041-17	OSC Coil (PB)	L10
03226-15	Trap Coil	L12, 13, 16, 17
V03101-5	Osc Coil (WB)	L15
V03071-64	Bar Ant Ass'y	L19, 20 (AM & SW)
V03084-1	Osc Coil (AM)	L21
V03106-2	〃 (SW)	L22
V03046-10	IFT (FM 1st)	L23, 24
V03068-7	〃	L25, C53
V03046-16	〃 (FM 2nd)	L26, 27, C85, 90
V03046-17	〃 (FM 3rd)	L28, 29, C96, 105
V03046-18	〃 (FM Det)	L30, 31, C11
V03029-20	〃 (AM 1st)	L32, C89
V03029-21	〃 (AM 2nd)	L33, C86
V03029-22	〃 (AM 3rd)	L34, C95
03226-12	Trap Coil	L35
03128-10	Input Transformer	T1
Q03342-2	Output Transformer	T2
VP41A3N-60A	Power Transformer	T3
Semiconductors		
2SC461 (B)	Transistor	X1, 2
2SC1047 (B)	〃	X3~6
2SC829 (C)	〃	X7
2SC829 (B)	〃	X8~12
2SC829 (A)	〃	X13, 14
2SB175 (B)	〃	X15
2SB175 (A)	〃	X16
2SB324 (B)	〃	X17, 18
1S80	Germanium Diode	D1, 4~7, 11~13
1S85 (WT)	Variable Capacitor Diode	D2
1N60	Germanium Diode	D3, 8
1N60 (P)	〃	D9, 10 (Pair)
10D1 or HR5A	Silicon Diode	D14, 15
HV23 (Y)	Varistor	Vr1
HV46	〃	Vr2, 3
Capacitors		
Q03463-1	Variable Capacitor	C6, 7, 14, 15, 65, 74, 77, 79
Q03460-001	〃	C28, 29, 38, 39
Q03638-1	Trimmer Capacitor	C67, 68
Q00402-10CH	Ceramic Capacitor	C18
Q00402-300CH	〃	C75
Q04062-1000	〃	C114, 115
〃 -01	〃	C25, 48, 52, 55, 58, 140
〃 -02	〃	C4, 5, 23, 34, 35, 56, 80, 82, 88, 92, 108, 118 119, 138
〃 -04	〃	C100, 109

Parts No.	Parts Name	Description
※Q04305-	Ceramic Capacitor	C1, 2, 8, 9, 12, 16, 20~22, 27, 31, 37, 41 44~47, 49, 50, 54, 57, 59, 60~64, 66, 69, 73, 78, 91, 103, 104, 111, 139
※Q04335-	〃	C10, 17, 30, 32, 40, 42, 112
Q03264-332	Polystyrol Capacitor	C76
Q03266-471	〃	C11, 51
〃 -821	〃	C33
〃 -102	〃	C84, 94
Q03246-102	〃	C24
〃 -332	Mylar Capacitor	C72
〃 -472	〃	C70, 97
〃 -103	〃	C13, 19, 36, 43, 71, 101, 106, 131, 132
〃 -223	〃	C3, 26, 99, 133, 134
〃 -473	〃	C83, 93, 117
〃 -823	〃	C126
Q03104-10	Electrolytic Capacitor	C102, 107, 116, 121, 137
〃 -30	〃	C125
〃 -100	〃	C128, 136
〃 -200	〃	C130
〃 -500	〃	C129
〃 -1000	〃	C110
Q03106-2000	〃	C135
Q03108-3	〃	C87, 98, 122~124, 127
Q03112-0.5	〃	C81, 120
Resistors		
Q30303-002M	Slide Variable Resistor	R68, 80
〃 -003M	〃	R73
04091 -1.5M	Composition Resistor	R90
※Q04802-	Carbon Resistor	Other Resistors

Note : When you require capacitors or resistors marked ※, write the required capacitance or resistance value in addition to the right-hand of hyphen.

Parts No.	Parts Name	Description
Others		
Q02720-001S	Push Switch	G1~5 BAND SELECTOR
Q03756-2	Slide Switch	S1 AFC
Q03781-5	Push Switch	S2 POWER SWITCH
Q03746-2	Slide Switch	S3 AC-BATTERY SELECTOR
V42121-A	Switch Ass'y	S4
D1E (BK)	Thermistor	Th
A04063-57.750	Crystal	Y1
V41948-2	Rod Antenna	
V03020-9	Tuning Indicator	
EAS10P70S	Speaker	4″ 8Ω
V40136-B	Antenna Jack	J1 AUTO ANTENNA
41537-E	Jack Ass'y	J2 TAPE REC
Q03936-1	Earphone Jack Ass'y	J3
Q03056-14	Power Cord with Plug	
V40976-2	Power Cord Bushing	
A41096-1	Tab	
Q03893-001	Magnetic Earphone	

COMPLETE ALIGNMENT PROCEDURE

The receiver has been fully aligned at the factory before shipment to you and does not normally require further adjustment. When necessary, however, the receiver may be aligned by performing the alignment in the order stated below.

IMPORTANT

Only qualified technicians with proper test equipment should undertake realignment of this receiver or damage may occur.

1. Remove the top cover of the receiver by removing the five [5] Philips head screws located on the rear of the top cover.

2. Refer to Figure 2. PF-300 Component Location, for the interior layout of the receiver for location of all adjustment points. A Functional Block Diagram and Schematic are also provided.

   CAUTION: Most of the coils in this unit have been sealed with wax. Be sure to melt the wax before attempting any adjustment.

VHF RECEIVER ALIGNMENT

ALIGNMENT OF VHF LOW BAND CRYSTAL OSCILLATOR [TR-3]

1. Insert a low band crystal into the X1 position in the low band section of the receiver, preferably one for operation in the middle of the band. Set the SELECTOR switch to X1 position for VHF LO band.

   NOTE: Alignment of this crystal oscillator is not necessary if the unit is to be operated in the VHF manual tuning mode only. Although, this alignment is necessary if UHF crystal control function is required.

2. Connect a VTVM between the emitter of TR-3 and ground [Use a DC range on the VTVM].

3. Turn the core of L7 counter-clockwise to the fully extended position [minimum inductance].

4. Turn the core of L7 clockwise until the oscillator begins oscillating. This can be confirmed by a reading on the VTVM.

5. Turn the core one turn clockwise past the point where oscillation began. The VTVM should indicate approximately 0.9 to 1.2 volts. Switch between VHF LO [tunable] position and X1 [crystal controlled] position and check that oscillation always takes place in the X1 position.

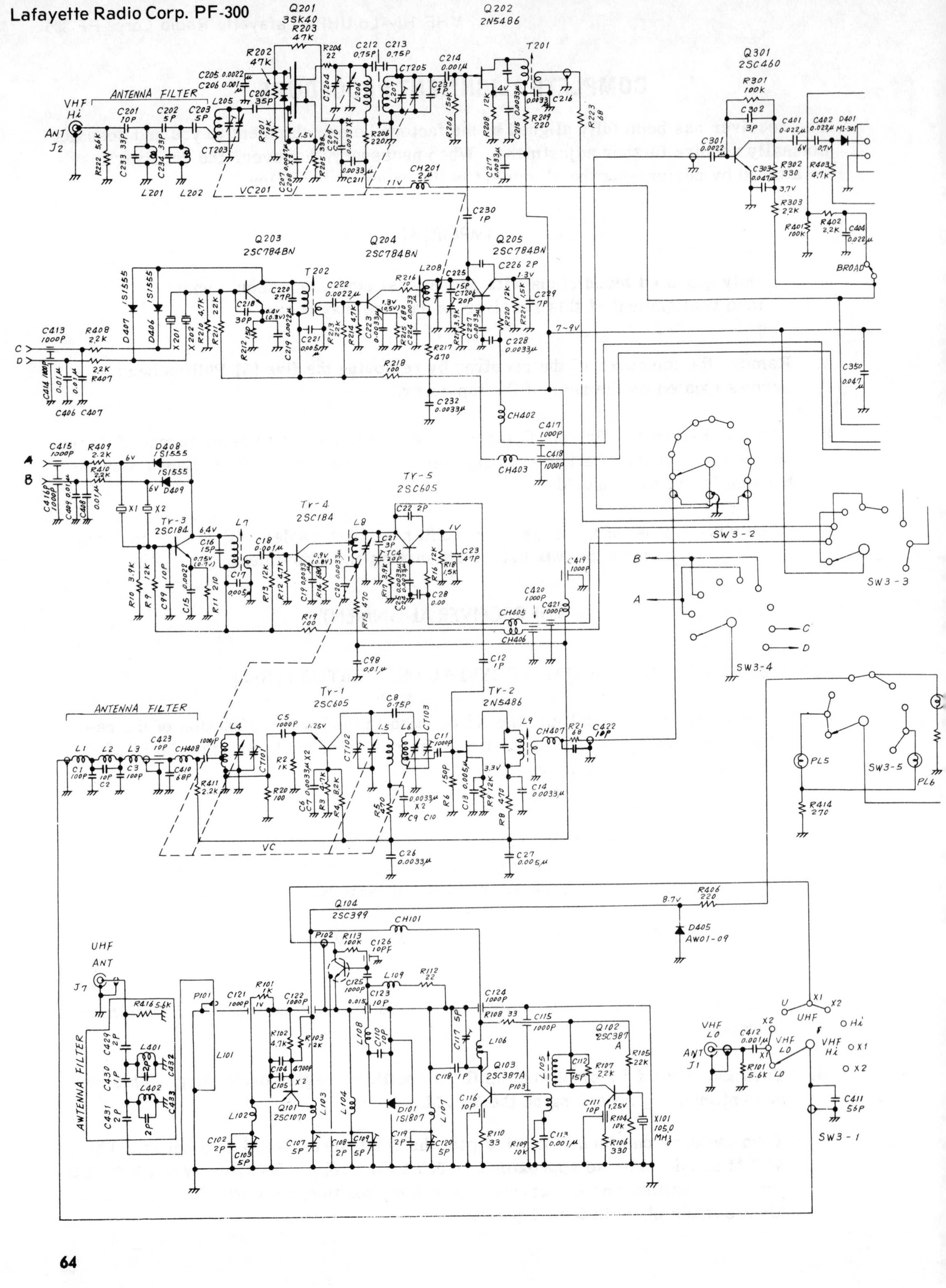
Q201
3SK40
Q202
2N5486
Q301
2SC460
VHF Hi
ANT J2
ANTENNA FILTER
T201
VC201
Q203
2SC784BN
Q204
2SC784BN
Q205
2SC784BN
T202
CH201
CH402
CH403
BROAD
D407 1S1555
D406 1S1555
D408 1S1555
D409
Tr-3
2SC184
Tr-4
2SC184
Tr-5
2SC605
SW3-2
SW3-3
SW3-4
SW3-5
PL5
PL6
CH405
CH406
ANTENNA FILTER
Tr-1
2SC605
Tr-2
2N5486
VC
CH407
CH408
Q104
2SC399
CH101
D405
AW01-09
UHF
ANT
J7
AWTENNA FILTER
P101
L101
Q101
2SC1070
D101
1S1807
Q103
2SC387A
Q102
2SC387A
X101
105.0 MHz
VHF LO
ANT J1
VHF LO
VHF Hi
UHF
SW3-1

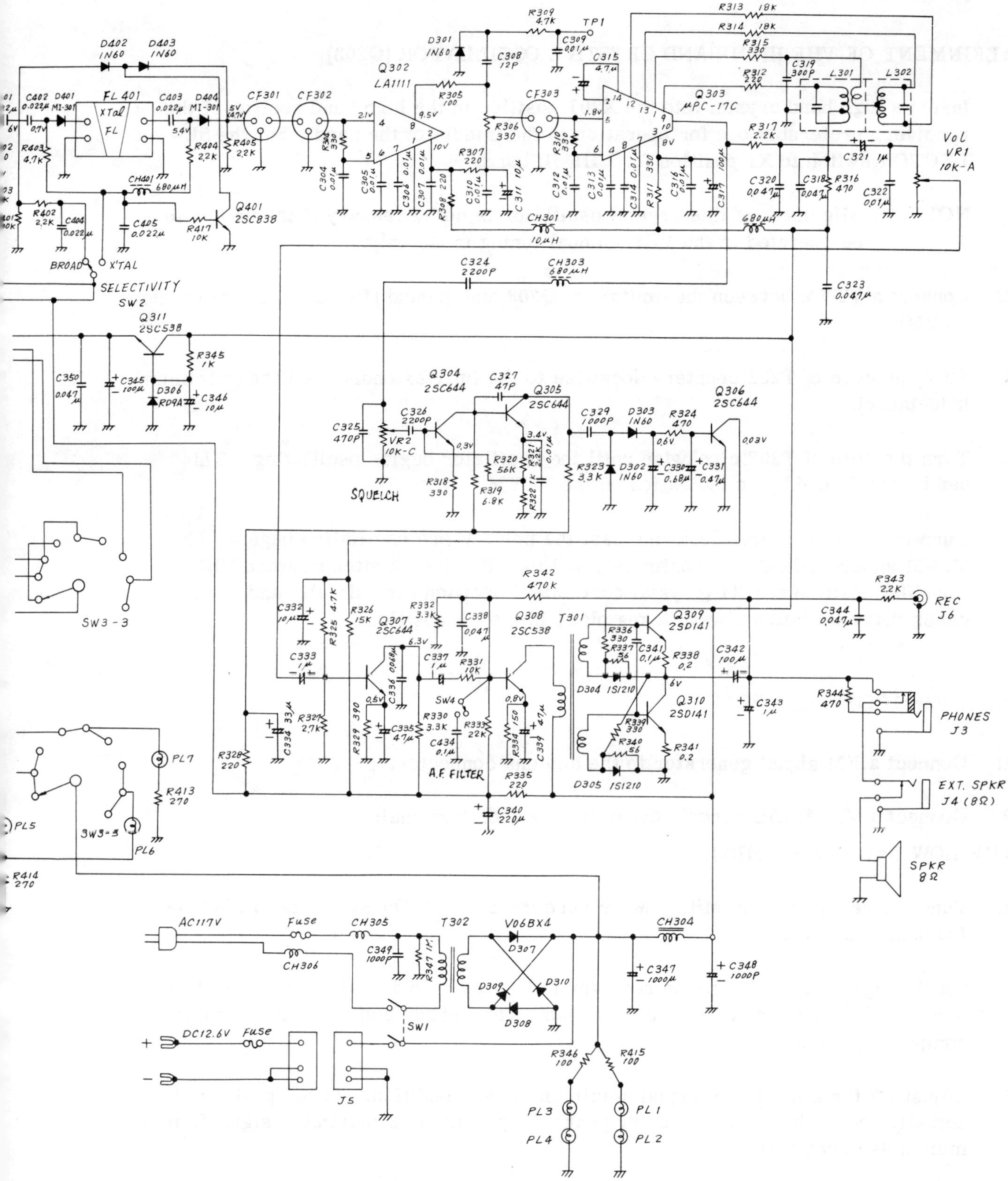
R313 18K
R314 18K
R309 4.7K
TP1
D301 1N60
C309 0.01μ
R315 330
C315 4.7μ
C308 12P
D402 1N60
D403 1N60
Q302 LA1111
R312 220
C319 300P
L301
L302
C402 0.022μ
D401 MI-301
FL 401 XTal FL
C403 0.022μ
D404 MI-301
CF301
CF302
R305 100
CF303
Q303 μPC-17C
R306 330
R317 2.2K
Vol VR1 10K-A
R403 4.7K
R405 2.2K
R404 2.2K
R307 220
R310 330
C321 1μ
C320 0.047μ
C318 0.047μ
R316 470
C322 0.01μ
CH401 680μH
C304 0.01μ
C305 0.01μ
C306 0.01μ
C307 0.01μ
R308 220
C310 0.01μ
C311 10μ
C312 0.01μ
C313 0.01μ
C314 0.01μ
R311 330
C316 0.01μ
C317 100μ
R402 2.2K
C404 0.022μ
C405 0.022μ
Q401 2SC838
R417 10K
CH301 10μH
680μH
BROAD
X'TAL
SELECTIVITY SW2
C324 2200P
CH303 680μH
C323 0.047μ
Q311 2SC538
R345 1K
C350 0.047μ
C345 100μ
D306 RD9A
C346 10μ
Q304 2SC644
C327 47P
Q305 2SC644
Q306 2SC644
C326 2200P
C325 470P
VR2 10K-C
C329 1000P
D303 1N60
R324 470
R320 56K
R321 2.2K
R323 3.3K
D302 1N60
C330 0.68μ
C331 0.47μ
SQUELCH
R318 330
R319 6.8K
R322 1K
SW3-3
R342 470K
R343 2.2K
REC J6
C344 0.047μ
C332 10μ
R325 4.7K
R326 15K
R332 3.3K
C338 0.047μ
Q307 2SC644
Q308 2SC538
T301
Q309 2SD141
R336 330
R337 56
C341 0.1μ
R338 0.2
C342 100μ
C333 1μ
C336 0.068μ
C337 1μ
R331 10K
SW4
D304 1S1210
Q310 2SD141
C343 1μ
R344 470
PHONES J3
C334 33μ
R327 2.7K
R329 390
R330 3.3K
C335 4.7μ
R333 22K
C434 0.1μ
R334 150
C339 47μ
R339 330
R340 56
R341 0.2
R328 220
A.F. FILTER
R335 220
D305 1S1210
EXT. SPKR J4 (8Ω)
C340 220μ
PL7
R413 270
PL5
SW3-3
PL6
SPKR 8Ω
R414 270
AC117V
Fuse
CH305
T302
V06BX4
CH304
CH306
C349 1000P
R347 1K
D307
C347 1000μ
C348 1000P
D309
D310
D308
SW1
DC12.6V
Fuse
R346 100
R415 100
J5
PL3
PL1
PL4
PL2

ALIGNMENT OF VHF HIGH BAND CRYSTA L OSCILLATOR [Q203]

1. Insert a high band crystal into the X201 position in the high band section of the receiver, preferably one for operation in the middle of the band. Set the SELECTOR switch to X1 position for VHF HI band.

   NOTE: Alignment of this crystal oscillator is not necessary if the unit is to be operated in the VHF manual tuning mode only.

2. Connect a VTVM between the emitter of Q203 and ground [Use a DC range on the VTVM].

3. Turn the core of T202 counter-clockwise to the fully extended position [minimum inductance].

4. Turn the core of T202 clockwise until the oscillator begins oscillating. This can be confirmed by an indication on the VTVM.

5. Turn the core one turn clockwise past the point where oscillation began. The VTVM should indicate approximately 0.9 to 1.2 volts. Switch between VHF HI [tunable] position and X1 [crystal controlled] position for VHF HI band, and check that oscillation always takes place in the X1 position.

ALIGNMENT OF RF SECTION

1. Connect a FM signal generator to the antenna connector.

2. Connect a VTVM [AC range] across the speaker terminals.

VHF LOW BAND [30-50 MHz]

3. Tune the receiver to 30 MHz. Make sure the SELECTOR switch is in the VHF LO position.

4. Set the signal generator to the receiver frequency with a modulated output level low enough [approximately 1-3uV] to enable you to hear both the signal and accompanying noise.

5. Adjust L8 for a maximum signal/minimum noise condition. Three peaks will actually occur, but only the center peak will provide this maximum signal/minimum noise condition.

6. Adjust L4, L5 and L6 for maximum indication on the VTVM.

7. Reduce the generator output level if necessary, and adjust L9 for a maximum signal/minimum noise condition.

8. Tune the receiver to 50 MHz and set the generator to the same frequency.

9. Adjust TC-4 for a maximum signal/minimum noise condition [as before, the middle peak is the correct one].

10. Adjust CT-101, CT-102 and CT-103 for a maximum signal/minimum noise condition. Reduce generator output as low as is necessary to hear both signal and noise.

VHF HIGH BAND [144-174 MHz]

11. Tune the receiver to 144 MHz. Make sure the SELECTOR switch is in the VHF HI position.

12. Set the FM generator to the receiver frequency with a modulated output level low enough [approximately 1-3uV] to enable you to hear both the signal and accompanying noise.

13. Adjust L208 for a maximum signal/minimum noise condition. Three peaks will actually occur, but only the center peak will provide this maximum signal/minimum noise condition.

14. Adjust L205, L206 and L207 for maximum on the VTVM.

15. Tune the receiver to 174 MHz and set the generator to the same frequency.

16. Adjust CT-206 for a maximum signal/minimum noise condition [as before, the middle peak is the correct one].

17. Adjust CT-203, CT-204 and CT-205 for a maximum signal/minimum noise condition. Reduce the generator output as low as is necessary to hear both signal and noise.

UHF RECEIVER ALIGNMENT

ALIGNMENT OF CRYSTAL OSCILLATOR [TR-3]

Alignment of Crystal Oscillator [TR-3] for the UHF Band is not necessary since this alignment has already been performed [see "ALIGNMENT OF VHF LOW BAND CRYSTAL OSCILLATOR TR-3]".

ALIGNMENT OF RF SECTION

This is a factory aligned converter, and requires no alignment in the field.

IMPORTANT

Do not under any circumstance attempt to align the RF converter for the UHF band. If malfunction of converter occurs, send receiver to Lafayette Service Dept. [see "Returning the Unit For Service"]. Converter may be installed only by a qualified technician.

ALIGNMENT OF FM DISCRIMINATOR

1. Connect a VTVM across the speaker terminals [Use AC range]. Make sure the squelch control is at minimum.

2. The adjustment of L301 and L302 is carried out using noise as an indication.

   Turn out the cores of L301 and L302 [counter-clockwise rotation] to completely detune the coils. Turn up volume so that noise will provide an indication on the VTVM. As the noise increases, you will have to gradually reduce the volume.

   Slowly, rotate the core of L302 clockwise, observing the VTVM. At a certain point, the noise will begin to increase and you will reach the first peak. This will be followed by a slight reduction in noise, then you will reach a second peak. This second peak is the point of proper adjustment.

   Now, rotate the core of L301 clockwise. The noise will increase to a peak and then decline. Adjust to the peak [maximum noise indication].

MINIMUM DISTORTION ADJUSTMENT

If a distortion analyzer is available proceed as follows:

With a signal generator output level of 1uV, modulated by 1 KHz, deviate the frequency no higher than 15 KHz while readjusting L302 for maximum audio output with minimum distortion on the distortion analyzer.

ALIGNMENT INSTRUCTIONS

For indication of output level, connect 8 ohm resistor (or speaker) and probe of VTVM to earphone jack. Standard signal Generator should be applied 1000Hz 30% Modulation.
The loop antenna should be placed 24 inch of distance from antenna coil in the unit and connected to the signal generator, by a 50 ohm cable.

BAND	STEP	GENERATOR	DIAL SETTING	ADJUST	REMARKS
AM SECTION	1	455KHz	Tuning gang T-10F	T-5 T-8 T-10	Adjust for maximum output. Repeat until no furthe improvement can be made.
	2	525KHz	Tuning gang Fully closed	L-20	Adjust for maximum output.
	3	1650KHz	Tuning gang Fully open	TC-5	Adjust for maximum output. Repeat steps 2 and 3.
	4	600KHz	Tune to signal	L-16	Adjust coil on ferrite core for maximum output.
	5	1400KHz	Tune to signal	TC-1	Adjust for maximum output. Repeat steps 4 and 5.
MB SECTION	1	1.55MHz	Tuning gang Fully closed	L-21	Adjust for maximum output.
	2	4.7MHz	Tuning gang Fully open	TC-6	Adjust for maximum output. Repeat steps 1 and 2.
	3	1.6MHz	Tune to signal	L-17	Adjust coil on ferrite core for maximum output.
	4	4.7MHz	Tune to signal	TC-2	Adjust for maximum output. Repeat steps 3 and 4.
SW-1 SECTION	1	4.3MHz	Tuning gang Fully closed	L-21	Adjust for maximum output.
	2	12.5MHz	Tuning gang Fully open	TC-6	Adjust for maximum output. Repeat steps 1 and 2.
	3	4.5MHz	Tune to signal	L-18	Adjust coil on ferrite core for maximum output.
	4	12MHz	Tune to signal	TC-3	Adjust for maximum output. Repeat steps 3 and 4.
SW-2 SECTION	1	11MHz	Tuning gang Fully closed	L-23	Adjust for maximum output.
	2	23MHz	Tuning gang Fully open	TC-8	Adjust for maximum output. Repeat stops 1 and 2.
	3	11.5MHz	Tune to signal	L-19	Adjust coil on jerrile core for maximum output.
	4	22MHz	Tune to signal	TC-4	Adjust for maximum output. Repeat steps 3 and 4.

For alignment of FM IFT and RATIO DETECTOR. connect the FM IF (10.7MHz) SWEEP GENERATOR to test point A and B. OSCILLOSCOPE to test point C and D. Then, adjust 5pcs. of IFT's (T-1, T-6, T-7, T-9, T-13) so as to indicate maximum level on oscilloscope. And repeat until no further improvement can be made. After the above alignment, adujst FM IFT T-12 (pink) for symmetrical responce-centered 10.7MHz on oscilloscope. These IFT's are provided for 4-band of FM, PB-L, PB-H and AIR, so, if above alignment done on one of these 4 bands, it is not necessory to adjust on other 3 bands.

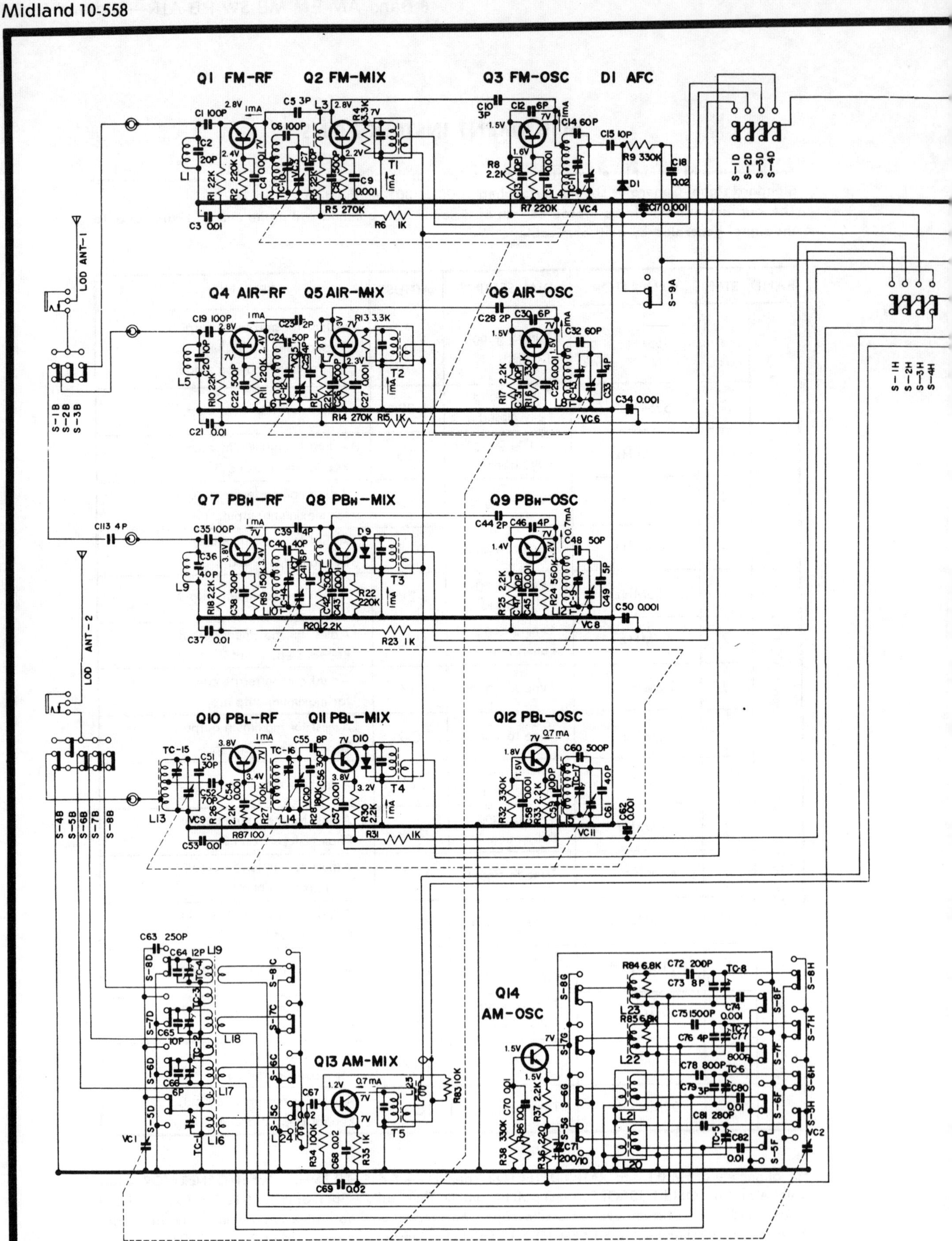
Q1 FM-RF
Q2 FM-MIX
Q3 FM-OSC
D1 AFC
Q4 AIR-RF
Q5 AIR-MIX
Q6 AIR-OSC
Q7 PBH-RF
Q8 PBH-MIX
Q9 PBH-OSC
Q10 PBL-RF
Q11 PBL-MIX
Q12 PBL-OSC
Q13 AM-MIX
Q14 AM-OSC
LOD ANT-1
LOD ANT-2
S-1B
S-2B
S-3B
S-4B
S-5B
S-6B
S-7B
S-8B
S-1D
S-2D
S-3D
S-4D
S-9A
S-1H
S-2H
S-3H
S-4H
R5 270K
R6 1K
R7 220K
R14 270K
R15 1K
R20 2.2K
R23 1K
R87 100
R31 1K
C113 4P
C63 250P
C64 12P
R83 10K
C69 0.02

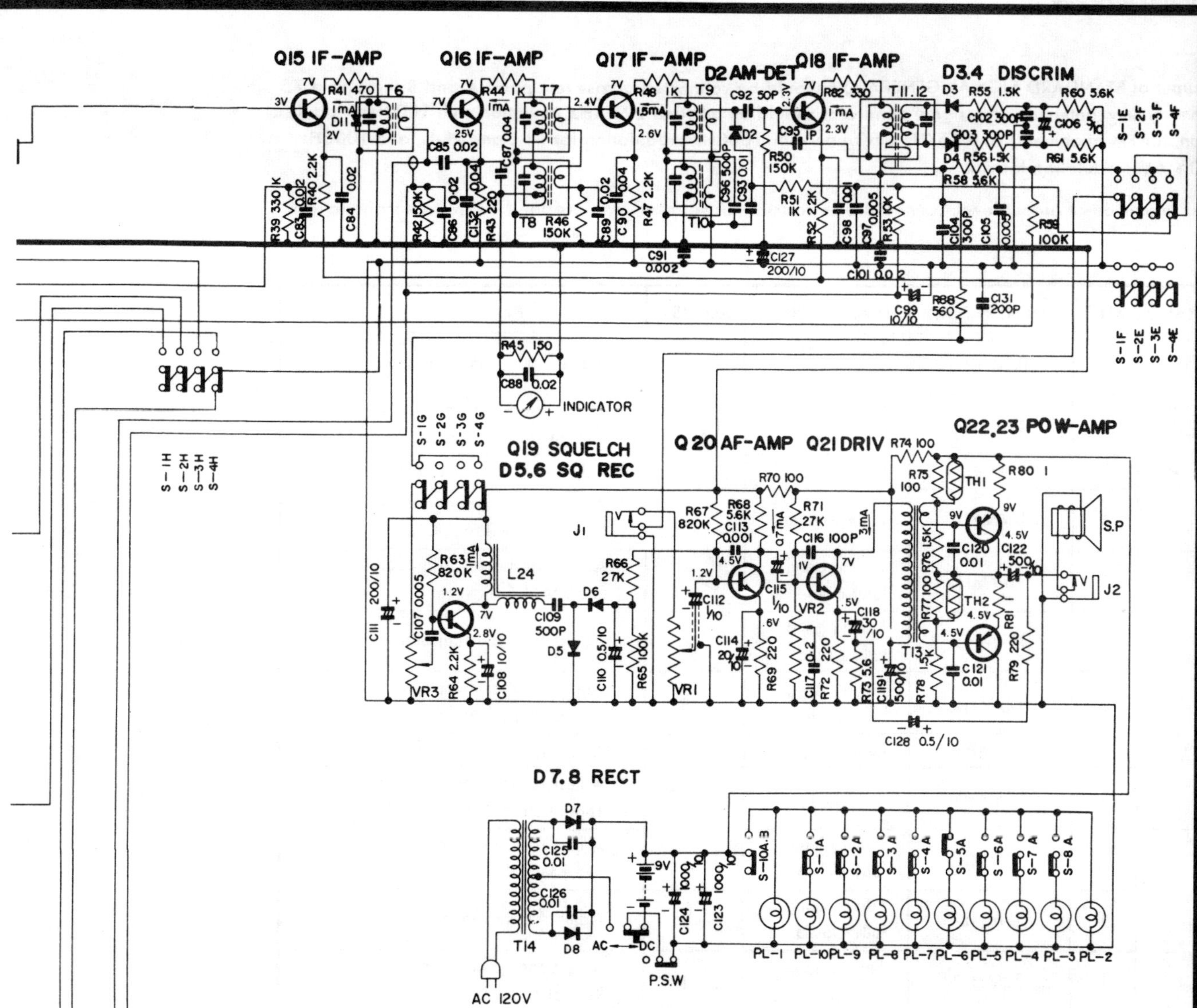

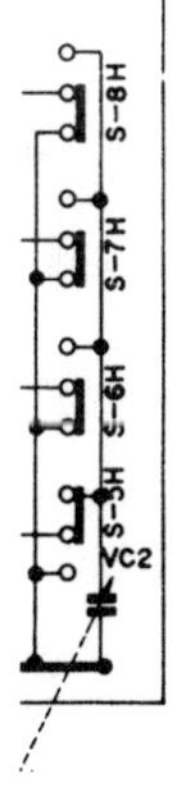

ALL RESISTANCE VALUES IN OHMS

ALL CAPACITANCE VALUES IN MICROFARADS

$K=10^3$ $P=10^{-6}$

(1) 9V BATTERY SIZE "D" UM-1

(2) CURRENT VALUES INDICATED ARE MEASURED AT NO SIGNAL.

(3) PUSH SWITCH IN AM

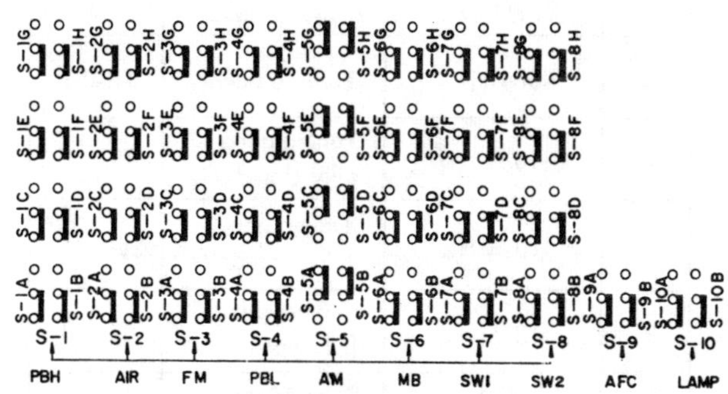

PUSH SWITCH TERMINAL LOCATION (BOTTOM VIEW)

Output of STANDARD SIGNAL GENERATOR should be connected across test point A and B through a 75 ohm dummy antenna. This signal generator should be applied 22.5KHz deviation with 1000Hz modulation, for FM alignment, and 15KHz deviation with 1000Hz modulation for PB-L and PB-H, and 1000Hz 30% AM modulation for AIR.

For indication of output level, connect 8 ohm resistor (or speaker) and probe of VTVM to earphone jack.

BAND	STEP	GENERATOR	DIAL SETTING	ADJUST	REMARKS
FM section	1	86MHz	Tuning gang fully closed		Adjust for maximum output.
	2	110MHz	Tuning gang fully open	TC-11	Adjust for maximum output. Repeat steps 1 and 2.
	3	90MHz	Tune to signal	L-2	Adjust for maximum output.
	4	106MHz	Tune to signal	TC-10	Adjust for maximum output. Repeat steps 3 and 4.
PB-H section	1	10.7MHz	Tuning gang fully closed	T-3	Adjust for maximum output. See FM IFT alignment.
	2	144MHz	Tuning gang fully closed	L-12	Adjust for maximum output.
	3	177MHz	Tuning gang fully open	TC-9	Adjust for maximum output. Repeat steps 2 and 3.
	4	145MHz	Tune to signal	L-10	Adjust for maximum output.
	5	175MHz	Tune to signal	TC-10	Adjust for maximum output. Repeat steps 4 and 5.
PB-L section	1	10.7MHz	Tuning gang fully closed	T-4	Adjust for maximum output. See FM IFT alignment.
	2	27MHz	Tuning gang fully closed	L-15	Adjust for maximum output
	3	52MHz	Tuning gang fully open	TC-17	Adjust for maximum output. Repeat steps 2 and 3.
	4	30MHz	Tune to signal	L-14	Adjust for miaxmum output.
	5	50MHz	Tune to signal	TC-16	Adjust for maximum output. Repeat steps 4 and 5.
AIR section	1	10.7MHz	Tuning gang fully closed	T-2	Adjust for maximum output. See FM IFT alignment.
	2	107MHz	Tuning gang fully closed	L-8	Adjust for maximum output.
	3	137MHz	Tuning gang fully open	TC-13	Adjust for maximum output. Repeat steps 2 and 3.
	4	108MHz	Tune to signal	L-6	Adjust for maximum output.
	5	135MHz	Tune to signal	TC-12	Adjust for maximum output.

PARTS LIST AND PRICES

COILS & TRANSFORMERS:

REF. NO.	DESCRIPTION	MIDLAND PART NUMBER	LIST PRICE	REMARKS
L24	Coil, Choke	10-178007	1.44	
T1,2,3,4	IFT-A, FM	10-090187	1.20	
T5	IFT-A, AM	10-090188	1.20	
T6,7	IFT-BC, FM	10-090189	1.20	
T8	IFT-B AM	10-090190	1.20	
T9	IFT-F FM	10-090191	1.20	
T10	IFT-C AM	10-090192	1.20	
T11	LFT-D FM	10-090193	1.20	
T12	IFT-E FM	10-090194	1.20	
R13	Transformer, Audio Input	10-096057	1.66	
T14	Transformer, AC Power	10-098013	4.70	

REF. NO.	DESCRIPTION	MIDLAND PART NUMBER	LIST PRICE	REMARKS
L1,5	Coil, VHF, AIR, ANT, FM ANT.	10-176042	.74	
L2	Coil, VHF, FM RF	10-176043	.74	
L3,7,11 24	Coil, IF Trap.	10-176044	.74	
L4	Coil, VHF Osc.	10-170128	.74	
L6	Coil, VHF, Air RF	10-176045	.74	
L8,12	Coil, VHF, Air, PB2 Osc.	10-170129	.74	
L9	Coil, VHF, PB2, Ant.	10-176046	.74	
L10	Coil, VHF, PB2, RF	10-176047	.74	
L13	Coil, VHF, PB1 Ant.	10-176048	1.20	
L14	Coil, VHF, PB1 RF	10-176049	1.20	
L15	Coil, VHF, PB1 Osc.	10-170130	1.20	
L16,17	Antenna Ferrite, AM	10-050067	2.64	
L18,19	Antenna, Ferrite SW1, SW2	10-050068	2.54	
L20	Coil, AM Osc.	10-170131	.96	
L21	Coil, MB Osc.	10-170132	.96	
L22	Coil, SW1 Osc.	10-170133	1.20	
L23	Coil, SW2 Osc.	10-170134	1.20	

RESISTORS: ALL RESISTORS ARE CARBON TYPE, ¼ WATT, REFER TO SCHEMATIC FOR VALUES.

TRANSISTORS:

REF. NO.	DESCRIPTION	MIDLAND PART NUMBER	LIST PRICE	REMARKS
Q1,2,4, 5,7,8,10, 11,13	2SC-668	U9-302009	2.55	
Q3,6,9 12,14,16 17,18	2SC-772	09-302010	2.55	
Q15,19, 20,21	2SC-537	09-302007	3.00	
Q22,23	2SB-405	09-301014	1.86	

MYLAR CAPACITORS:

REF. NO.	DESCRIPTION	MIDLAND PART NUMBER	LIST PRICE	REMARKS
C117	.02MFD, 50V	78-651203	.48	
C120,121	.01MFD, 50V	78-651103	.48	
C107	.005MFD, 50V.	78-651502	.48	

ELECTROLYTIC CAPACITORS:

REF. NO.	DESCRIPTION	MIDLAND PART NUMBER	LIST PRICE	REMARKS
C123,124	1000MFD, 10V	77-336108	.96	
C119,122	470MFD, 10V	77-336477	.96	
C71,111 127	220MFD, 10V	77-336227	.86	
C118	33MFD, 10V	77-336336	.76	
C114	22MFD, 10V	77-336226	.76	
C108,99	10MFD, 10V	77-336106	.76	
C106	4.7MFD, 10V	77-336475	.76	
C112,115	1MFD, 10V	77-336105	.76	
C110,128	.47 MFD, 10V	77-336474	.76	

POLYSTYRENE CAPACITORS:

REF. NO.	DESCRIPTION	MIDLAND PART NUMBER	LIST PRICE	REMARKS
C75	1500PF, 50V	78-751152	.46	
C77, 78	800PF, 50V	78-751801	.46	
C81	280PF, 50V	78-751281	.46	

CAPACITORS: ALL CAPACITORS NOT SHOWN ON THIS PARTS LIST ARE STANDARD DISC TYPE, 500 W.V. REFER TO SCHEMATIC. FOR SPECIFIC VALUES:

DIODES:

REF. NO.	DESCRIPTION	MIDLAND PART NUMBER	LIST PRICE	REMARKS
D1	IS-553	09-306014	1.40	
D2,3,4	IS-188 FM	09-306010	.80	
D5,6	36D32	09-306035	.68	
D7,8	DS-16E	09-306036	.80	
D9,10,11	IS-188 AM	09-306037	.80	

THERMISTORS:

REF. NO.	DESCRIPTION	MIDLAND PART NUMBER	LIST PRICE	REMARKS
TH1,2	M-60	09-307011	.92	

Midland 10-558

PARTS LIST AND PRICES

MISC.

REF. NO.	DESCRIPTION	MIDLAND PART NUMBER	LIST PRICE	REMARKS
	Antenna, Telescope (SW)	10-040035	4.90	
	Antenna, Telescope (FM)	10-040036	5.16	
	Spring, Tuning Knob	10-152017	.30	
	Spring, Case Lid	10-152019	.30	
VC7, 8,9,10,11	Hi/Lo Police Capacitor, Tuning	10-120052	5.92	
VC1,2,3, 4,5,6	AM/MB/SW1/SW2/FM /AIR Capacitor, Tuning	10-120053	7.32	
	10-Key (Front) Switch, Band Selector	10-183027	7.36	
	8-Key (Rear) Switch, Band Selector	10-183028	9.20	
	Switch, Power	10-183026	.96	
	Washer, Felt, Tuning Shaft	10-157065	.30	
	Spring, Fine Tuning Knob	10-152018	.30	
VR1	Control, Volume (Slide Type)	10-160060	2.54	
VR2	Control, Tone (Slide Type)	10-163011	2.54	
VR3	Control, Squelch	10-160061	1.10	

MISC.

REF. NO.	DESCRIPTION	MIDLAND PART PNUMBER	LIST PRICE	REMARKS
	Holder, Battery	10-030108	1.66	
TC1 thru 8	Capacitor, Trimmer	10-123008	.74	
TC9	Capacitor, Trimmer	10-123009	.74	
SP	Speaker, 4" × 6" 8ohm	10-060039	4.30	
	Meter, Battery/S	10-200006	5.46	
	Switch, AC-DC (Slide Type)	10-183029	.74	
	Cord, AC Power	10-034013	1.44	
	Clamp, AC Power Cord	10-157066	.30	
	Earphone	10-038019	.96	
PL1	Lamp, Dial Illumination (Right)	10-201004	.96	
PL2	Lamp, Dial Illumination (Left)	10-201005	.96	
PL3 thru 10	Lamp, Band Indicator	10-201006	.76	
	Plug, Ext. Ant.	10-159009	.74	
	Mount, Ferrite Antenna (Poly)	10-157067	.46	
	Bracket, Ferrite Antenna (Metal)	10-158093	.74	

ALIGNMENT

Equipment Required
1. RF Signal Generator
2. Electronic Voltmeter V.T.V.M.

General
1. Signal input must be as low as possible to avoid overload and clipping. (Use highest sensitivity of output indicator)
2. Standard modulation is 400 Hz at 30% amplitude.
3. Connect 8 ohm dummy load across voice coil.

NOTE: Perform alignment in following sequence.
1. AM band alignment.
2. MB band alignment.
3. PB2 band alignment.
4. PB1 band alignment.
5. AIR band alignment.
6. FM band alignment.
7. When trimmer (TC9) is realigned for the AM band, realignment of the MB and PB1 bands is also required as TC9 is common to the AM, MB and PB1 bands.
8. When trimmer (TC1) is realigned for the PB2 band, realignment of the PB1, AIR and FM bands is also required as TC1 is common to the PB1, PB2, AIR and FM bands.

Step	Connect Signal Source To-	Connect Output Indicator To-	Set Signal Generator To-	Set Radio Dial To-	Adjust -	Adjust For -	Step
			Set Function Switch to AM				
1	Loop of several turns of wire connected across gen. leads. Place loop close to the AM antenna.	A.C.V.T.V.M. across MUX OUT Jack	455 kHz	Tuning gang closed	T6, T7, T8	Maximum output on V.T.V.M.	1
2			520 kHz	Tuning gang closed	L19		2
3			1650 kHz	Tuning gang open	TC11		3
4	Repeat Steps 2 and 3 for optimum sensitivity						4
5	Same as above	A.C.V.T.V.M. across MUX OUT Jack	600 kHz	Tune for signal	L17	Maximum output	5
6			1400 kHz		TC9 (See Note 7)		6
7	Repeat Steps 5 and 6 for optimum sensitivity						7
			Set Function Switch to MB				
1	Same as above	A.C.V.T.V.M. across MUX OUT Jack	1.55 MHz	Tuning gang closed	L20	Maximum output on V.T.V.M.	1
2			3.55 MHz	Tuning gang open	TC12		2
3	Repeat Steps 1 and 2 for optimum sensitivity						3
4	Same as above	A.C.V.T.V.M. across MUX OUT Jack	1.8 MHz	Tune for signal	L18	Maximum output	4
5			3.2 MHz		TC10		5
6	Repeat Steps 4 and 5 for optimum sensitivity						6

			Set Function Switch to FM (IF Alignment)				
1	Place gen. leads across telescopic antenna	A.C.V.T.V.M. across voice coil	10.7 MHz	Tuning gang closed	T1, T2, T3, T4	Maximum output on V.T.V.M.	1
2		DC probe across MUX OUT Jack	10.7 MHz (Unmod.)		T5	Zero reading, a positive or negative reading will be obtained on either side of correct setting.	2
3	Repeat Steps 1 and 2 for optimum sensitivity						3

Montgomery Ward GEN-1463A

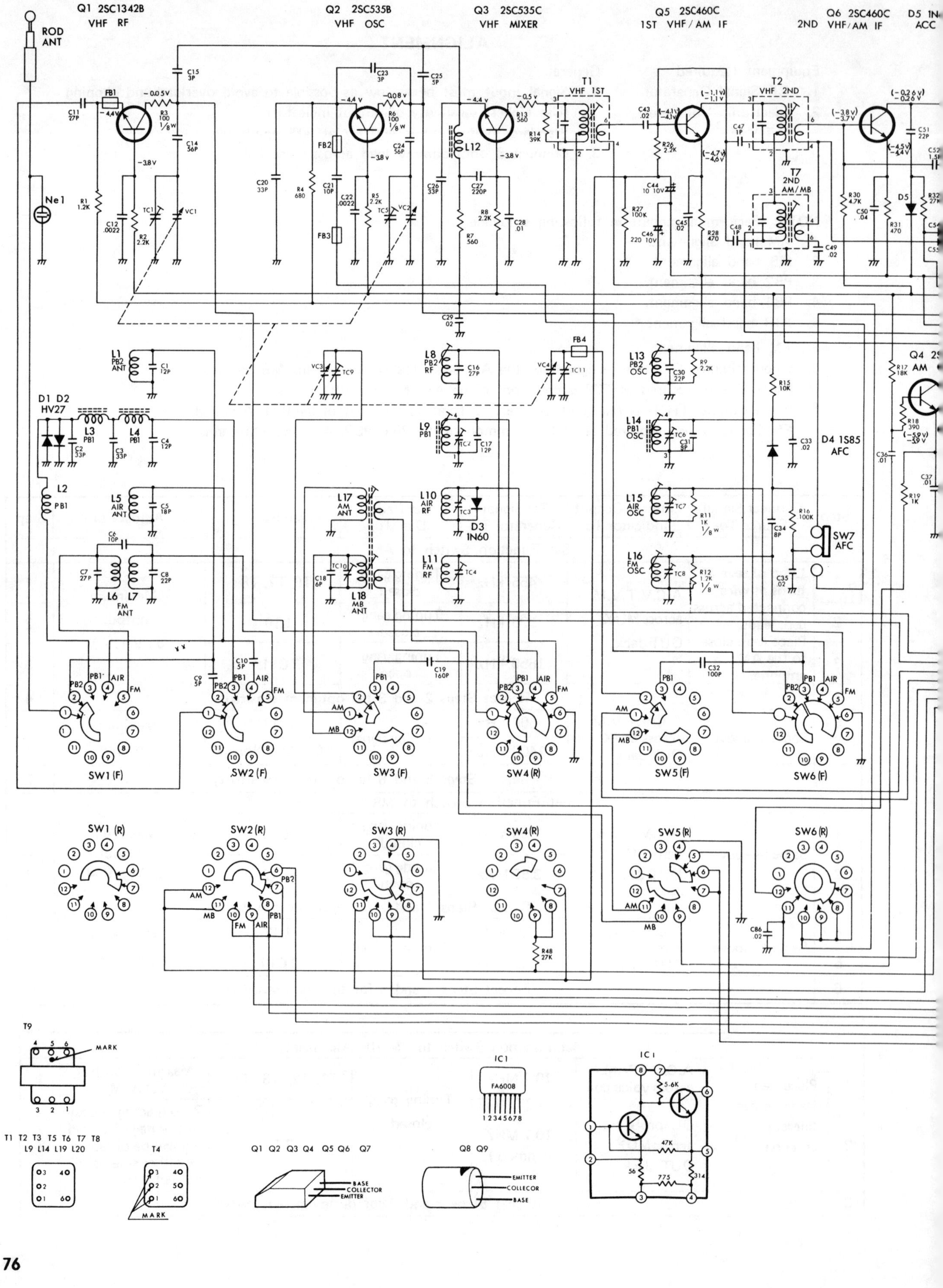

Montgomery Ward GEN-1463A

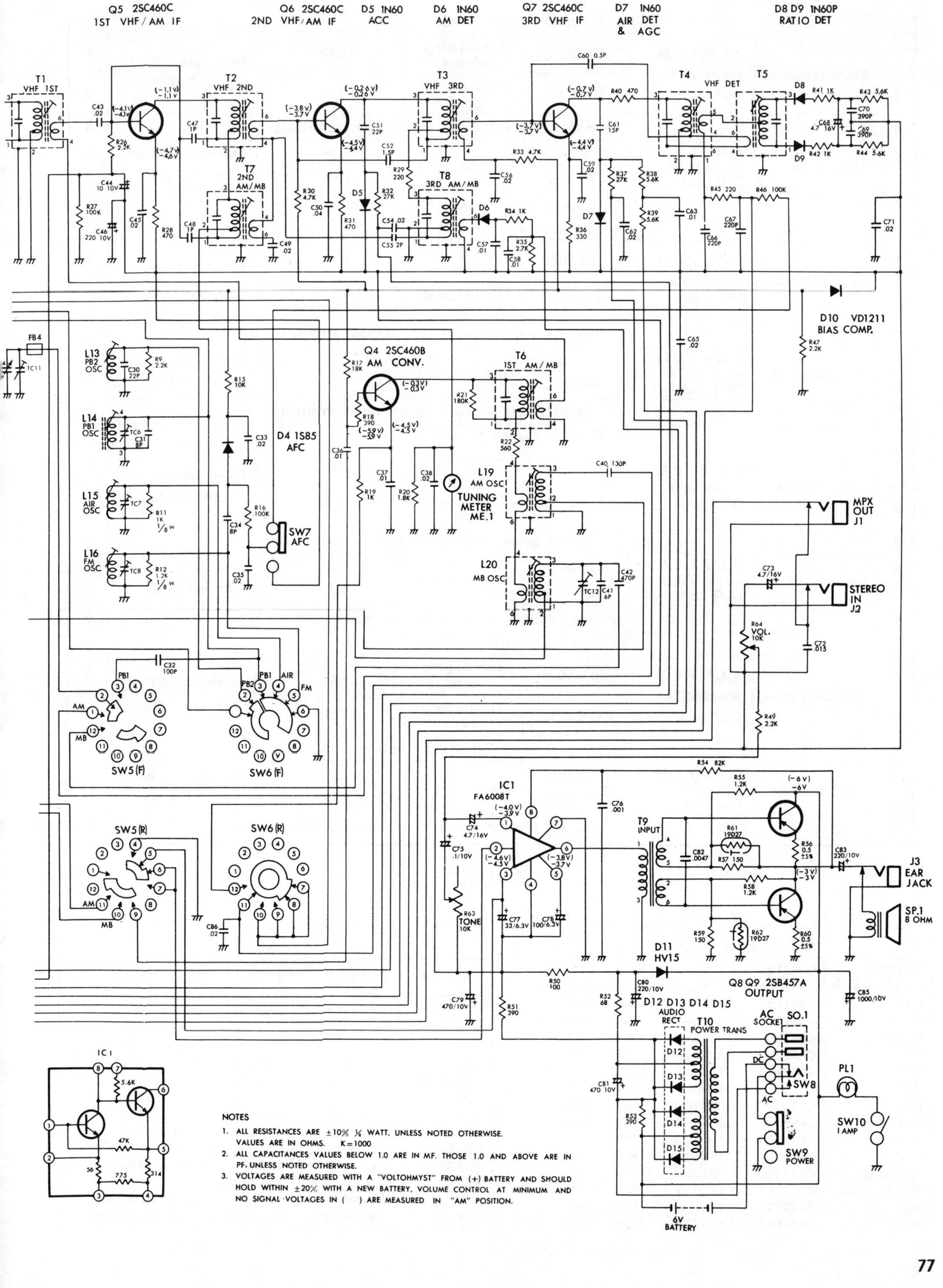

NOTES

1. ALL RESISTANCES ARE ±10% ¼ WATT, UNLESS NOTED OTHERWISE. VALUES ARE IN OHMS. K=1000
2. ALL CAPACITANCES VALUES BELOW 1.0 ARE IN MF. THOSE 1.0 AND ABOVE ARE IN PF. UNLESS NOTED OTHERWISE.
3. VOLTAGES ARE MEASURED WITH A "VOLTOHMYST" FROM (+) BATTERY AND SHOULD HOLD WITHIN ±20% WITH A NEW BATTERY, VOLUME CONTROL AT MINIMUM AND NO SIGNAL. VOLTAGES IN ( ) ARE MEASURED IN "AM" POSITION.

Set Function Switch to PB2							
1	Same as above	A.C.V.T.V.M. across MPX OUT Jack	145 MHz	Tuning gang closed	L13	Maximum output on V.T.V.M.	1
2			180 MHz	Tuning gang open	TC5		2
3	Repeat Steps 1 and 2 for optimum sensitivity						3
4	Same as above	A.C.V.T.V.M. across MPX OUT Jack	148 MHz	Tune for signal	L8	Maximum output on V.T.V.M.	4
5			175 MHz		TC1 (See Note 8)		5
6	Repeat Steps 4 and 5 for optimum sensitivity						6
Set Function Switch to PB1							
1	Same as above	A.C.V.T.V.M. across MPX OUT Jack	29 MHz	Tuning gang closed	L14	Maximum output on V.T.V.M.	1
2			51.5 MHz	Tuning gang open	TC6		2
3	Repeat Steps 1 and 2 for optimum sensitivity						3
4	Same as above	A.C.V.T.V.M. across MPX OUT Jack	31 MHz	Tune for signal	L9	Maximum output on V.T.V.M.	4
5			48 MHz		TC2		5
6	Repeat Steps 4 and 5 for optimum sensitivity						6
Set Function Switch to AIR							
1	Same as above	A.C.V.T.V.M. across MPX OUT Jack	106 MHz	Tuning gang closed	L15	Maximum output on V.T.V.M.	1
2			140 MHz	Tuning gang open	TC7		2
3	Repeat Steps 1 and 2 for optimum sensitivity						3
4	Same as above	A.C.V.T.V.M. across MPX OUT Jack	110 MHz	Tune for signal	L10	Maximum output on V.T.V.M.	4
5			135 MHz		TC3		5
6	Repeat Steps 4 and 5 for optimum sensitivity						6
Set Function Switch to FM							
1	Same as above	A.C.V.T.V.M. across MPX OUT Jack	86.5 MHz	Tuning gang closed	L16	Maximum output on V.T.V.M.	1
2			109.5 MHz	Tuning gang open	TC8		2
3	Repeat Steps 1 and 2 for optimum sensitivity						3
4	Same as above	A.C.V.T.V.M. across MPX OUT Jack	88 MHz	Tune for signal	L11	Maximum output on V.T.V.M.	4
5			108 MHz		TC4		5
6	Repeat Steps 4 and 5 for optimum sensitivity						6

PARTS LIST

REF NO.	PART NO.	DESCRIPTION
RESISTORS		
R1,55,58	CA1212	1.2K ohm, 1/4W ±10%
R2,5,8,26,47,49	CA1222	2.2K ohm, 1/4W ±10%
R3,6	CA1110	100 ohm, 1/8W ±10%
R4	CA1168	680 ohm, 1/4W ±10%
R7,13,22	CA1156	560 ohm, 1/4W ±10%
R9	CA1222	2.2K ohm, 1/8W ±10%
R11	CA1210	1K ohm, 1/8W ±10%
R12	CA1212	1.2K ohm, 1/8W ±10%
R14	CA1339	39K ohm, 1/4W ±10%
R15	CA1310	10K ohm, 1/4W ±10%
R16,27,46	CA1410	100K ohm, 1/4W ±10%
R17	CA1318	18K ohm, 1/4W ±10%
R18,51,53	CA1139	390 ohm, 1/4W ±10%
R19,34,41,42	CA1210	1K ohm, 1/4W ±10%
R20	CA1218	1.8K ohm, 1/4W ±10%
R21	CA1418	180K ohm, 1/4W ±10%

REF NO.	PART NO.	DESCRIPTION
RESISTORS		
R28,31,40	CA1147	470 ohm, 1/4W ±10%
R29,45	CA1122	220 ohm, 1/4W ±10%
R30,33	CA1247	4.7K ohm, 1/4W ±10%
R32,37,48	CA1327	27K ohm, 1/4W ±10%
R35	CA1227	2.7K ohm, 1/4W ±10%
R36	CA1133	330 ohm, 1/4W ±10%
R38,39,43,44	CA1256	5.6K ohm, 1/4W ±10%
R50	CA1110	100 ohm, 1/4W ±10%
R52	CA1068	68 ohm, 1/4W ±10%
R54	CA1382	82K ohm, 1/4W ±10%
R56,60	23MW382	0.5 ohm, 1/4W ±5%
R57,59	CA1115	150 ohm, 1/4W ±10%
R61,62	24MW1045	Thermistor
R63	25MW434	10K ohm, ±20% Tone Control
R64	25MW435	10K ohm, ±20% Volume Control

REF NO.	PART NO.	DESCRIPTION
CAPACITORS		
C1,4,17	DD-120	12PFD ±10% 50V Ceramic
C2,3,20,26	DD-330	33PFD ±10% 50V Ceramic
C5	DD-180	18PFD ±10% 50V Ceramic
C6,21	DD-100	10PFD ±0.5PFD 50V Ceramic
C7,11,16	DD-270	27PFD ±10% 50V Ceramic
C8,30	DD-220	22PFD ±10% 50V Ceramic
C9,10,25	DD-050	5PFD ±0.5PFD 50V Ceramic
C12,22	33MW920	0.0022MFD ±20% 50V Ceramic
C14,24	DD-560	56PFD ±10% 50V Ceramic
C15,23	33MW921	3PFD ±0.5PFD 50V Ceramic
C18,41	33MW922	6PFD ±0.5PFD 50V Ceramic
C19	31MW745	160PFD ±5% 50V Stylor
C27,66,67	DD-221	220PFD ±10% 50V Ceramic
C28,57,58	CK-103	0.01MFD +80% -20% 25V Ceramic
C29,65,71	33MW913	0.02MFD ±20% 50V Ceramic
C31,34	33MW914	8PFD ±0.5 PFD 50V Ceramic
C32	31MW741	100PFD ±5% 50V Stylor
C33,35,38,43,45, 49,54,56,59,62,86	CK-203	0.02MFD +80% -20% 25V Ceramic
C36,37,63	DD-1032	0.01MFD ±20% 50V Ceramic
C40	31MW742	130PFD ±5% 50V Stylor
C42	31MW572	470PFD ±5% 50V Stylor
C44	32MW871	10MFD +100%-10% 10V Electrolytic
C46,80,83	32MW872	220MFD +100%-10% 10V Electrolytic
C47,48	33MW915	1PFD ±0.5PFD 50V Ceramic
C50	DD-403	0.04MFD +80% -20% 25V Ceramic
C51	DD-220	22PFD ±10% 50V Ceramic
C52	33MW916	1.5PFD ±0.25PFD 50V Ceramic
C55	33MW917	2 PFD ±0.5PFD 50V Ceramic
C60	33MW918	0.5PFD ±0.25PFD 50V Ceramic
C61	DD-150	15PFD ±10% 50V Ceramic
C68,74	32MW873	4.7MFD +100%-10% 16V Electrolytic
C69,70	DD 391	390PFD ±20% 50V Ceramic
C72	31MW743	0.015MFD ±20% 50V Mylar
C73	32MW874	4.7MFD +100% -10% 16V Electrolytic
C75	32MW875	0.1MFD +40% -20% 10V Al Solid
C76	DD-102	0.001MFD ±20% 50V Ceramic
C77	32MW710	33MFD +100% -10% 6.3V Electrolytic
C78	32MW706	100MFD +100%-10% 6.3V Electrolytic
C79,81	32MW708	470MFD +100%-10% 10V Electrolytic
C82	DD-472	0.0047MFD ±20% 50V Ceramic
C85	32MW709	1000MFD +100% -10% 10V Electrolytic
VC1-4	35MW346	Variable Tuning Gang with Trimmers (TC1,5,9,11)
TC-1	N.A. Separately	Trimmer, Part of VC1
TC-2	35MW388	Trimmer, PB1 RF
TC-3	35MW388	Trimmer, AIR RF
TC-4	35MW388	Trimmer, FM RF
TC-5	N.A. Separately	Trimmer, Part of VC2
TC-6	35MW388	Trimmer, PBI Osc.
TC-7	35MW388	Trimmer, AIR Osc.
TC-8	35MW388	Trimmer, FM Osc.
TC-9	N.A. Separately	Trimmer, Part of VC3
TC-10	35MW388	Trimmer, MB Ant.
TC-11	N.A. Separately	Trimmer, Part of VC4
TC-12	35MW388	Trimmer, MB Osc.

SEMICONDUCTORS

REF NO.	PART NO.	DESCRIPTION
IC1	24MW1036	Integrated Circuit, FA6008T Audio Output
Q1	24MW1037	Transistor, VHF RF Amp.
Q2	24MW812	Transistor, VHF Osc.
Q3	24MW1038	Transistor, VHF Mix.
Q4	24MW815	Transistor, AM Conv.
Q5	24MW1039	Transistor, 1st AM/VHF IF Amp.
Q6	24MW1039	Transistor, 2nd AM/VHF IF Amp.
Q7	24MW1039	Transistor, 3rd VHF IF Amp.
Q8.9	**24MW1040	Transistor, Audio Output (matched pair)
D1,2	24MW1041	Diode, PB1 Limiter
D3	1N60	Diode, AIR RF Over load
D4	24MW1042	Diode, FM AFC
D5	1N60	Diode, AGC
D6	1N60	Diode, AM/MB Detector
D7	1N60	Diode, AIR Det & AGC
D8,9	**24MW1043	Diode, Ratio Detector (matched pair)
D10	24MW1044	Diode, Bias Comp.
D11	24MW860	Diode, DC Switch
D12-15	24MW862	Selenium Rectifier

** Replace with matched pairs only

COILS AND TRANSFORMERS

REF NO.	PART NO.	DESCRIPTION
L1	61MW929	Coil, PB2 Ant.
L2	61MW930	Coil, PB1 Loading
L3,4	61MW931	Coil, PB1 Filter
L5	61MW932	Coil, AIR Ant.
L6,7	61MW933	Coil, FM Ant.
L8	61MW934	Coil, PB2 RF
L9	61MW935	Coil, PB1 RF
L10	61MW936	Coil, AIR RF
L11	61MW937	Coil, FM RF
L12	61MW931	Coil, Emitter
L13	61MW934	Coil, PB2 Osc.
L14	61MW938	Coil, PB1 Osc.
L15	61MW939	Coil, AIR Osc.
L16	61MW936	Coil, FM Osc.
L17	61MW940	Coil, AM Ant.
L18	61MW940	Coil, MB Ant.
L19	61MW941	Coil, AM Osc.
L20	61MW942	Coil, MB Osc.
T1	62MW563	Transformer, 1st VHF IF
T2	62MW569	Transformer, 2nd VHF IF
T3	62MW569	Transformer, 3rd VHF IF
T4	62MW663	Transformer, Ratio Det. (Primary)
T5	62MW664	Transformer, Ratio Det. (Secondary)
T6	62MW568	Transformer, 1st AM/MB IF
T7	62MW564	Transformer, 2nd AM/MB IF
T8	62MW665	Transformer, 3rd AM/MB IF
T9	11MW369	Transformer, Input
T10	11MW370	Transformer, Power

MISCELLANEOUS

REF NO.	PART NO.	DESCRIPTION
FB1,2,3,4	61MW943	Ferrite Bead
ME1	13MW30	Tuning Meter
PL1	18MW344	Pilot Lamp
NE1	18MW417	Neon Lamp
SP1	10MW138	Speaker, 3½" PM 8 ohm 1.26oz. Mag. Weight
SW1-6	12MW353	Switch, Selector
SW7	12MW278	Switch, AFC
SW8	N.A. Separately	Switch, AC/DC Changeover, Part of A.C. Power Socket (SO1)
SW9	12MW354	Switch, Power
SW10	12MW282	Switch, Lamp
SO1	26MW428	Socket, AC Power with SW8
	56MW125	P.C. Board Complete, Tuner

REF NO.	PART NO.	DESCRIPTION
		MISCELLANEOUS
	56MW126	P.C. Board Complete, IF
	56MW127	P.C. Board Complete, VHF Osc. Coil
	56MW128	P.C. Board Complete, RF
	43MW506	Pointer Dial
	43MW507	Drum
	43MW412	Dual Pulley
	43MW508	Pulley
	50MW723	Knob, Band Selector
	50MW724	Knob, Tuning
	50MW725	Knob, Tone, Volume
	50MW726	Knob, Pilot Lamp SW.
	50MW727	Knob, AFC SW.
	50MW728	Knob, Power SW.
	98MW683	Cabinet Front (Assembly)
	98MW684	Cabinet Back (Assembly)
	60MW231	Rod Antenna
	43MW509	Cord, Dial
	43MW510	Cord, Band Change
	43MW511	Spring, Band Change Cord
J1	63MW255	Jack, MPX Output
J2	63MW255	Jack, Stereo Input
J3	63MW255	Jack, Earphone
	43MW511	Spring, Dial Cord

■ ALIGNMENT INSTRUCTIONS

READ CAREFULLY BEFORE ATTEMPTING ALIGNMENT

Notes:

1. Volume control——Maximum, Minimum (FM-IF)
2. Treble control——Maximum
3. Bass control——Maximum
4. Band selector switch——AM, MB, SW, FM, VHF, AIR
5. Squelch switch——OFF
6. AC-BATT. selector switch——BATT
7. AFC switch——OFF (FM-IF & RF)
8. Fine Tuning——Center
9. PHONO-RADIO selector switch——RADIO
10. Power source voltage——DC 6 volts
11. Output of signal generator should be no higher than necessary to obtain an output reading.

Notes:

1. When alignment step 2~4, unsolder lead between test point **TP6** and point [A] before alignment and resolder it after alignment.
2. When alignment step 6~11, connect 12mmf between test point **TP4** and chassis before alignment and remove it after alignment.

	SIGNAL GENERATOR or SWEEP GENERATOR: CONNECTIONS	SIGNAL GENERATOR or SWEEP GENERATOR: FREQUENCY	RADIO DIAL SETTING 〔DISTANCE〕	INDICATOR (VTVM or SCOPE)	ADJUSTMENT	REMARKS
	AM, MB & SW IF ALIGNMENT					
1	Fashion loop of several turns of wire and radiate signal into loop of receiver.	455 kHz 30% Mod. with 400Hz.	Point of non-interference. (on/about 600 kHz)	Output meter across voice coil.	T_4 (1st IFT) T_6 (2nd IFT) T_{10} (3rd IFT)	Adjust for maximum output.
	FM-IF ALIGNMENT					
2	High side thru. 0.001 mfd to point **TP3**, Common to chassis.	10.7 MHz (400 kHz SWP.)	Point of non-interference. (on/about 90 MHz).	Connect vert. amp. of scope to point **TP6**, Common to chassis.	T_3 (1st FM IFT) T_5 (2nd FM IFT) T_7 (3rd FM IFT) T_8 (4th FM IFT) (Primary)	Adjust for maximum amplitude and proper linearity between ±100 kHz markers. (Refer to fig. 1)
	VHF-IF ALIGNMENT					
3	High side thru. 0.001 mfd to point **TP1**, Common to chassis.	10.7 MHz (400 kHz SWP.)	Point of non-interference. (on/about 152 MHz)	Connect vert. amp. of scope to point **TP6**, Common to chassis.	T_1 (1st VHF IFT)	Adjust for maximum amplitude and proper linearity between ±100 kHz markers. (Refer to fig. 1)
	AIR-RF ALIGNMENT					
4	High side thru. 0.001 mfd to point **TP2**, Common to chassis.	10.7 MHz (400 kHz SWP.)	Point of non-interference.	Connect vert. amp. of scope to point **TP6**, Common to chassis.	T_2 (1st AIR IFT)	Adjust for maximum amplitude and proper linearity between ±100 kHz markers. (Refer to fig. 1)
	FM-DET ALIGNMENT					
5	High side thru. 0.001 mfd to point **TP3**, Common to chassis.	10.7 MHz (400 kHz SWP.)	Point of non-interference.	Connect vert. amp. of scope to point **TP7**, Common to chassis.	T_9 (4th FM IFT) (Secondary)	Adjust **T9** so that 10.7 MHz marker appears at the center (Refer to fig. 2)
	AM-RF ALIGNMENT					
6	Fashion loop of several turns of wire and radiate signal into loop of receiver.	550 kHz	550 kHz 〔15/32″〕	Output meter across voice coil.	L_{24} (OSC Coil) (*)L_{19} (ANT Coil)	Adjust for maximum output. Adjust L_{19} by moving coil bobbin along ferrite core.
7	〃	1500 kHz	1500 kHz 〔3 7/8″〕	〃	C_{77} (OSC Trimmer) C_{63} (ANT Trimmer)	Adjust for maximum output. Repeat steps (6) and (7).
	MB-RF ALIGNMENT					
8	Fashion loop of several turns of wire and radiate signal into loop of receiver.	1.6 MHz	1.6 MHz 〔7/32″〕	Output meter across voice coil.	L_{23} (OSC Coil) (*)L_{20} (ANT Coil)	Adjust for maximum output. Adjust L_{20} by moving coil bobbin along ferrite core.
9	〃	4.5 MHz	4.5 MHz 〔4 1/32″〕	〃	C_{74} (OSC Trimmer) C_{64} (ANT Trimmer)	Adjust for maximum output. Repeat steps (8) and (9).
	SW-RF ALIGNMENT					
10	Fashion loop of several turns of wire and radiate signal into loop of receiver.	5.9 MHz	5.9 MHz 〔5/32″〕	Output meter across voice coil.	L_{22} (OSC Coil) (*)L_{21} (ANT Coil)	Adjust for maximum output. Adjust L_{21} by moving coil bobbin along ferrite core.
11	〃	18 MHz	18 MHz 〔4 3/32″〕	〃	C_{65} (OSC Trimmer)	Adjust for maximum output. Repeat steps (10) and (11).

*Cement antenna bobbin with wax after completing alignment.

Schematic Diagram

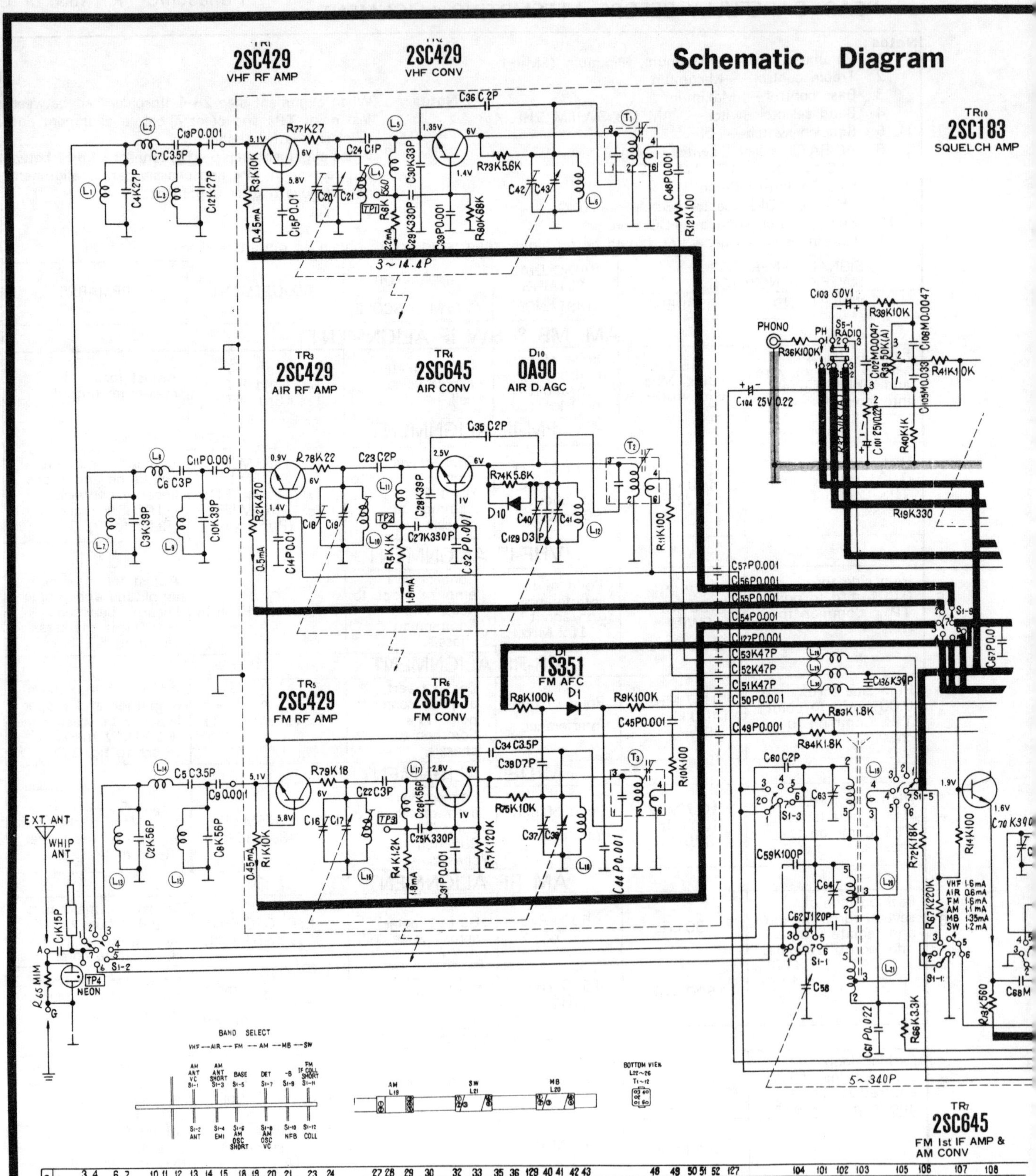

Notes :

1. S_{1-1}~S_{1-12} : Band selector switch in "VHF" position.
2. S_{2-1}, S_{2-2} : AC-Battery selector switch in "BATTERY" position.
3. S_{3-1}, S_{3-2} : Power source switch in "OFF" position.
4. S_4 : AFC switch in "ON" position.
5. S_{5-1}, S_{5-2} : Phono-Radio selector switch in "RADIO" position.
6. S_6 : Dial light switch in "OFF" position.
7. S_7 : Squelch control in "OFF" position.
8. DC voltage measurements are taken with circuit tester 10KΩ/V from negative terminal of battery.
   TR1 & TR2......VHF position TR5 & TR6...... FM position
   TR3 & TR4......AIR position TR7~TR14 ... AM position
9. Battery current : No signal FM & AM...... 22mA
   Maximum output FM & AM......385mA

Model RF-1600 or C

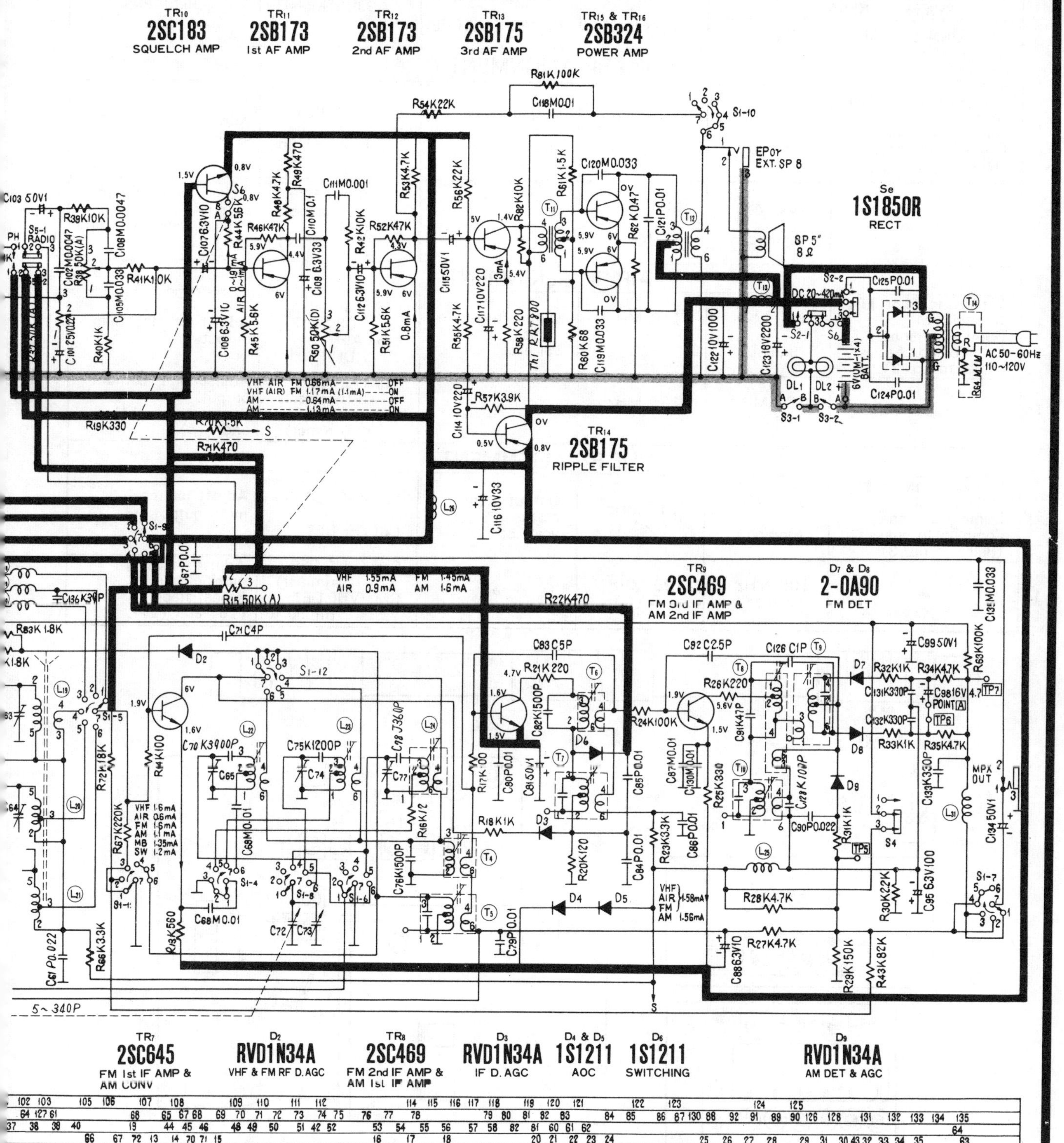

Positive Voltage Line

Negative Voltage Line

	SIGNAL GENERATOR or SWEEP GENERATOR		RADIO DIAL SETTING (DISTANCE)	INDICATOR (VTVM or SCOPE)	ADJUSTMENT	REMARKS
	CONNECTIONS	FREQUENCY				
	FM-RF ALIGNMENT					
12	Connect to point **TP4** through FM Dummy antenna Common to chassis. (Refer to fig. 3)	90 MHz (unmod.)	90 MHz (1 3/32")	Output meter across voice coil.	L18 (OSC Coil) L16 (FM DET Coil)	Adjust for maximum output.
13	〃	106 MHz (unmod.)	106 MHz (3 9/16")	〃	C37 (FM OSC Trimmer) C16 (FM Trimmer)	Adjust for maximum output. Repeat steps (12) and (13)
	AIR-RF ALIGNMENT					
14	Connect to point **TP4** through FM Dummy antenna, Common to chassis. (Refer to fig. 3)	110 MHz (unmod.)	110 MHz (27/32")	Output meter across voice coil.	L12 (AIR OSC Coil) L10 (AIR DET Coil)	Adjust for maximum output.
15	〃	133 MHz (unmod.)	133 MHz (3 19/32")	〃	C40 (AIR OSC Trimmer) C18 (AIR DET Trimmer)	Adjust for maximum output. Repeat steps (14). and (15).
	VHF-RF ALIGNMENT					
16	Connect to point **TP4** through FM Dummy antenna, Common to chassis. (Refer to fig. 3)	140 MHz (unmod.)	140 MHz (27/32")	Output meter across voice coil.	L6 (VHF OSC Coil) L4 (VHF DET	Adjust for maximum output.
17	〃	106 MHz (unmod.)	106 MHz (3 13/32")	〃	C42 (VHF OSC Trimmer) C20 (VHF DET Trimmer)	Adjust for maximum output. Repeat steps (16) and (17).

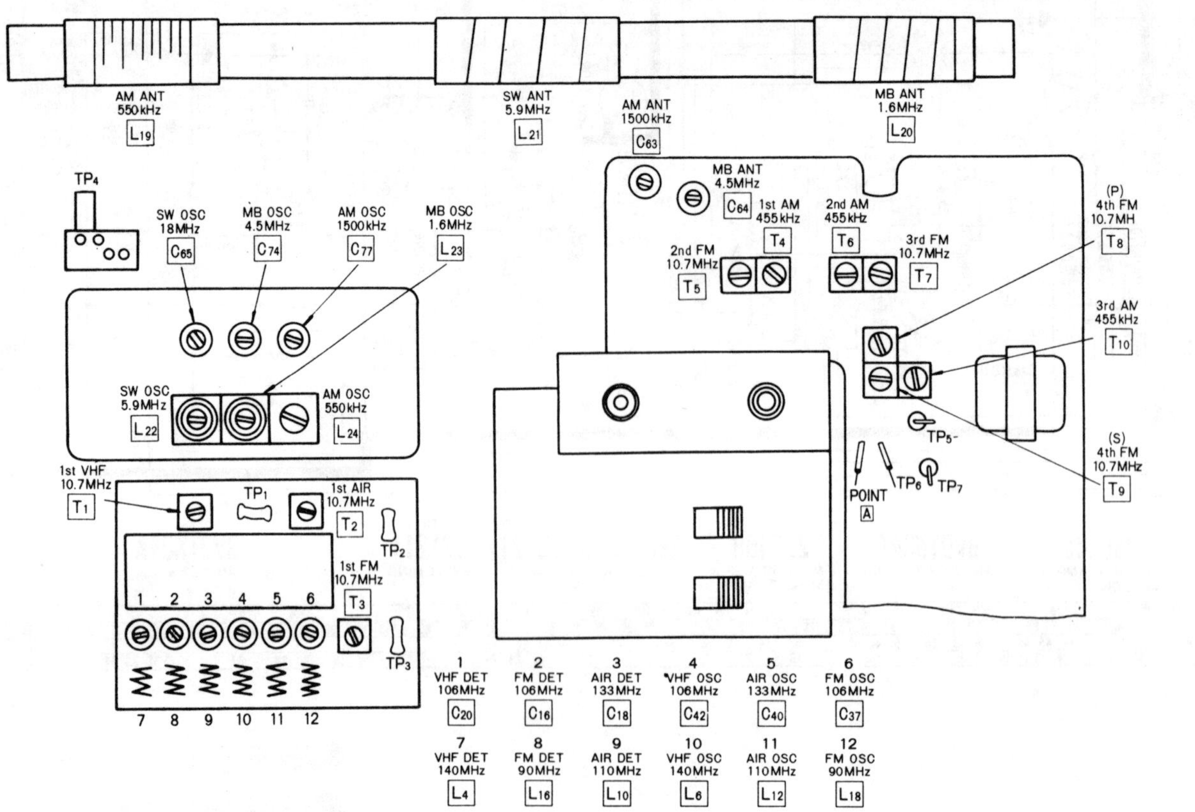

Alignment Points

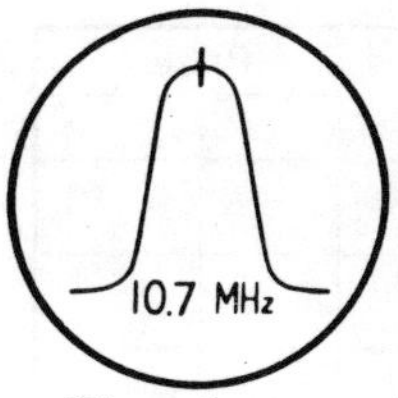

Fig. 1

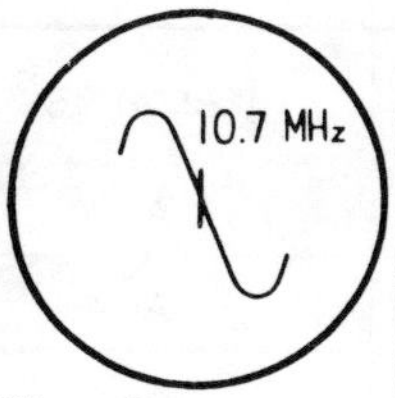

Fig. 2

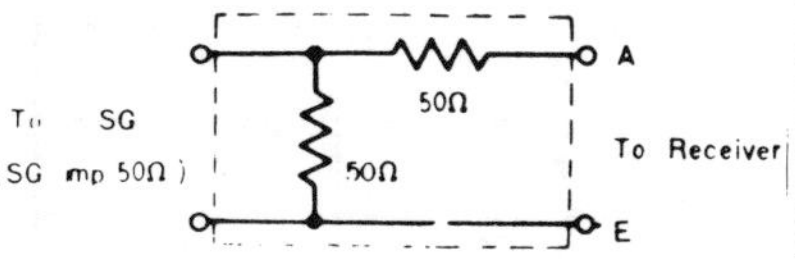

Fig. 3 FM Dummy Antenna

NOTES:
1. Part numbers are indicated on most mechanical parts. Please use this part number, for parts orders.
2. ISO metric thread screws & parts which employ ISO metric thread screws are identified by ISO marking (ISO).
3. Ⓝ indicates the New Parts.
4. **A—C** rank: **A** rank parts will cover 80% of repair needs.
   A+B rank parts will cover 95% of repair needs.
   C rank parts are less necessary.
5. Old part No. is shown in ( ).
6. Please use the "Price" column as desired.
7. These parts are illustrated on prior pages acording to their reference numbers.

Ref.No.	Part No.	Description	Per Set (Pcs.)	Remarks	Price	
TRANSISTORS AND DIODES						
$TR_{1,2,3,5}$	2SC429	VHF RF Amp., VHF Converter, AIR RF Amp., FM RF Amp.	4	A		
$TR_{4,6,7}$	2SC645	AIR Converter,.FM Converter, 1st FM IF Amp. & AM Converter	3	A		
$TR_{8,9}$	2SC469	2nd FM IF Amp. & AM 1st Amp., 3rd FM IF Amp. & AM 2nd Amp.	2	A		
TR_{10}	2SC183	Squelch Amplifier	1	A		
$TR_{11,12}$	2SB173	1st, 2nd AF Amplifier	2	A		
$TR_{13,14}$	2SB175	3rd AF Amplifier, Ripple Filter	2	A		
$TR_{15,16}$	2SB324	Power Amplifier (push-pull)	2	A		
D_1	1S351	FM AFC	1	A		
$D_{2,3,9}$	RVD1N34A (1N34A)	FM, VHF RF D.AGC, IF D.AGC, AM Detector & AGC	3	A		
$D_{4,5,6}$	1S1211	Operation Compensator, Switching	3	A		
$D_{7,8}$	2-0A90	FM Ratio Detector	1 pair	A		
D_{10}	0A90	AIR D.AGC	1	A		
THERMISTOR						
Th_1	RRT800 (MT-080)	Temperature Compensator	1	A		
Se	1S1850R	Rectifier	1	A		
COILS AND TRANSFORMERS						
$L_{1,3}$	RLQY05S-5	VHF Antenna Coil	2	B		
L_2	RLQY25S-5	VHF Antenna Coil	1	B		
L_4	RLD4Y53	VHF Detector Coil	1	A		
$L_{5,11,14,17,28,29,30}$	RLQX121-1	Choke Coil	7	B		
L_6	RLD4Y43	VHF Oscillator Coil	1	A		
$L_{7,9,13,15}$	RLQY10S-5	AIR & FM Antenna Coil	4	B		
L_8	RLQY50S-5	AIR Antenna Coil	1	B		
L_{10}	RLD4Y54	AIR Detector Coil	1	A		

Ref.No.	Part No.	Description	Per Set (Pcs.)	Remarks	Price	
L_{12}	RLD4Y44	AIR Oscillator Coil	1	A		
L_{16}	RLD4Y43	FM Detector Coil	1	A		
L_{18}	RLD4Y45	FM Oscillator Coil	1	A		
$L_{19,20,21}$	RLF9H11	AM.SW & MB Antenna Coil	1	A		
L_{22}	RLO3R26-T	SW Oscillator Coil	1	A		
L_{23}	RLO3P60	MB Oscillator Coil	1	A		
L_{24}	RLO2C44	AM Oscillator Coil	1	A		
$L_{25,26,31}$	RLQX121-1	Choke Coil	3	B		
$T_{1,2,3}$	RLI4B152	1st VHF, AIR, FM IF Transformer	3	A		
$T_{4,6}$	RLI2B156	1st & 2nd AM IF Transformer	2	A		
$T_{5,7}$	RLI4B351	2nd & 3rd FM IF Transformer	2	A		
T_8	RLI4B508	4th FM IF Transformer, Primary	1	A		
T_9	RLI4B552	4th FM IF Transformer, Secondary	1	A		
T_{10}	RLI2B457	3rd AM IF Transformer	1	A		
T_{11}	RLT3G10-S	Input Transformer, P=5KΩ:S=2.5KΩ	1	A		
T_{12}	RLT2I12-S	Output Transformer, P=50Ω:S=8Ω	1	A		
T_{13}	RLT6E1-S	Choke Coil	1	A		
T_{14}	RLT5J31-W	Power Transformer	1	A		
		RESISTORS				
R_{62}	ERW12LKR47	0.47Ω, ½Watt, ±10%, Wire	1	B		
R_{16}	ERD14VK120	12Ω, ¼Watt, ±10%, Carbon	1	B		
R_{60}	ERD14VK680	68Ω, ¼Watt, ±10%, Carbon	1	B		
$R_{10,14,17,24}$	ERD14VK101	100Ω, ¼Watt, ±10%, Carbon	4	B		
R_{20}	ERD14VK121	120Ω, ¼Watt, ±10%, Carbon	1	B		
$R_{21,26,58}$	ERD14VK221	220Ω, ¼Watt, ±10%, Carbon	3	B		
$R_{19,25}$	ERD14VK331	330Ω, ¼Watt, ±10%, Carbon	2	B		
$R_{22,49,71}$	ERD14VK471	470Ω, ¼Watt, ±10%, Carbon	3	B		
$R_{6,13}$	ERD14VK561	560Ω, ¼Watt, ±10%, Carbon	2	B		
$R_{18,32,33,40}$	ERD14VK102	1KΩ, ¼Watt, ±10%, Carbon	4	B		
R_4	ERD14VK122	1.2KΩ, ¼Watt, ±10%, Carbon	1	B		
$R_{61,70}$	ERD14VK152	1.5KΩ, ¼Watt, ±10%, Carbon	2	B		
$R_{83,84}$	ERD14VK182	1.8KΩ, ¼Watt, ±10%, Carbon	2	B		
R_{30}	ERD14VK222	2.2KΩ, ¼Watt, ±10%, Carbon	1	B		
$R_{23,66}$	ERD14VK332	3.3KΩ, ¼Watt, ±10%, Carbon	2	B		
R_{57}	ERD14VK392	3.9KΩ, ¼Watt, ±10%, Carbon	1	B		
$R_{27,28,34,48,55,59}$	ERD14VK472	4.7KΩ, ¼Watt, ±10%, Carbon	6	B		
$R_{44,45,51}$	ERD14VK562	5.6KΩ, ¼Watt, ±10%, Carbon	3	B		

Ref.No.	Part No.	Description	Per Set (Pcs.)	Remarks	Price	
$R_{3,39,41,42}$	ERD14VK103	10KΩ, ¼Watt, ±10%, Carbon	4	B		
R_{72}	ERD14VK183	18KΩ, ¼Watt, ±10%, Carbon	1	B		
$R_{54,56}$	ERD14VK223	22KΩ, ¼Watt, ±10%, Carbon	2	B		
$R_{46,52}$	ERD14VK473	47KΩ, ¼Watt, ±10%, Carbon	2	B		
R_{43}	ERD14VK823	82KΩ, ¼Watt, ±10%, Carbon	1	B		
$R_{63,81}$	ERD14VK104	100KΩ, ¼Watt, ±10%, Carbon	2	B		
R_{7}	ERD14VK124	120KΩ, ¼Watt, ±10%, Carbon	1	B		
R_{29}	ERD14VK154	150KΩ, ¼Watt, ±10%, Carbon	1	B		
R_{67}	ERD14VK224	220KΩ, ¼Watt, ±10%, Carbon	1	B		
R_{79}	ERD14TK180	18Ω, ¼Watt, ±10%, Carbon	1	B		
R_{78}	ERD14TK220	22Ω, ¼Watt, ±10%, Carbon	1	B		
R_{77}	ERD14TK270	27Ω, ¼Watt, ±10%, Carbon	1	B		
$R_{11,12}$	ERD14TK101	100Ω, ¼Watt, ±10%, Carbon	2	B		
R_{2}	ERD14TK471	470Ω, ¼Watt, ±10%, Carbon	1	B		
$R_{5,31}$	ERD14TK102	1KΩ, ¼Watt, ±10%, Carbon	2	B		
R_{35}	ERD14TK472	4.7KΩ, ¼Watt, ±10%, Carbon	1	B		
$R_{73,74}$	ERD14TK562	5.6KΩ, ¼Watt, ±10%, Carbon	2	B		
$R_{1,75,82}$	ERD14TK103	10KΩ, ¼Watt, ±10%, Carbon	3	B		
R_{80}	ERD14TK683	68KΩ, ¼Watt, ±10%, Carbon	1	B		
$R_{8,9,36}$	ERD14TK104	100KΩ, ¼Watt, ±10%, Carbon	3	B		
$R_{64,65}$	ERC12GM105	1MΩ, ½Watt, ±20%, Solid	2	B		
		VARIABLE RESISTORS				
$R_{37,38,50}$	EVAG0AA01A54	50KΩ (A), Treble, Bass Volume Control	3	A		
R_{15}	EVJA0BT12A54	50KΩ (A), Squelch Control W/ON-OFF Switch(S_7)	1	A		
		CAPACITORS				
$C_{24,126}$	ECCD05010C	1mmf, 50WV, ±0.25mmf, Ceramic	2	C		
$C_{23,35,36,60}$	ECCD05020C	2mmf, 50WV, ±0.25mmf, Ceramic	4	C		
C_{92}	ECCD052R5C	2.5mmf, 50WV, ±0.25mmf, Ceramic	1	C		
C_{22}	ECCD05030C	3mmf, 50WV, ±0.25mmf, Ceramic	1	C		
$C_{7,34}$	ECCD053R5C	3.5mmf, 50WV, ±0.25mmf, Ceramic	2	C		
$C_{5,6,71}$	ECCD05040C	4mmf, 50WV, ±0.25mmf, Ceramic	3	C		
C_{83}	ECCD05050C	5mmf, 50WV, ±0.25mmf, Ceramic	1	C		
$C_{131,132,133}$	ECCD05331K	330mmf, 50WV, ±10%, Ceramic	3	C		
C_{136}	ECCD05390K	39mmf, 50WV, ±10%, Ceramic	1	C		

Ref.No.	Part No.	Description	Per Set (Pcs.)	Remarks	Price	
C_{39}	ECCD05070DX	7mmf, 50WV, ±0.5mmf, Ceramic	1	C		
C_{129}	ECCU5030DY	3mmf, 500WV, ±0.5mmf, Ceramic	1	C		
C_{1}	ECMS05150K-H	15mmf, 50WV, ±10%, Mica	1	C		
$C_{4, 12}$	ECMS05270K-H	27mmf, 50WV, ±10%, Mica	2	C		
C_{30}	ECMS05330K-H	33mmf, 50WV, ±10%, Mica	1	C		
$C_{3, 10, 28}$	ECMS05390K-H	39mmf, 50WV, ±10%, Mica	3	C		
C_{91}	ECMS05470K-H	47mmf, 50WV, ±10%, Mica	1	C		
$C_{2, 8, 26}$	ECMS05560K-H	56mmf, 50WV, ±10%, Mica	3	C		
$C_{59, 128}$	ECMS05101K-H	100mmf, 50WV, ±10%, Mica	2	C		
C_{62}	ECMS05121J-H	120mmf, 50WV, ±5%, Mica	1	C		
$C_{25, 27, 29}$	ECCU05331K	330mmf, 50WV, ±10%, Ceramic	3	C		
$C_{9, 11, 13, 31, 32, 33, 44, 45, 48}$	ECKD05102P	0.001mfd, 50WV, $\pm^{100}_{0}\%$, Ceramic	9	C		
$C_{68, 69, 87, 130}$	ECKE05103MY	0.01mfd, 50WV, ±20%, Ceramic	4	C		
$C_{14, 15, 67, 79, 80, 84, 85, 86, 121}$	ECKE05103P	0.01mfd, 50WV, $\pm^{100}_{0}\%$, Ceramic	9	C		
$C_{124, 125}$	ECKD5103P	0.01mfd, 500WV, $\pm^{100}_{0}\%$, Ceramic	2	C		
$C_{61, 90}$	ECKE05223P	0.022mfd, 50WV, $\pm^{100}_{0}\%$, Ceramic	2	C		
$C_{51, 52, 53}$	ECCL05470K-D	47mmf, 50WV, ±10%, Ceramic	3	C		
$C_{49, 50, 54, 55, 56, 57, 127}$	ECKL05102P-D	0.001mfd, 50WV, $\pm^{100}_{0}\%$, Ceramic	7	C		
C_{111}	ECQG05102MZ-N	0.001mfd, 50WV, ±20%, Polyester	1	C		
$C_{102, 106}$	ECQG05472MZ-N	0.0047mfd, 50WV, ±20%, Polyester	2	C		
C_{118}	ECQG05103MZ-N	0.01mfd, 50WV, ±20%, Polyester	1	C		
$C_{105, 119, 120, 135}$	ECQG05333MZ-N	0.033mfd, 50WV, ±20%, Polyester	4	C		
C_{110}	ECQG05104MZ-N	0.1mfd, 50WV, ±20%, Polyester	1	C		
C_{78}	ECQS1361JZ	360mmf, 125WV, ±5%, Styrol	1	C		
C_{75}	ECQS1122KZ	1200mmf, 125WV, ±10%, Styrol	1	C		
$C_{76, 82}$	ECQS1152KZ	1500mmf, 125WV, ±10%, Styrol	2	C		
C_{70}	ECQS02392KZ	3900mmf, 25WV, ±10%, Styrol	1	C		
$C_{101, 104}$	ECAF25VR22	0.22mfd, 25WV, Electrolytic	2	B		
$C_{81, 99, 103, 115, 134}$	ECEA50V1	1mfd, 50WV, Electrolytic	5	B		
C_{98}	ECEA16V4R7	4.7mfd, 16WV, Electrolytic	1	B		
$C_{88, 107, 108, 112}$	ECEA6V10	10mfd, 6.3WV, Electrolytic	4	B		
C_{109}	ECEA6V33	33mfd, 6.3WV, Electrolytic	1	B		
C_{116}	ECEA10V33	33mfd, 10WV, Electrolytic	1	B		
C_{95}	ECEA6V100	100mfd, 6.3WV, Electrolytic	1	B		
$C_{114, 117}$	ECEA10V220	220mfd, 10WV, Electrolytic	2	B		
C_{123}	ECEB16V2200	2200mfd, 16WV, Electrolytic	1	B		

Ref.No.	Part No.	Description	Per Set (Pcs.)	Remarks	Price	
C_{122}	ECEA10V1000	1000mfd, 10WV, Electrolytic	1	B		
		VARIABLE CAPACITORS				
C_{73}	ECV1YW02D59A	Fine Tuning Gang	1	A		
$C_{17,19,21,38,41,43}$	ECV6EW02X50A	Tuning Gang, FM, AIR, VHF	1	A		
$C_{58,72}$	PVC2R-3	Tuning Gang, AM, MB, SW	1	A		
$C_{16,18,20,37,40,42,64}$	ECV1ZW10P12	Trimmer	7	A		
$C_{63,65,74,77}$	ECV1ZW20P12	Trimmer	4	A		
		SWITCHES				
$S_{1-1} \sim S_{1-11}$	ESRE6C6L50A	Band Selector Switch	1	A		
S_{2-1}, S_{2-2} & S_{5-1}, S_{5-2}	RSS153	AC-BATT, RADIO-PHONO Selector Switch	2	A		
$S_{3-1} \sim S_{3-2}$	RSH62	Power Source Switch	1	A		
S_4	RSS91-1	AFC Switch	1	A		
		SPEAKER				
SP	EAS12P70SA	5″ PM Dynamic Speaker, 8Ω	1	A		

6-Band Portable AM-FM-MB-AIR-PB—Sharp FV-1500

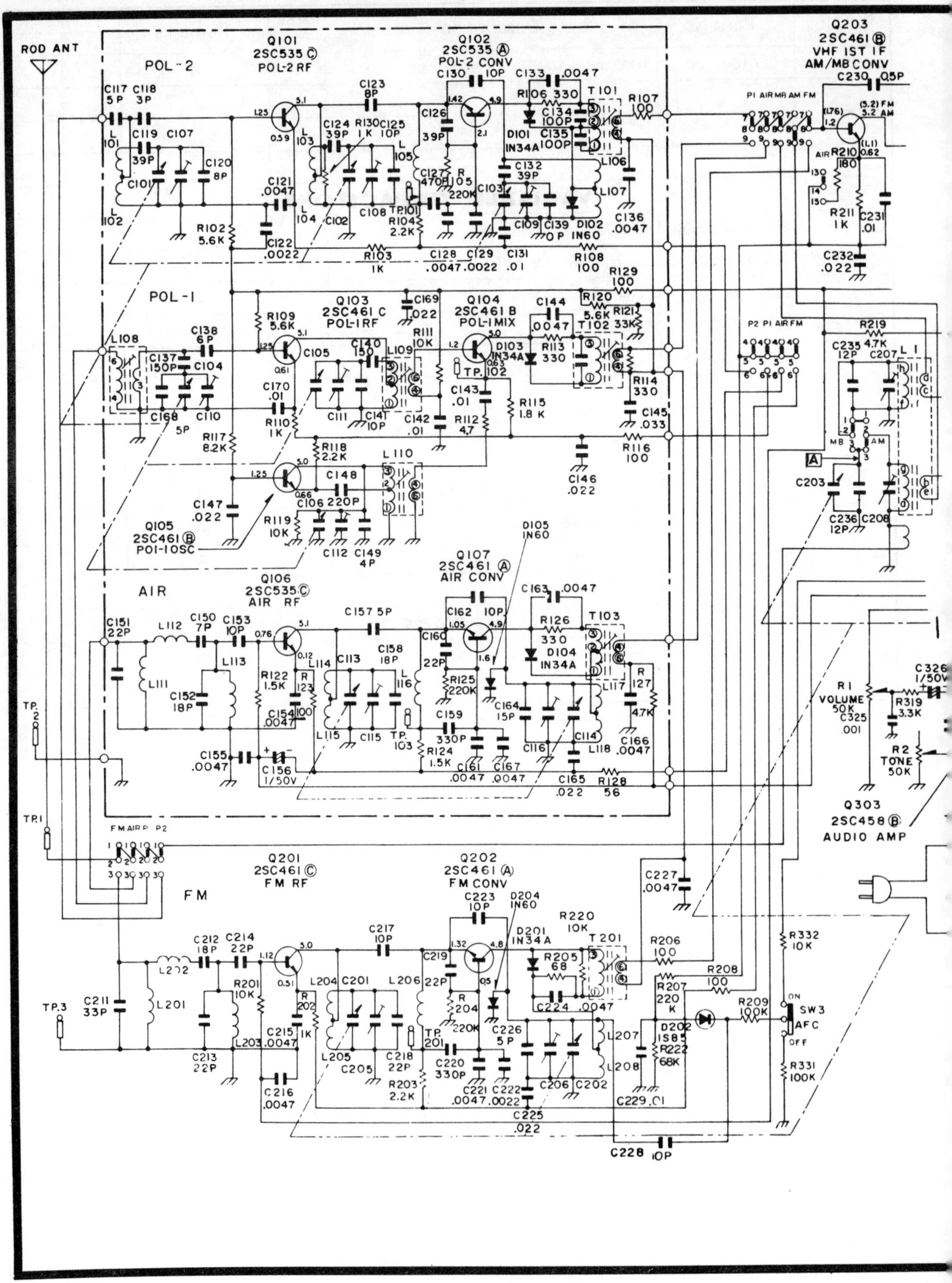

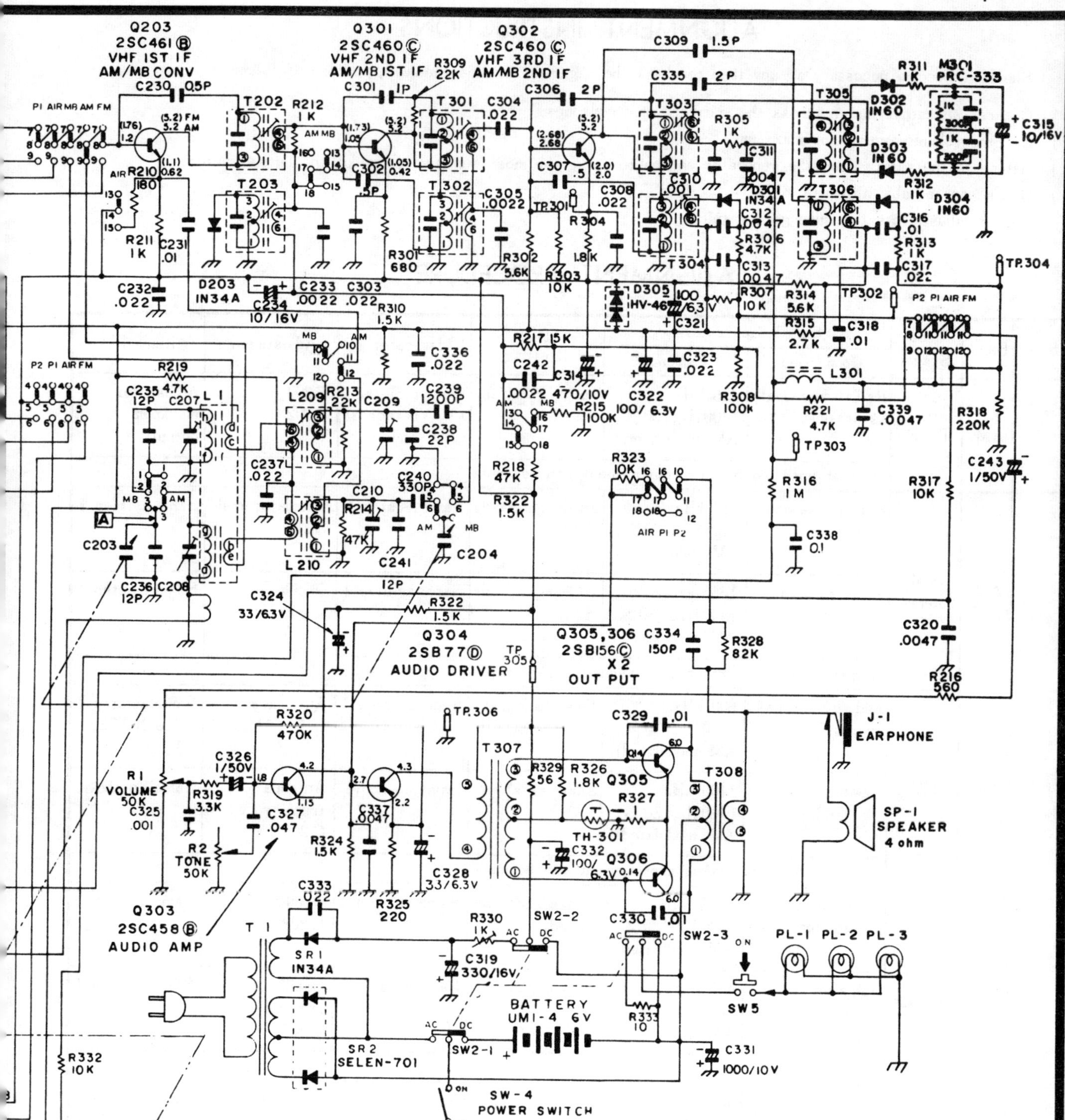

NOTES

1) BAND SELECTOR SWITCH IS IN "AM" POSITION.
2) FREQ RANGE

POL2	148 — 176 MHz.	POL1	30 — 50 MHz.
AIR	108 — 136 MHz.	MB	1.6 — 4.3 MHz.
AM	520 — 1620 KHz.	FM	88 — 108 MHz.

3) RESISTANCE VALUES ARE IN OHM K=1000
4) CAPACITANCE VALUES ARE IN MFD P=MMFD
5) VOLTAGE READINGS ARE MEASURED WITH VTVM WITH NO SIGNAL AND VOLUME CONTROL AT MINIMUM.
Q101 TO Q107, Q201 TO Q203, AND Q301 TO Q303 ARE POSITIVE TO CIRCUIT POINT ※B, Q304 TO Q306 ARE. NECATIVE TO CHASSIS CROUND.
6) VOLTAGES MARKED ( ) ARE MEASURED IN "FM" POSITON.
7) C120 MAY BE ANY VALUE FROM 3P TO 10P.
8) C125 MAY BE ANY VALUE FROM 3P TO 10P.
9) C158 MAY BE ANY VALUE FROM 13P TO 22P.
10) C164 MAY BE ANY VALUE FROM 10P TO 15P.
11) C168 MAY BE ANY VALUE FROM 0 TO 5P.
12) C238 MAY BE ANY VALUE FROM 10P TO 22P.
13) C241 MAY BE ANY VALUE FROM 8P TO 12P.
14) C301 MAY BE ANY VALUE FROM 0 TO 2P.
15) C306 MAY BE ANY VALUE FROM 0.5P TO 2P.
16) C339 MAY BE ANY VALUE FROM .0022 TO .0047.
17) R125 MAY BE ANY VALUE FROM 100K TO 270K.
18) R130 MAY BE ANY VALUE FROM 1K TO ∞.
19) R204 MAY BE ANY VALUE FROM 100K TO 270K.
20) R221 MAY BE ANY VALUE FROM 0 TO 10K.
21) R323 MAY BE ANY VALUE FROM 10K TO 22K.
22) R217 MAY BE ANY VALUE FROM 10K TO 22K.

ALIGNMENT INSTRUCTIONS

Should it become necessary at any time to check the alignment of this receiver, proceed as follows;

1) Connect an output meter across the speaker voice coil lugs.
2) Set the volume control for maximum.
3) Attenuate the signals from the generator enough to swing the most sensitive range of the output meter.
4) Use a non-metallic alignment tool.
5) Repeat adjustments to insure good results.

ALIGNMENT PROCEDURE

Step	Band	Test Stage	Signal Generator		Dial Setting	Indicator	Adjustment	Remarks
			Connection	Frequency				
1	AM	I F	Radiated as small as possible.	455 kHz (400 Hz, 30%, AM modulated)	Tuning gang fully closed.	Output meter (VTVM) across voice coil.	T 203 T 302 T 304	Adjust for maximum output.
			Repeat until no further improvement can be made.					
2	AM	Band Coverage	Radiated as small as possible.	510 kHz (400 Hz, 30%, AM modulated).	Tuning gang fully closed.	Same as step 1	Oscillator coil L 210.	Same as step 1
			Same as above.	1650 kHz (400 Hz, 30%, AM modulated).	Tuning gang fully open.	Same as step 1	Oscillator trimer C 210.	Same as step 1
			Repeat until no further improvement can be made.					
3	AM	Tracking	Radiated as small as possible.	600 kHz (400 Hz, 30%, AM modulated).	Tune to signal.	Same as step 1	Antenna coil L 1 (AM)	Same as step 1 (See NOTE A)
			Same as above.	1400 kHz (400 Hz, 30%, AM modulated).	Same as above.	Same as above.	Antenna trimmer C 208.	Same as above.
			Repeat until no further improvement can be made.					
4	MB	Band Coverage	Radiated as small as possible.	1.58 MHz (400 Hz, 30%, AM modulated).	Tuning gang fully closed.	Same as step 1	Oscillator coil L 209.	Same as step 1
			Same as above.	4.5 MHz (400 Hz, 30%, AM modulated).	Tuning gang fully open.	Same as step 1	Oscillator trimmer C 209.	Same as step 1
			Repeat until no further improvement can be made.					
5	MB	Tracking	Radiated as small as possible.	1.7 MHz (400 Hz, 30%, AM modulated).	Tune to signal.	Same as step 1	Antenna coil L 1 (MB).	Same as step 1 (See NOTE A)
			Same as above.	3.7 MHz (400 Hz, 30%, AM modulated).	Same as above.	Same as step 1	Antenna trimmer C 207.	Same as above.
			Repeat until no further improvement can be made.					

NOTE A

Check the alignment of the receiver antenna coil by bringing a piece of ferrite (such as coil slug) near the antenna loop stick, then a piece of brass. If ferrite increases output, loop requires more inductance. If brass increases output, loop requires less inductance. Change loop inductance by sliding the bobbin toward the center of ferrite core to increase inductance, or away to decrease inductance.

ALIGNMENT PROCEDURE

Step	Band	Test Stage	Signal Generator Connection	Signal Generator Frequency	Dial Setting	Indicator	Adjustment	Remarks
1	FM	IF	Through 5 PF capacitor to TP 201 and ground.	10.7 MHz unmodulated.	Tuning gang fully closed.	DC probe of VTVM to point TP301 and ground.	T 201 T 202 T 301	Adjust for maximum indication.
			Repeat until no further improvement can be made.					
2	FM	Ratio Detector	Same as step 1	Same as step 1	Same as step 1	DC probe of VTVM to point TP303 and ground.	T 303 T 305	Adjust for zero reading. A positive or negative reading will be obtained on either side of the correct setting.
			Repeat until no further improvement can be made.					
3	FM	Band Coverage	Through 75 ohm matching pad to FM antenna terminal.	87 MHz 400Hz, 30%, (22.5 kHz Dev), FM modulated.	Tuning gang fully closed.	Output meter (VTVM) across voice coil.	Oscillator coil L 207, L 208.	Adjust for maximum output.
			Same as above.	110 MHz 400Hz, 30%, (22.5 kHz Dev), FM modulated.	Tuning gang fully open.	Same as above.	Oscillator trimmer C 206.	Same as above.
			Repeat until no further improvement can be made.					
4	FM	Tracking	Same as step 3	88 MHz 400Hz, 30%, (22.5 kHz Dev), FM modulated.	Tune to signal.	Same as step 3.	RF coil L 204, L 205.	Same as step 3.
			Same as above.	108 MHz 400Hz, 30%, (22.5 kHz Dev), FM modulated.	Same as above.	Same as above.	RF trimmer C 205.	Same as above.
			Repeat until no further improvement can be made.					
5	AIR	IF	Through 5PF capacitor to TP 103 and ground.	10.7 MHz unmodulated.	Tuning gang fully closed.	DC probe of VTVM to point TP301 and ground.	T 103 T 306	Adjust for maximum indication.
			Repeat until no further improvement can be made.					
6	AIR	Band Coverage	Through 75 ohm matching pad to AIR antenna terminal.	107 MHz 400Hz, 30%, AM modulated.	Tuning gang fully closed.	Same as step 3.	Oscillator coil L 117, L 118.	Adjust for maximum output.
			Same as above.	138 MHz 400Hz, 30%, AM modulated.	Tuning gang fully open.	Same as above.	Oscillator trimmer C 116.	Same as above.
			Repeat until no further improvement can be made.					
7	AIR	Tracking	Same as step 6	108 MHz 400Hz, 30%, AM modulated.	Tune to signal	Same as step 3.	RF coil L 114, L 115.	Same as step 6.
			Same as above.	136 MHz 400Hz, 30%, AM modulated.	Same as above.	Same as above.	RF trimmer C 115.	Same as above.
			Repeat until no further improvement can be made.					
8	PB-1	IF	Through 5 PF capacitor to TP 102 and ground.	10.7 MHz unmodulated.	Tuning gang fully closed.	DC probe of VTVM to point TP 301 and ground.	T 102	Adjust for maximum indication
			Repeat until no further improvement can be made.					

9	PB-1	Band Coverage	Through 75ohm matching pad to POL-1 antenna terminal.	29.5 MHz 400Hz, 30%, (5 kHz Dev), FM modulated.	Tuning gang fully closed.	Output meter (VTVM) across voice coil.	Oscillator coil L 110.	Adjust for maximum output.
			Same as above.	51 MHz 400Mz, 30%, (5 kHz Dev), FM modulated.	Tuning gang fully open.	Same as above.	Oscillator trimmer C 112.	Same as above.
			Repeat until no	further improvement can be made.				
10	PB-1	Tracking	Same as step 2.	32 MHz 400Hz, 30%, (5 kHz Dev), FM modulated.	Tune to signal.	Same as step 2.	Antenna coil L 108. RF coil L 109.	Same as step 2.
			Same as step 2.	48 MHz 400Hz, 30%, (5 kHz Dev), FM modulated.	Same as above.	Same as step 2.	Antenna trimmer C 110. RF trimmer C 111.	Same as step 2.
			Repeat until no	further improvement can be made.				
11	PB-2	I F	Through 5 PF capacitor to TP101 and ground	10.7 MHz unmodulated.	Tuning gang fully closed.	DC prode of VTVM to point TP 301 and ground.	T 101	Adjust for maximum indication.
			Repeat until no	further improvement can be made.				
12	PB-2	Band Coverage	Through 75 ohm matching pad to POL-2 antenna terminal.	147 MHz 400Hz, 30%, (5 kHz Dev), FM modulated.	Tuning gang fully closed.	Output meter (VTVM) across voice coil.	Oscillator coil L 106, L 107.	Adjust for maximum output.
			Same as above.	178 MHz 400Hz, 30%, (5 kHz Dev), FM modulated.	Tuning gang fully open.	Same as above.	Oscillator trimmer C 109.	Same as above.
			Repeat until no	further improvement can be made.				
13	PB-2	Tracking	Same as step 5	148 MHz 400Hz, 30%, (5 kHz Dev), FM modulated.	Tune to signal.	Same as step 5.	Antenna coil L 101, L 102. RF coil L 103, L 104.	Same as step 5.
			Same as step 5	176 MHz 400Hz, 30%, (5 kHz Dev), FM modulated.	Same as above.	Same as step 5.	Antenna trimmer C 107. RF trimmer C 108.	Same as step 5.
			Repeat until no further improvement can be made.					

AM ALIGNMENT BY USING BROADCASTING STATION (If generator is not available)

Step	Band	Test Stage	Dial pointer Setting	Indicator	Adjustment	Remarks
1	A M	I F	Tune to maximum broadcasting signal around 1600kHz. Then rotate set slightly away from the maximum signal direction.	Output meter (VTVM) across voice coil.	T 203 T 302 T 304	Adjust for maximum output.
			Repeat until no further improvement can be made.			
2	A M	Tracking	Tune to maximum broadcasting signal around 600kHz. Then rotate set slightly away from the maximum signal direction.	Same as step 1	Antenna coil L 1 (AM).	Slide and adjust antenna winding for maximum output (See NOTE A)
			Tune to maximum broadcasting signal around 1300kHz. Then rotate set slightly away from the maximum signal direction.	Same as step 1	Antenna trimmer C 208.	Same as step 1 (See NOTE A)
			Repeat until no further improvement can be made.			

FM IF ALIGNMENT BY USING BROADCASTING STATION (If generator is not available)

Step	Band	Test Stage	Dial pointer setting	Indicator	Adjustment	Remarks
1	F M	I F	Tune to maximum any broadcasting signal.	Connect VTVM to point TP301 and ground. Adjust antenna for the weakest signal as possible, while still retaining a positive indication on VTVM.	T 201 T 202 T 301	Adjust for maximum indication.
			Repeat until no further improvement can be made.			
2	F M	Tracking	Tune to maximum any broadcasting signal.	Output meter across voice coil.	T 303	Adjust for maximum output.
			Set the pointer to low end of dial.	Same as above.	T 305	Adjust for maximum rushing noise.
			Repeat until no further improvement of maximum output and mimimum rushing noise can be made.			

SERVICE PARTS PRICE LIST

REF. NO.	PART NO.	DESCRIPTION	LIST PRICE
		SEMICONDUCTOR COMPLEMENT	
Q101	2SC-535	Transistor, PB-2 Amp.	3.00
Q102	2SC-535	Transistor, PB-2 Converter	3.00
Q103	2SC-461	Transistor, PB-1 RF Amp.	3.00
Q104	2SC-461	Transistor, PB-1 Mixer	3.00
Q105	2SC-461	Transistor, PB-1 Osc.	3.00
Q106	2SC-535	Transistor, AIR RF Amp.	3.00
Q107	2SC-461	Transistor, AIR Converter	3.00
Q201	2SC-461	Transistor, FM RF Amp.	3.00
Q202	2SC-461	Transistor, FM Converter	3.00
Q203	2SC-461	Transistor, AM/MB Converter 1st FM IF Amp.	3.00
Q301	2SC-460	Transistor, 1st AM/MB, 2nd FM IF Amp.	3.00
Q302	2SC-460	Transistor, 2nd AM/MB, 3rd FM IF Amp.	3.00
Q303	2SC-458	Transistor, Audio Amp.	3.00
Q304	2SB-77	Transistor, Audio Driver	2.50
Q305,306	2SB-156	Transistor, Audio Output (Matched Pair)	2.50
D101	1N-34A	Diode, PB-2 Overload	1.25
D102	1N-60	Diode, PB-2 Osc. Stabilizer	1.25
D103	1N-34A	Diode, PB-1 Overload	1.25
D104	1N-34A	Diode, AIR Overload	1.25
D105	1N-60	Diode, AIR Osc. Stabilizer	1.25
D201	1N-34A	Diode, FM Overload	1.25
D202	1S-85	Diode, FM AFC	2.50
D203	1N-34A	Diode, AM/MB Overload	1.25
D204	1N-60	Diode, FM Osc. Stabilizer	1.25
D301	1N-34A	Diode, AM/MB Detector	1.25
D302,303	1N-60	Diode, FM Detector (Matched Pair)	1.25
D304	1N-60	Diode, AIR Detector	1.25
D305	HV-46	Diode, Bias Regulator	1.00
SR1	1N-34A	Diode, Bias	1.25
SR2	SELEN-701	Rectifier, Power	1.25
TH301	D-1B	Thermistor, AOC	1.00
		COILS AND TRANSFORMERS	
L1	8L-275A10	Coil, AM/MB Antenna	1.60
L101,102	7L-598C	Coil, PB-2 Antenna	.25
L103,104	7L-599C	Coil, PB-2 RF	.25
L105	7L-954	Coil, PB-2 Emitter Choke	1.25
L106,107	7L-598C	Coil, PB-2 Osc.	.25
L108	8L-276A	Coil, PB-1 Antenna	.80
L109	7L-651C	Coil, PB-1 RF	.80
L110	8L-353B	Coil, PB-1 Osc.	.80
L111,113	7L-597C	Coil, AIR Antenna	.25
L112	7L-604C	Coil, AIR Antenna	.25
L114,115	7L-598C	Coil, AIR RF	.25
L116	7L-954	Coil, AIR Emitter Chcke	1.25
L117,118	7L-606C	Coil, AIR Osc.	.25
L201,203	7L-597C	Coil, FM Antenna	.25
L202	7L-604C	Coil, FM Antenna	.25
L204,205	7L-623C	Coil, FM RF	.25
L206	7L-954	Coil, FM Emitter Choke	1.25
L207,208	7L-598C	Coil, FM Osc.	.25
L209	7L-424B	Coil, MB Osc.	1.25
L210	7L-045B	Coil, AM Osc.	.60
L301	7L-946	Coil, Choke	1.25
T1	5T-823	Transformer, Power	5.00
T101	8IF-703	Transformer, 1st PB-2 IF	1.00
T102	7IF-840	Transformer, 1st PB-1 IF	1.25
T103	7IF-840	Transformer, 1st AIR IF	1.25
T201	7IF-840	Transformer, 1st FM IF	1.25
T202	7IF-842	Transformer, 2nd FM IF	1.25
T203	7IF-871	Transformer, 1st AM/MB IF	1.25
T301	7IF-842	Transformer, 3rd FM IF	1.25
T302	7IF-872	Transformer, 2nd AM/MB IF	1.25
T303	7IF-820	Transformer, 4th FM IF	1.25
T304	8IF-704	Transformer, AM/MB Detector	1.00
T305	7IF-821	Transformer, PB-1/PB-2, FM Detector	1.25
T306	7IF-811	Transformer, AIR Detector	1.00
T307	6T-801	Transformer, Audio Driver	2.00
T308	7T-837	Transformer, Audio Output	1.50
		CONTROLS	
R1	8V-884	50K OHM, Pot, Volume	1.50
R2	8V-884	50K OHM, Pot, Tone	1.50
R330	8V-889	1K OHM, Pot, AC Bias	.50
		CAPACITORS	
C101-107 110,111,112	VC-426	Variable Capacitor, PB-1/PB-2 Tuning, Trimmer Capacitors	5.00
C108,109	TO-701	Trimmer Capacitor, PB-2 RF, PB-2 Osc.	.50
C113-C116	VC-424	Variable Capacitor, AIR Tuning, Trimmer Capacitors	2.50

Chapter 4

Schematics, Alignment Procedures, and Parts Lists for Representative AM-FM-Multiplex Radios and Stereo-8 Tape Players

Portable AM-FM-STEREO 8—Realistic 14-916 (Allied Radio Shack, A Tandy Corp. Co.)

■REPLACEMENT PARTS LIST

NOTES : 1. Part numbers are indicated on most mechanical parts. Please use this Part number for parts orders.
2. Ⓝindicates the New Parts.
3. Gothic part No indicates the Service Standard Parts. Production part No is indicated in ( ).
4. **X−Z** rank: **X** rank parts will cover 80% of repair needs. **X+Y** rank parts will cover 95% of repair needs. **Z** rank parts are less necessary.

Ref.No.	Part No.	Description	Per Set (Pcs.)	Remarks
		INTEGRATED CIRCUITS, TRANSISTORS AND DIODES		
IC1	AN217	FM & AM IF Amplifier	1	X
IC2	AN211	FM MPX Amplifier	1	X
IC3, 4	RV1UPC22C	Power Amplifier	2	X
TR1	2SK37	FM RF Amplifier	1	X
TR2, 3	2SC920	FM Oscillator, FM Mixer	2	X
TR4	2SC945	Meter Amplifier	1	X
TR5	2SB175B	Eye Switching	1	X
TR6	2SC945	Eye Switching	1	X
TR7, 8	2SC900	AF Amplifier	2	X
TR9, 10	2SC945	AF Amplifier	2	X
D1, 2, 4, 8, 9	0A90	FM D. AGC, AM DET, Meter RECT	5	X
D3	RVDSC-15	FM AFC	1	X
D6, 7	2-0A90	FM Detector	1 pair	X
D10	**OA90** (OA95)	Eye Switching	1	X
		THERMISTOR AND RECTIFIER		
Se1	RVD10DC1R	Rectifier	1	X
Se2	RVD10DC1	Rectifier	1	X
		COILS AND TRANSFORMERS		
L1	RLA4P7	FM ANT Coil	1	X
L2	RLD4N21	FM DET Coil	1	X
L3	RLQY15G-5	FM Choke	1	X
L4	RL04N41	FM OSC Coil	1	X
L5	RLF2D89-0	AM ANT Coil	1	Ⓝ X
L6	RL02B83-M	AM OSC Coil	1	X
L7	RLM1×1-Q	MPX Coil	1	X
L8	RLM1C16	MPX Coil	1	X
T1	RLI4B152	FM IF Transformer	1	X
T2	RLI7W104-T	AM IF Transformer	1	X
T3	RLI2B451-T	AM IF Transformer	1	X
T4	RLI4B551	FM IF Transformer	1	X
T5	RLI4B552	FM IF Transformer	1	X
T6	RLM1C17	MPX Coil	1	X
T7	RLT5L66-W	Power Transformer	1	Ⓝ X
CF1, 2	RVFCF10S12CR	FM Ceramic Filter	2	X
		COMPONENT COMBINATION		
M1	EXAF203Z471	0.01 μF×2, 470Ω	1	Y
M2	EXA5DL04C	330pF×2, 4.7kΩ×2	1	Y
M3, 4	RXAF103P22HD	0.01pF×2	2	Y
		SWITCHES		
S3-1～S3-6	RSR41A	Band Selector Switch	1	Ⓝ X
S2-1～S2-2	RST13A	Loudness Switch	1	Ⓝ X
S4-1～S4-2	RST13A	AFC Switch	1	Ⓝ X
		SPEAKER		
SP		PM Dynamic Speaker, Ω	1	X
		RESISTORS		
R53, 54	ERC12GM100	10Ω, ½W, ±10%, Carbon	2	Y
R46, 108	ERC12GM221	220Ω, ½W, ±10%, Carbon	2	Y
R101, 102	ERC12GM331	330Ω, ½W, ±10%, Carbon	2	Y
R95, 96	ERC12GM105	1MΩ, ½W, ±10%, Carbon	2	Y
R97, 98, 52	ERC12GM335	3.3MΩ, ½W, ±10%, Carbon	3	Y
R55, 56	ERC12GM475	4.7MΩ, ½W, ±10%, Carbon	2	Y
R15, 44, 48, 74, 76, 92, 94	ERD14VJ101	100Ω, ¼W, ±10%, Carbon	7	Y
R14	ERD14VJ221	220Ω, ¼W, ±10%, Carbon	1	Y
R51, 86	ERD14VJ331	330Ω, ¼W, ±10%, Carbon	2	Y
R41	ERD14VJ471	470Ω, ¼W, ±10%, Carbon	1	Y
R9, 25, 26, 58, 60, 87, 88	ERD14VJ102	1kΩ, ¼W, ±10%, Carbon	7	Y
R40, 57, 59, 17, 63, 66	ERD14VJ152	1.5kΩ, ¼W, ±10%, Carbon	6	Y
R3, 42, 45	ERD14VJ222	2.2kΩ, ¼W, ±10%, Carbon	3	Y
R73, 75	ERD14VJ272	2.7kΩ, ¼W, ±10%, Carbon	2	Y
R20, 49, 50, 80, 81, 82, 83, 84, 85, 109	ERD14VJ332	3.3kΩ, ¼W, ±10%, Carbon	10	Y
R43	ERD14VJ472	4.7kΩ, ¼W, ±10%, Carbon	1	Y
R67, 68	ERD14VJ562	5.6kΩ, ¼W, ±10%, Carbon	2	Y
R47, 99, 100	ERD14VJ682	6.8kΩ, ¼W, ±10%, Carbon	3	Y
R28	ERD14VJ822	8.2kΩ, ¼W, ±10%, Carbon	1	Y
R10, 61, 64, 18, 106	ERD14VJ103	10kΩ, ¼W, ±10%, Carbon	5	Y

Ref.No.	Part No.	Description	Per Set (Pcs.)	Remarks
R6, 91, 93	ERD14VJ223	22kΩ, ¼W, ±10%, Carbon	3	Y
R5, 8	ERD14VJ333	33kΩ, ¼W, ±10%, Carbon	2	Y
R24, 107	ERD14VJ473	47kΩ, ¼W, ±10%, Carbon	2	Y
R34	ERD14VJ183	18kΩ, ¼W, ±10%, Carbon	1	Y
R7, 31, 32, 37, 111, 112	ERD14VJ104	100kΩ, ¼W, ±10%, Carbon	6	Y
R1, 11, 22, 29	ERD14VJ224	220kΩ, ¼W, ±10%, Carbon	4	Y
R23	ERD14VJ474	470kΩ, ¼W, ±10%, Carbon	1	Y
R110	ERD14VJ124	120kΩ, ¼W, ±10%, Carbon	1	Y
R71, 72	ERD14VJ824	820kΩ, ¼W, ±10%, Carbon	2	Y
R4	ERD14VJ100	10Ω, ¼W, ±10%, Carbon	1	Y
R13	ERD14VJ330	33Ω, ¼W, ±10%, Carbon	1	Y
R2, 12, 16, 19, 21	ERD14VJ470	47Ω, ¼W, ±10%, Carbon	5	Y
R89, 90	ERM14PR33	0.33Ω, ½W, ±10%, Wire	2	Y
R27	ERD14TJ102	1kΩ, ¼W, ±10%, Carbon	1	Y
R30	ERD14TJ222	2.2kΩ, ¼W, ±10%, Carbon	1	Y
R105	ERD14TJ104	100kΩ, ¼W, ±10%, Carbon	1	Y
R113	ERD14TJ333	33kΩ, ¼W, ±10%, Carbon	1	Y
R33	ERD14TJ183	18kΩ, ¼W, ±10%, Carbon	1	Y
R114, 115	ERD14TJ103	10kΩ, ¼W, ±10%, Carbon	2	Y
		VARIABLE RESISTORS		
R62, 65, 69, 70	EVFN0AU40A54	50K (A), Bass, Treble Control	4	Ⓝ X
R77	EVF475U40G14	10K (G), Balance Control	1	Ⓝ X
R78, 79	EVFS2AU40G14	10K (G), Volume Control	2	Ⓝ X
R38	EVLS3AA00B14	10K (B), Separation Control	1	Ⓝ X
		CAPACITORS		
C16	ECCD051R5C	1.5PF, 50WV, Ceramic	1	Z
C14	ECCD05040C	4PF, 50WV, Ceramic	1	Z
C4	ECCD05050C	5PF, 50WV, Ceramic	1	Z
C11, 21	ECCD05070DC	7PF, 50WV, Ceramic	2	Z
C43	ECCD05470K	47PF, 50WV, Ceramic	1	Z
C116	ECCD05101K	100PF, 50WV, Ceramic	1	Z
C12, 96, 97	ECCD05181K	180PF, 50WV, Ceramic	3	Z
C52	ECCD05560K	56PF, 50WV, Ceramic	1	Z
C114	ECKE05120KC	12PF, 50WV, Ceramic	1	Z
C41	ECKE05020C	2PF, 50WV, Ceramic	1	Z
C112, 113	ECKE05030C	3PF, 50WV, Ceramic	2	Z
C47	ECCD05331K	330PF, 50WV, Ceramic	1	Z
C56, 17	ECKE05102MY	0.001μF, 50WV, Ceramic	3	Z
C8	ECKD05102MA	0.001μF, 50WV, Ceramic	1	Z
C35	ECKE05103MY	0.01μF, 50WV, Ceramic	1	Z
C23, 25, 26, 30, 36, 44, 45, 48, 71, 72	ECKE05223P	0.022μF, 50WV, Ceramic	10	Z
C64	ECKD14101P	100PF, 1000WV, Ceramic	1	Z
C119	ECKE05472MY	4700PF, 50WV, Ceramic	1	Z
C109, 110	ECKE05222MY	0.0022PF, 50WV, Ceramic	2	Z
C7, 15	ECMS05100K-H	10PF, 50WV, Mica	2	Z
C3, 20, 22	ECMS05120K-H	12PF, 50WV, Mica	3	Z
C13	ECMS05220K-H	22PF, 50WV, Mica	1	Z
C34	ECMS05361J-H	360PF, 500WV, Mica	1	Z
C51, 98, 99	ECQS1471JZ	470PF, 100WV, Styrol	3	Z
C54	ECQS05562KZ	5600PF, 50WV, Styrol	1	Z
C55	ECQS1152KZ	1500PF, 100WV, Styrol	1	Z
C108	ECQG05152MZ-N	0.0015μF, 50WV, Polyester	1	Z
C80, 83	ECQG05222MZ-N	2200PF, 50WV, Polyester	2	Z
C38, 89, 91	ECQG05103MZ-N	0.01μF, 50WV, Polyester	3	Z
C75, 77	ECQG05153MZ-N	0.015μF, 50WV, Polyester	2	Z
C24, 39, 58, 63, 81, 84	ECQG05223MZ-N	0.022μF, 50WV, Polyester	6	Z
C102, 103	ECQG05224MZ-N	0.22μF, 50WV, Polyester	2	Z
C115	ECQG05333MZ-N	0.033μF, 50W, Polyester	1	Z
C76, 78, 87, 88	ECAG16ER22	0.22μF, 16WV, Electrolytic	4	Y
C59	ECEA6V47	47μF, 6.3WV, Electrolytic	1	Y
C42, 49, 57, 60, 106	ECEA10V33	33μF, 10WV, Electrolytic	5	Y
C62, 104, 105	ECEA10V47	47μF, 10WV, Electrolytic	3	Y
C90, 94, 95, 96	ECEA10V100	100μF, 10WV, Electrolytic	4	Y
C37, 107, 117, 118	ECEA16V10	10μF, 16WV, Electrolytic	4	Y
C73, 74, 111	ECEA25V3R3	3.3μF, 25WV, Electrolytic	3	Y
C40, 50, 79, 82, 85, 86, 92, 93, 100, 101	ECEA50V1	1μF, 50WV, Electrolytic	10	Y
C66	ECEA25V4R7	4.7μF, WV, Electrolytic	1	Y
C69, 70	ECEB16V2200	2200μF, 16WV, Electrolytic	2	Y
C46	ECEB25V4R7	4.7μF, 25WV, Electrolytic	1	Y
		VARIABLE CAPACITORS		
C1, 2, 10, 18, 28, 29, 32, 33	ECV5MD34×21A	Tuning Gang, W/Trimmer (C2, 10, 29, 33)	7	X
C19	ECV1ZW10P12	Trimmer, FM Oscillator	1	X
		CHASSIS		
CH1	RJJ41A	Jack, Headphone	1	Y

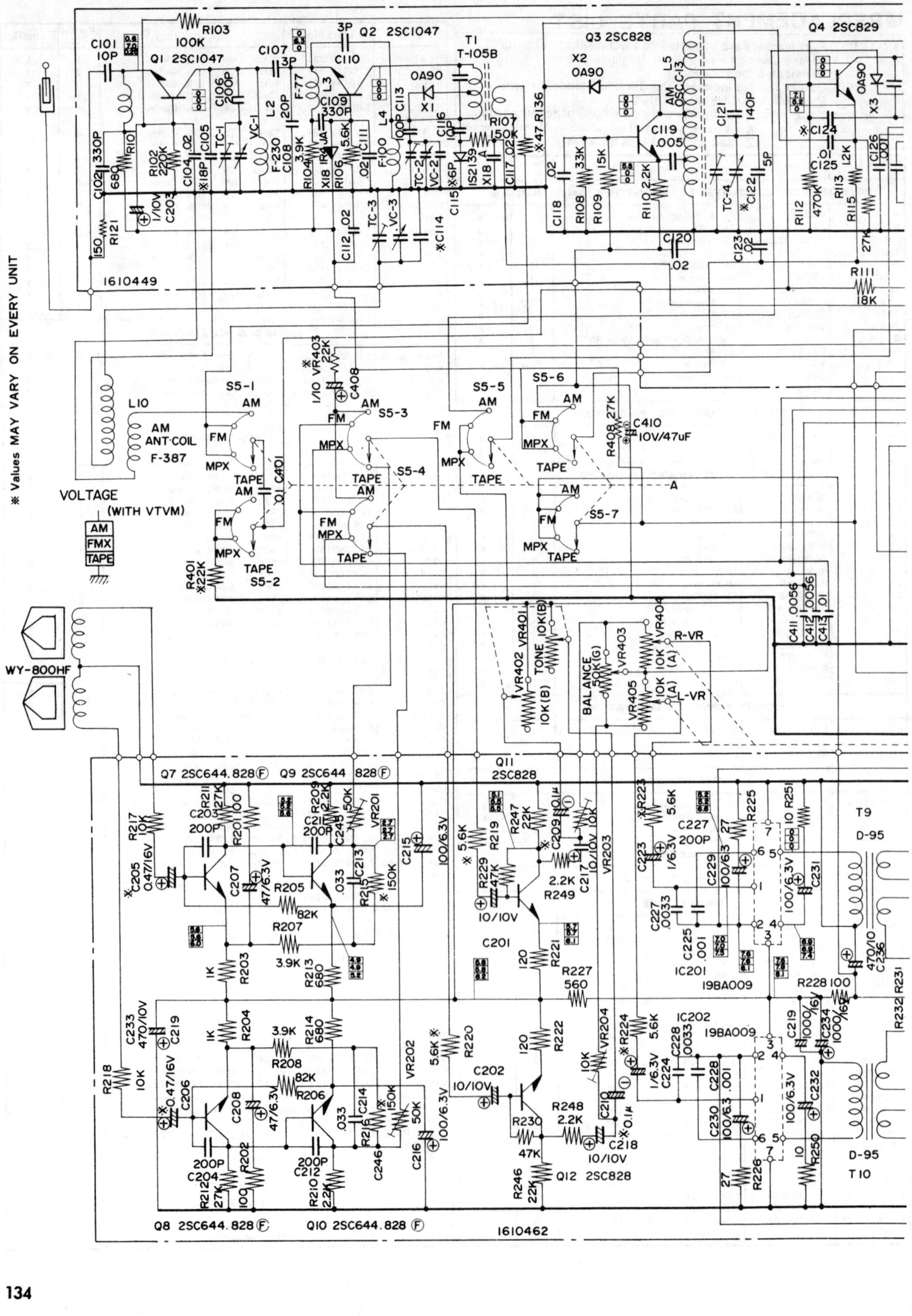
※ Values MAY VARY ON EVERY UNIT
VOLTAGE (WITH VTVM)
AM
FMX
TAPE
Q1 2SC1047
Q2 2SC1047
Q3 2SC828
Q4 2SC829
T1 T-105B
L5 AM OSC C-13
L10 AM ANT·COIL F-387
1610449
WY-800HF
S5-1
S5-2
S5-3
S5-4
S5-5
S5-6
S5-7
TONE 10K(B)
BALANCE 50K(G)
R-VR
L-VR
Q7 2SC644.828 Ⓕ
Q9 2SC644 828 Ⓕ
Q11 2SC828
Q8 2SC644.828 Ⓕ
Q10 2SC644.828 Ⓕ
Q12 2SC828
IC201 19BA009
IC202 19BA009
T9 D-95
D-95 T10
1610462

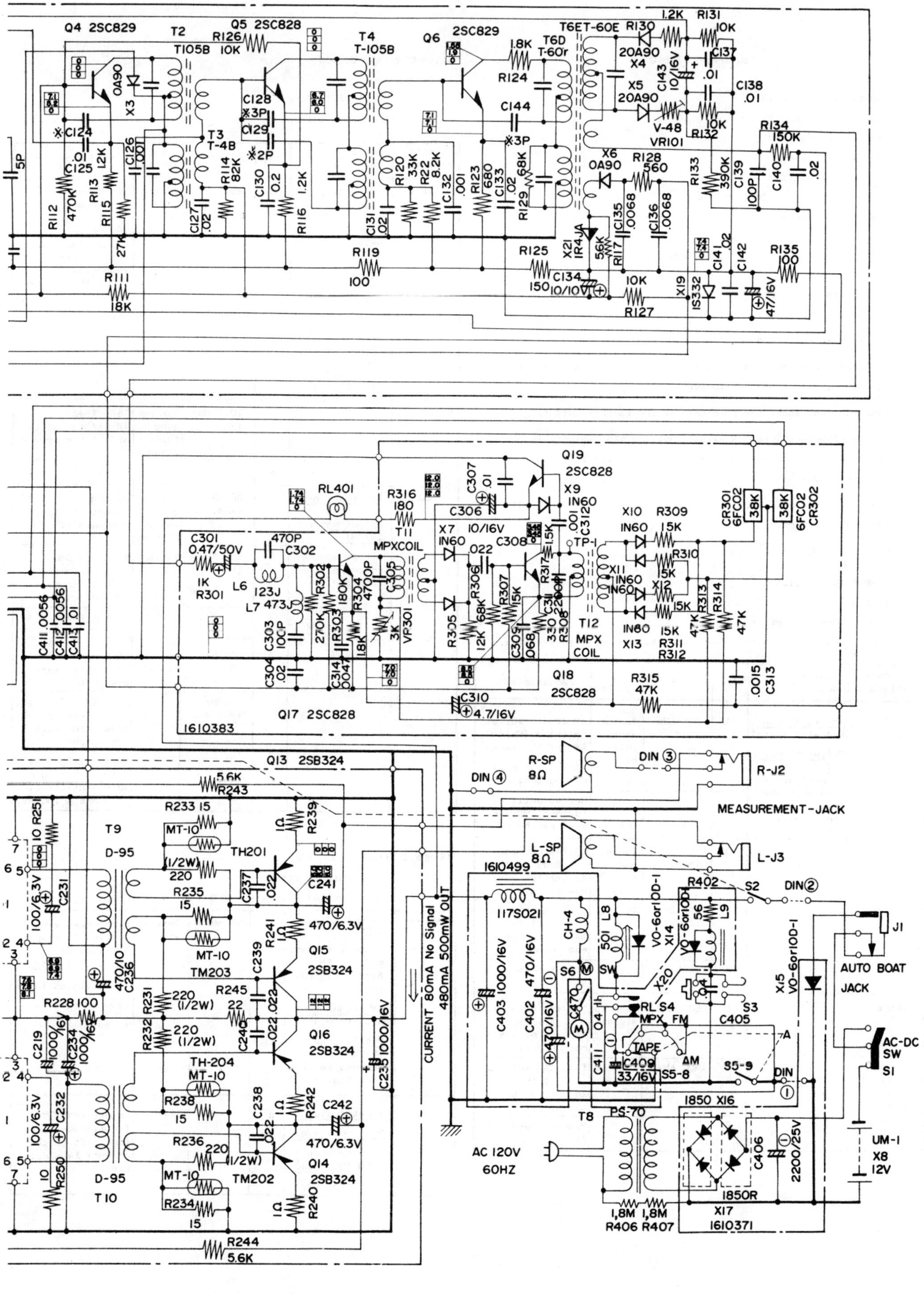

EQUIPMENT NEEDED

1. AM Signal Generator
2. FM Signal Generator
3. IF Sweeper (AM, FM)
4. Marker generator
5. Output Meter (VTVM)
6. Dummy Load 8 ohm (2W more than)
7. FM Stereo Generator

IMPORTANT

1. For FM AM Alignment, apply 1KHz modulation on the signal generator and radiation signal.
2. Dummy Load must be 8 ohm.
3. Connect the Dummy Load instead of the speaker, and connect a VTVM or circuit tester to it.
4. You must not short the terminals of the speaker output.
5. IF the Tape mechanism is aligned outside the cabinet, it will have to be re-alligned after re-insertion into the cabinet.
6. Use only non-metalic alignment tools to insure proper alignment at Radio section.

ALIGNMENT INSTRUCTION

AM Section

Control setting

Function switch ----- AM, Volume control ----- Max.

Circuit Alignment	Equipment connection	Step	Gen. Freq.	Dial Setting	Adjustment
IF	AM signal gen. Radiated signal Output meter. (VTVM) Connect across speaker voice coil. (See Fig.-1)	1	455 KHz (Mod.)	Tuning gang fully closed	AM IFT 1 (White) AM IFT 2 (White) AM IFT 3 (Black) Adjust for maximum output.
		2	"	"	Repeat until no further improvement can be made.
BAND	"	3	525 KHz (Mod.)	Tuning gang fully closed	L5 (Am OSC coil, Red) Adjust for maximum output.
		4	1650 KHz (Mod.)	Tuning gang fully opened.	Tc-4 (AM OSC Trim Trimmer) Adjust for maximum output.
		5	"	"	Repeat steps 3 & 4.
TRACKING	"	6	600 KHz	Tune to signal.	L10 (AM antenna trimmer) Adjust for maximum output.
		7	1400 KHz	Tune to signal.	Tc-3 (AM antenna trimmer) Adjust for maximum output.
		8	"	"	Repeat steps 6 & 7.

ALIGNMENT INSTRUCTION

FM Section

Control setting

Function switch ----- FM Volume control ----- Max.

Circuit Alignment	Equipment connection	Step	Gen. Freq.	Dial Setting	Adjustment
IF	IF sweep gen. high & low side to FM antenna terminal. Marker gen. The same as sweep generator. Oscilloscope Across point VR404 (VR405) (Pre amp. out and GND.) (See Fig. 2)	1	10.7 MHz (100-200 KHz sweep)	Tuning gang fully closed.	T1 (Blue) T2 (Blue) T4 (Blue) T6D (Blue) Adjust for maximum symmetrical response. (10.7 MHz at the center point.)
		2	"	"	Repeat step 1.
Ratio Det.	"	3	10.7 MHz (Mod.)	Tuning gang fully closed.	T6E (Red) Adjust for suitable "S" curve.
		4	"	"	Repeat step 3.
BAND	FM signal gen. Output meter (VTVM). Across speaker voice coil. (See fig.4)	5	86 MHz (Mod.)	Tuning gang fully closed.	L4 (FM OSC Coil) Adjust for maximum output.
		6	110 MHz (Mod.)	Tuning gang fully closed.	Tc-2 (FM OSC trimmer) Adjust for maximum output.
		7	"	"	Repeat steps 5 & 6.
RF	" "	8	90 MHz (Mod.)	Tune to signal.	L2 (FM RF Coil) Adjust for maximum output.
		9	106MHz (Mod.)	"	Tc-1 (FM RF trimmer) Adjust for maximum output.
		10	"	"	Repeat steps 8 & 9 to obtain suitable sensitivity at 90MHz & 106MHz.

FM MPX Section

Control setting

Function switch ---- FM MPX, Volume control ---- Max.

T11, T12	FM Stereo Gen. Composite out connect to ext. mod. of FM Signal generator. FM signal gen. Ant. terminal. Osciloscope (VTVM) Connect to TP1 (See Fig. 3)	1	98 MHz	Tune to signal.	First make sure section is properly aligned; Adjust T11, T12 to obtain suitable wave on osciloscope.
	FM stereo gen. FM signal gen. Same as step 1 VTVM (O.Scope) (See Fig. 4)	2	"	"	Adjust VR301,T12 for minimum output from left (Right channel is modulated.

MECHANICAL PARTS FOR MODEL 14-916

Ref. No.	Parts No.	RS Part No.	Description
L - 1	SRM2204		M 2 X 4 m/m Slotted Head Screw
L - 2	SOC1208		M 2 X 8 m/m Closs Head Screw
L - 3	SPM1904		M 2.6 X 4 m/m Closs Head Screw
L - 4	SPM1906		M 2.6 X 6 m/m Closs Head Screw
L - 5	SRM1908		M 2.6 X 8 m/m Closs Head Screw
L - 6	SBM1910		M 2.6 X 10 m/m Closs Head Screw
L - 7	SPM3306		M 3 X 6 m/m Closs Head Screw
L - 8	SPC3306		M 3 X 6 m/m Closs Head Screw
L - 9	SDM3306		M 3 X 6 m/m Close Head Screw
L - 10	SPM3308		M 3 X 8 m/m Closs Head Screw
L - 11	SPC3310		M 3 X 10 m/m Closs Head Screw
L - 12	SPM3310		M 3 X 10 m/m Closs Head Screw
L - 13	SDC3310		M 3 X 10 m/m Closs Head Screw
L - 14	SPM3408		M 4 X 8 m/m Closs Head Screw
L - 15	SDM3408		M 4 X 8 m/m Closs Head Screw
L - 16	STM3308		M 3 X 8 m/m Closs Head Screw
L - 17	SRP3306		M 3 X 6 m/m Closs Head Screw (color)
L - 18	SRP3310		M 3 X 10 m/m Closs Head Screw (color)
L - 19	SZM1310		M 3 X 10 m/m Tapping Screw
L - 20	SZM1308		M 3 X 8 m/m
L - 21	NHB0026		M 2.6 m/m Nut
L - 22	NHM3030		M 3 m/m Nut
L - 23	NHB3040		M 4 m/m Nut
L - 24	WSK9066		M 2.6 m/m Spring Lock Washer
L - 25	WSK3067		M 3 m/m Spring Lock Washer
L - 26	WSK4081		M 4 m/m Spring Lock Washer
L - 27	WPB9085		M 2.6 m/m Plain Washer
L - 28	WPM3108		M 3 m/m Plain Washer
L - 29	EEU0030		3 m/m E Pin
L - 30	RFD0Z03		2.5 X 3 m/m Rivet
L - 31	BQS0030		3 m/m Steel Ball
L - 32	HE03000		M 3 m/m Ground Lug
L - 33	HT03000		M 3 m/m Hold Lug
L - 34	HHR2003		2 X 3 m/m Eyelet Pipe
L - 35	WPB3063		3 m/m Plain Washer
L - 36	EEK0040		4 m/m E Pin
L - 37	WAU3075		M 3 m/m Toothed Lock
A - 1	21C8500	Z -0098	Front Cabinet (Left)
A - 2	21C8505	Z -0099	Rear Cabinet (Left)
A - 3	21C8501	Z -0100	Front Cabinet (Right)
A - 4	21C8504	Z -0101	Rear Cabinet (Right)
A - 5	21B8500	DB-0017	Battery Door
A - 6	22H7000	H -6008	Handle
A - 7	21H8500	H -6009	Handle Grip
A - 8	23M8510		Control Metal (Left)
A - 9	23M8511		Control Metal (Right)
A - 10	24T8505		Salan Net
A - 12	21N7067	K -0760	Volume Knob
A - 13	23M7132		Volume Knob Metal
A - 14	21N850'	K -0761	Channel Selector Button
A - 15	23M7136	H -0667	Channel Indicator Plate
A - 16	23M8517	DA-0012	Cartridge Door
A - 17	23M8516	J -4162	Jack Plate
A - 18	21W7018	RA-1095	Cap
A - 19	23U7005	H -0669	Hinge (Left)
A - 20	23U0006	H -0670	Hinge (Right)
A - 21	27U7000	H -0671	Latch (Right)
A - 22	23U7004	H -0672	Latch (Left)
A - 23	24L7043		Caution Label
A - 24	24L8511		Serial No. Label
A - 25	24L8507		Model No. Label
A - 27	24L7058		UL Label
A - 29	21N7066	K -0762	Tuning Knob
A - 30	23M7131	H -0701	Tuning Knob Metal
A - 31	21D8500	G -0071	Dial Window
A - 32	23P7004	D -1020	Dial Pointer
A - 33	23D8500	D -5019	Backscreen
A - 34	27A7000	A -0094	Telescopic Antenna
A - 35	21U8501		Stereo Lamp Window
A - 36	24W7067		Hinge Spacer
A - 37	24T8506		Sealed Board

MECHANICAL PARTS FOR MODEL 14-916

Ref. No.	Parts No.	RS Part No.	Description
B - 1	23S7012		Main Chassis
B - 2	23S7004	H -0673	Cartridge Support
B - 3	23W7023	RA-4667	Tape Guide
B - 4	23W7350	S -8026	Switch (A)
B - 5	23W7351	S -8027	Switch (B)
B - 6	24W7035		Switch Fiber (A)
B - 7	24W7039		Switch Fiber (B)
B - 8	23W7061	RA-5227	(B) Guide Arm
B - 9	21W0054		(B) Guide Roller
B - 10	25W0443	RA-4668	(B) Guide Roller Shaft
B - 11	21W7059	RA-0072	Guide Roller
B - 12	25W0444		Guide Roller Shaft
B - 13	23W0360	RA-4669	Motor Switch Plate (A)
B - 14	23W0361	RA-4690	Motor Switch Plate (B)
B - 15	24W0340		Switch Bake
B - 16	23W7020	RA-4671	Head Setting Plate
B - 17	23W7022	RA-4672	Head Arm
B - 18	23W7359	RA-4673	Solenoid Arm
B - 19	23W7357	RA-4674	Cam Arm (A)
B - 20	23W7358	RA-4675	Cam Arm (B)
B - 21	25W446		Both Rivet
B - 22	26W0048	RA-5228	"U" Spring
B - 23	22R7002	RA-3082	Fly Wheel
B - 24	22R7001	RA-7105	Fly Wheel Stand
B - 25	23W7053	RA-4676	Fly Wheel Holder (A)
B - 26	22R7003	RA-4677	Cam
B - 27	25W7017		Cam Shaft Holder
B - 28	25W7019	RA-4678	Guide Shaft (A)
B - 29	25R7021	RA-4679	Cam Shaft
B - 30	25W7011		Head Height Adjusting
B - 31	25W7018		Head Height Adjusting Screw Guide
B - 32	23W7370		22mm Washer
B - 33	26W7003	RA-5229	Cam Spring
B - 35	26W0104	RA-5230	Lock Spring
B - 36	21W0117	B -6144	Neoprene Belt
B - 37	21V7002	B -6145	Timing Belt
B - 38	25R0066	RA-2174	Motor Pulley
B - 39	21R7000	RA-2175	Tooth Pulley
B - 40	25W7072	RA-2172	Timing Belt Guide Pulley
B - 41	25W7071	H -0676	Timing Belt Guide Stud
B - 42	26W7025		Door Shaft
B - 43	26W0105	RA-5231	Door Spring
B - 44	23W7254		Door Shaft Holder
B - 45	26W7086	RA-5232	Solenoid Spring
B - 46	25W0743	RA-4680	Guide Shaft (B)
B - 47	23W1020	H -0677	Fly Wheel Holder (B)
B - 48	22W0002		Fly Wheel Bearing
B - 49	23W7405	H -0678	Volume Setting Bracket
B - 50	23W7060	H -0679	Channel Indicator Holder
B - 51	25W7039		Channel Indicator Shaft
B - 52	21W0086	H -0680	Channel Slector Shaft Holder
B - 53	25W7161	RA-4681	Channel Selector Shaft
B - 54	26W7064	RA-5233	Channel Selector Spring
B - 55	25W7058		P.C.B. Spacer
B - 56	21A0061		Insulator Pipe
B - 57	25W0565		Channel Switch (B)
B - 58	24W0687		Fiber Washer
B - 59	23W7425	H -0681	Hinge Setting Metal
B - 60	23W7343	H -0682	Rear Cabinet Setting Metal
B - 61	23W7125		Radiator Plate (A)
B - 62	23W7257		Radiator Plate (B)
B - 63	23B7000		Battery Terminal
B - 64	26B0033		Battery Terminal Spring
B - 65	25W1081		Hexagon Screw M 3 X 10 mm
B - 66	21W0069		AC. Cord Stopper
B - 67	23W1046		Cord Clamp (A)
B - 68	24L736		Battery Label
B - 69	25W7074		Chassis Setting Stud (B)
B - 70	21B7003	B -0063	Battery Case
B - 71	25W7080		Amp. P.C.B. Setting Stud
B - 72	25W0744		Channel Switch (A)
B - 73	23W7397		Speaker Setting Piece
B - 74	25W7075		Chassis Setting Stud (C)

MECHANICAL PARTS FOR MODEL 14-916

Ref. No.	Parts No.	RS Part No.	Description
B - 75	25W7164		Chassis Setting Stud (D)
B - 76	25W7166		Rear Cabinet Setting Stud (B)
B - 77	25W7167		Rear Cabinet Setting Stud (C)
B - 78	25W7168		Rear Cabinet Setting Stud (D)
B - 79	25W7256		Rear Cabinet Setting Stud (E)
B - 80	24W7063		Shaft Spacer
B - 81	24W0440		Cushion for Preventing Vibration (A)
B - 82			Vinyltube 3mm X 30mm
B - 83	25W8504		Chassis Setting Stud (E)
B - 84	25W7165		Rear Cabinet Setting Stud (A)
B - 85	23W7412		P.C.B. Setting Braket
B - 86	21P7009	D -0157	Dial Drive Pulley
B - 87	21A7001	H -0683	Antenna Holder
B - 88	23W7403		Dial String Pulley Holder (A)
B - 89	23W7404		Dial String Pulley Holder (B)
B - 90	25W0750		Dial String Pulley Stud
B - 91	25W7062		Dial Stringing Pulley Stud (B)
B - 92	27W7009	D -3015	Tuning Shaft
B - 93	21P7003	D -0158	Dial Stringing Pulley
B - 94	26W0122	RA-5234	Dial Stringing Spring
B - 95	24W7044		Tape for Dial Pointer
B - 96	25W7286		Dial Scale Setting Stud
B - 97	21W7012		Gum Spacer
B - 98	23W7235		Telescopic Antenna Setting Metal
B - 99	23A0021	A -4008	Telescopic Antenna Terminal
B - 100	25W0969		Hexagon Screw M 2.6 X 8 mm
B - 101	23W0132	H -0684	Lamp Holder
B - 102	25R0024		Button Spacer
B - 103	23W0886		Condencer Holder
B - 104			
B - 105	TOY0050		Dial Cord
B - 106	25W7082		Hexagon Screw M 3 X 16 mm
B - 107	23W7424		Jack Setting Metal
B - 108	24W7081		Insulator Board
B - 109			
B - 110	24W0668		Cloth (A) for Preventing Vibration
B - 111	24W7072		Cloth (B) for Preventing Vibration
B - 112	24W7087		M.P.X. Insulater Board
B - 113	23W0065		Cord Clamp (B)
B - 114	24W0308		Knob Spacer
B - 115	23W7398		P.C.B. Terminal
B - 116	24W7099		Cloth (C) for Preventing Vibration
B - 117	24W8502		Cushion for Preventing Vibration (B)
B - 118	24W7077		Dust Net (A)
B - 119	24W7078		Dust Net (B)
B - 120	23W1052		Cord Clamps (C)
B - 121	24W8505		Vinyl Washer
B - 122	24W8506		Gum Spacer
B - 123	21W8504		Tape Holder
B - 124	24L8514		Schematic Diagram

(D) ELECTRICAL PARTS LIST (AMP)

Ref. No.	Parts No.	RS Part No.	Description
HEAD	164M012	H -4118	Playback Head, Impedance 2.5K ohm at 1KHz
SP	1520167	S -4252	Speaker 8 ohm 2W
M	164M016	M -4158	DC Motor DC 12V
S5	1620131	S -1017	Rotary switch
S1	1623020	S -3004	AC DC Changeover switch
	1661016	S -9004	Solenoid 8.5 ohm
VR403	139N130	P -3012	Variable resistor 50K ohm (G)
VR404,405	139N208	P -2013	Variable resistor 10K ohm (B)
VR203,204	1390171	P -6018	Semi fixed resistor RV-10HG
VR201,202	1390182	P -6019	Semi fixed resistor RB-312
T8	118N070	TA-0242	Power Transformer (UL) AC120V;AC12V,13VA
T9, 10	1150095	TD-0015	Input Transformer Imp. 1K : 150ohm X 2
CH-3	117S021	CB-2087	Choke Coil 3 -- 6mH at 1KHz 400mA
CH-4	1170004	CA-2379	Choke Coil 110uH at 1KHz
PL401	1830021	L -0252	Pilot Lamp 6.3V 40mA
J1	1631040	J -0272	Ext. Battery Jack
J2, 3	1630047	J -0273	Measurement Jack
	1610462		Printed Curcuit Board
	1610371		Printed Curcuit Board (UL)
	1750052	W -1239	AC Cord (UL)
	1720041	H -0685	Lug Plate 1L4P

Ref. No.	Parts No.	RS Part No.	Description
	168C001	R -8026	Relay
VR401,402	139N231	P -1053	Variable Resistor 10K ohm (A) two sect.
	1720043	J -6068	7P Connector Plug
	1720044	J - .9	7P Connector Socket
	1720206	J -6070	4P DIN Connector Plug
	1720205	J -0270	4P DIN Connector Jack

(E) ELECTRICAL PARTS

Ref. No.	Parts No.	RS Part No.	Description	
Q7,8,9,10	C644		Transistor	2SC644 (MATSUSHITA) or 2SC828F (MATSUSHITA)
Q11,12	C828		Transistor	2SC828 (MATSUSHITA)
Q13,14,15, 16	B324		Transistor	2SB324 (MATSUSHITA)
IC201,202	19BA009		Integrated Curcuit	19BA009 (FUNAI) or LAD010 (MITSUMI)
X14,15,20	10D-1		Silicon Diode	10D-1 (STANLAY)
X16	1S-1850		Rectification Diode Silicon Compound Type	1S-1850 (SUNKEN)
X17	1S-1850R		Rectification Diode Silicon Compound Type	1S-1850R (SUNKEN)
TH-201,202 203,204	MT-10		Thermister	10 ohm at $25^{\circ}C$
C201,202 217,218	1203106		Electrolytic Capacitor	10μF/16V
C234,219	1202108		Electrolytic Capacitor	1000μF/10V
C406	1220045		Electrolytic Capacitor	2000μF/25V
C402,235	1203108		Electrolytic Capacitor	1000μF/16V
C403,407	1203477		Electrolytic Capacitor	470μF/16V
C213,236	1202477		Electrolytic Capacitor	470μF/10V
C241,242	1201477		Electrolytic Capacitor	470μF/6.3V
C215,216 229,230 231,232	1201107		Electrolytic Capacitor	100μF/6.3V
C207,208	1201476		Electrolytic Capacitor	47μF/6.3V
C223,224	1201105		Electrolytic Capacitor	1μF/6.3V
C409	1203306		Electrolytic Capacitor	30μF/16V
C205,206	1203474		Electrolytic Capacitor	0.47μF/16V
C225,226	1250102		Mylar Capacitor	0.001μF/50V (K)
C213,214	1250333		Mylar Capacitor	0.033μF/50V (K)
C237,238 239,240	1250223		Mylar Capacitor	0.022μF/50V (K)
C203,204 211,212	1270201		Ceramic Capacitor	200pF/50V (K)
C405	1275403		Ceramic Capacitor	0.04μF/50V (K)
C227,228	1250332		Mylar Capacitor	0.0033μF/50V(K)
C209,210	1211104		Aluminium Capacitor	0.1μF/10V (M)
R221,222	1326121		Carbon Resistor	120 ohm ¼W (K)
R201,202 228	1326101		Carbon Resistor	100 ohm ¼W (K)
R223,224 243,244 219,220	1326562		Carbon Resistor	5.6Kohm ¼W (K)
R211,212	1326273		Carbon Resistor	27K ohm ¼W (K)
R213,214	1326681		Carbon Resistor	680 ohm ¼W (K)
R217,218	1326103		Carbon Resistor	10K ohm ¼W (K)
R209,210 248,249	1326222		Carbon Resistor	2.2K ohm ¼W (K)
R207,208	1326392		Carbon Resistor	3.9K ohm ¼W (K)
R225,226	1326270		Carbon Resistor	27 ohm ¼W (K)
R403,246 247,401	1321223		Carbon Resistor	22K ohm ¼W (K)
R233,234 237,238 245,402	1326560		Carbon Resistor	56 ohm ¼W (K)
R205,206	1326823		Carbon Resistor	82K ohm ¼W (K)
R203,204	1326102		Carbon Resistor	1 K ohm ¼W (K)
R215,216	1326154		Carbon Resistor	150K ohm ¼W (K)
R229,230	1326473		Carbon Resistor	47K ohm ¼W (K)
R227	1326561		Carbon Resistor	560 ohm ¼W (K)
R250,251	1326100		Carbon Resistor	10 ohm ¼W (K)
R233,234 237,238	1326150		Carbon Resistor	15 ohm ¼W (K)
R245	1326220		Carbon Resistor	22 ohm ¼W (K)
R239,240 241,242	1330019		Winding Resistor	1 ohm ½W (K)
R252,253	1317185		Solid Resistor	1.8M ohm ½W (K)
R235,236 231,232	1317221		Solid Resistor	220 ohm ½W (K)

ELECTRICAL PARTS FPR MODEL 14-916

(D) ELECTRICAL PARTS LIST (TUNER)

Ref. No.	Parts No.	RS Part No.	Description
	1610449		Printed Circuit Board
VC-1,2,3,4	1290096	C -4220	Polyester Film Variable Capacitor VC-96
L5	1130135	CA-4226	AM OSC Coil 320uH ± 30uH C-135
L10	111B387	CA-2477	AM Ant. Coil 720uH ± 30uH F-387
T2,4	11LD105	CA-6300	FM IFT LD-105(10.7MHz) IMP.10K:2K PAT. C.50p
T6D	11DM060	CA-6750	FM IFT DD-060(10.7MHz) IMP.10K 30P
T6E	11ED060	CA-6796	FM IFT ED-060(10.7MHz) IMP.--- 50P
T3,5	1141004	CA-6716	AM IFT BD-004(455KHz) IMP.20K:350ohm 180P
T7	11CM204	CA-6912	AM IFT CD-204(455KHz) IMP.20K:5K 180P
L4	111B100	CA-4142	FM ANT Coil F-100
L2	111B230	CA-4176	FM RF Coil F-230 0.85uH ± 8%
L1,3	111B077	CA-2372	FM Trap Coil F-77 0.7uH ± 10%
VR101	1390048	P -0269	Semi-fixed Resistor V-48

(E) ELECTRICAL PARTS

Ref. No.	Parts No.	RS Part No.	Description
Q1,2	C1047		Transistor 2SC1047 (MATSUSHITA)
Q4,6	C829		Transistor 2SC829 (MATSUSHITA)
Q3,5	C828		Transistor 2SC828 (MATSUSHITA)
X1,2,3,6	OA90	R-7028	Diode OA90 (MATSUSHITA) Silicon
X4,5	20A90	R -7029	Diode 20A90 (MATSUSHITA) Silicon
X19	1S332	R -7030	Zener Diode 1S332 (JRC) Silicon
X22	1S2139A	R -7031	Variable Capacitor Diode 1S2139A (JRC)
X18,21	VD-1RAJA	R -7032	Variable VD-1RAJA (MATSUSHITA) (EYV-420DIR4JA)
C121	1235141		Styrol Capacitor 140pF/50V (J)
C135,136	1250682		Mylar Capacitor 0.0068µF/50V(K)
C104,111 112,117 118,120 123,127 131,130 134,140 141	1273202		Ceramic Capacitor 0.02µF/50V
C125,131 138,401	1272103		Ceramic Capacitor 0.01µF/50V (M)
C119	1272502		Ceramic Capacitor 0.005µF/50V (M)
C126,132	1272102		Ceramic Capacitor 0.001µF/50V (M)
C102,109	1271331		Ceramic Capacitor 330pF/50V (K)
C106	1271201		Ceramic Capacitor 220pF/50V (K)
C113,137	1271101		Ceramic Capacitor 100pF/50V (K)
C108	1271200		Ceramic Capacitor 20pF/50V (K),(D)
C105	1271180		Ceramic Capacitor 18pF/50V (K),(D)
C101,116	1270100		Ceramic Capacitor 10pF/50V (D)
C107,110 115,122	1270309		Ceramic Capacitor 3pF/50V (D)
C124,128 129,144	1270209		Ceramic Capacitor 2pF/50V, (D)
C103,408	1201105		Electrolytic Capacitor 1µF/6.3V
C124,410	1202476		Electrolytic Capacitor 47µF/10V
C134,143	1203106		Electrolytic Capacitor 10µF/16V
C411	1202227		Electrolytic Capacitor 220µF/10V
R119,135	1326101		Carbon Resistor 100 ohm ¼W (K)
R121,125	1326151		Carbon Resistor 150 ohm ¼W (K)
R128	1326561		Carbon Resistor 560 ohm ¼W (K)
R101,123	1326681		Carbon Resistor 680 ohm ¼W (K)
R113,116 130	1326122		Carbon Resistor 1.2K ohm ¼W (K)
R124	1326182		Carbon Resistor 1.8K ohm ¼W (K)
R110	1326222		Carbon Resistor 2.2K ohm ¼W (K)
R104	1326392		Carbon Resistor 3.9K ohm ¼W (K)
R122	1326822		Carbon Resistor 8.2K ohm ¼W (K)
R126,127 131,132	1326103		Carbon Resistor 10K ohm ¼W (K)
R109	1326153		Carbon Resistor 15K ohm ¼W (K)
R111	1326183		Carbon Resistor 18K ohm ¼W (K)
R401	1326223		Carbon Resistor 22K ohm ¼W (K)
R115,408	1326273		Carbon Resistor 27K ohm ¼W (K)
R108,120	1326333		Carbon Resistor 33K ohm ¼W (K)
R129	1326683		Carbon Resistor 68K ohm ¼W (K)
R114	1326823		Carbon Resistor 82K ohm ¼W (K)
R103	1326104		Carbon Resistor 100K ohm ¼W (K)
R107,134	1326154		Carbon Resistor 150K ohm ¼W (K)
R102	1326224		Carbon Resistor 220K ohm ¼W (K)
R133	1326394		Carbon Resistor 390K ohm ¼W (K)

ELECTRICAL PARTS FOE MODEL 14-916

Ref. No.	Parts No.	RS Part No.	Descritpion
R112	1326474		Carbon Resistor 470K ohm ¼W (K)
R106,117	1326562		Carbon Resistor 5.6K ohm ¼W (K)

(D) ELECTRICAL PARTS LIST (MPX)

Ref. No.	Parts No.	RS Part No.	Descritpion
T11	11PN001	CB-0069	MPX Multiplier Coil Inductance 14.8mH
T12	11QN001	CB-0070	MPX Tuning Coil Inductance 7mH
L6	1170017	CA-2339	Choke Coil 123J 12mH 50mA
L7	1170016	CB-2088	Choke coil 123J 47mH 80m
VR301	1390175	P -6022	Semi Fixed Resistor V-175 3K ohm
	1610383		Printed Circuit Board
	1720154	J -0271	10P Connector
CR301,302	1810013	CB-0071	Combined part (38KHz Filter) 6FC02

(E) ELECTRICAL PARTS

Ref. No.	Parts No.	RS Part No.	Descritpion
Q17,18,19	C828		Transistor 2SC828 (MATSUSHITA)
X7,8,9,10 11,12,13	1N60		Diode 1N60 (HITACHI) Silicon
C304	1273203		Ceramic Capacitor 0.02µF/50V (Z)
C307	1273103		Ceramic Capacitor 0.01µF/50V (Z)
C303	1230101		Styrol Capacitor 100pF/50V (J)
C302	1230471		Styrol Capacitor 470pF/50V (J)
C311	1230222		Styrol Capacitor 2200pF/50V (J)
C305	1230472		Styrol Capacitor 4700pF/50V (J)
C313	1250152		Mylar Capacitor 0.0015µF/50V (K)
C312	1250102		Mylar Capacitor 0.001µF/50V (K)
C308	1250223		Mylar Capacitor 0.022µF/50V (K)
C309	1250683		Mylar Capacitor 0.068µF/50V (K)
C314	1250472		Mylar Capacitor 0.0047µF/50V (K)
C411,412	1250562		Mylar Capacitor 0.0056µF/50V (K)
C413	1250103		Mylar Capacitor 0.01µF/50V
C301	1201474		Electrolytic Capacitor 0.47µF/6.3V
C310	1201475		Electrolytic Capacitor 4.7µF/10V
C306	1203106		Electrolytic Capacitor 10µF/16V
R316	1326181		Carbon Resistor 180 ohm ¼W (K)
R308	1326331		Carbon Resistor 330 ohm ¼W (K)
R301	1326102		Carbon Resistor 1K ohm ¼W (K)
R304	1326182		Carbon Resistor 1.8K ohm ¼W (K)
R305	1326123		Carbon Resistor 12K ohm ¼W (K)
R307,309 310,311 312	1326153		Carbon Resistor 15K ohm ¼W (K)
R315	1326473		Carbon Resistor 47K ohm ¼W (K)
R306	1326683		Carbon Resistor 68K ohm ¼W (K)
R303	1326184		Carbon Resistor 180K ohm ¼W (K)
R302	1326274		Carbon Resistor 270K ohm ¼W (K)
R318	1326150		Carbon Resistor 1.5K ohm ¼W (K)
R313,314	1321473		Carbon Resistor 47K ohm ¼W (J)

Chapter 5

Schematics, Alignment Instructions, and Parts Lists for Representative Automobile Stereo Tape Players

SCHEMATIC DIAGRAM (RADIO SECTION)

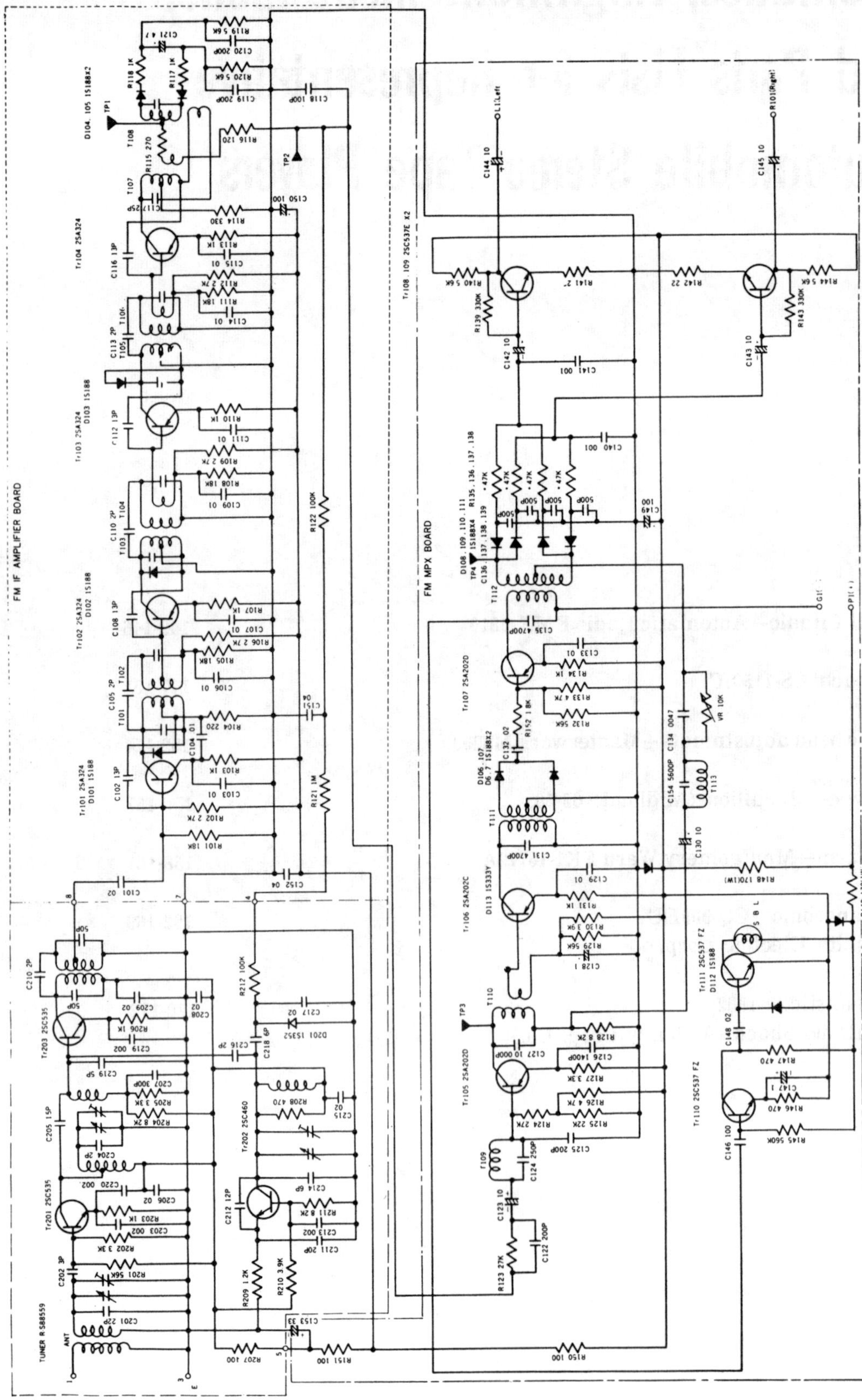

NORMAL VOTAGE OF EACH POINT

Transistor	Measurement point		
	Base	Emitter	Collector
Tr 101	6.3V	6.5V	162mV
Tr 102	6.3V	6.5V	0V
Tr 103	6.3V	6.5V	0V
Tr 104	6.3V	6.5V	235mV
Tr 105	7V	7.2V	2.8V
Tr 106	7.8V	7.9V	11.2V
Tr 107	7.7V	7.8V	5.1mV
Tr 108	0.62V	23.4mV	2.6V
Tr 109	0.63V	25.6mV	2.1V
Tr 110	1.9V	1.3V	7V
Tr 111	0.1mV	0.1mV	13.2V

DC voltage measured no signal, volume and tone is maximam position.

NOTE

1. All resistor values in OHMS unless otherwise noted. K 1000 OHMS
2. All capacitor values in MFD unless otherwise noted P Micro Micro FARAD
3. All resister 10%, ¼ Watts unless otherwise noted • – 5, ¼ Watts
4. Components in tape player numbered 1 thru 37
5. Components in FM radio numbered 101 thru 154
6. Components in FM tuner numbered 201 thru 218

SCHEMATIC DIAGRAM (TAPE PLAYER SECTION)

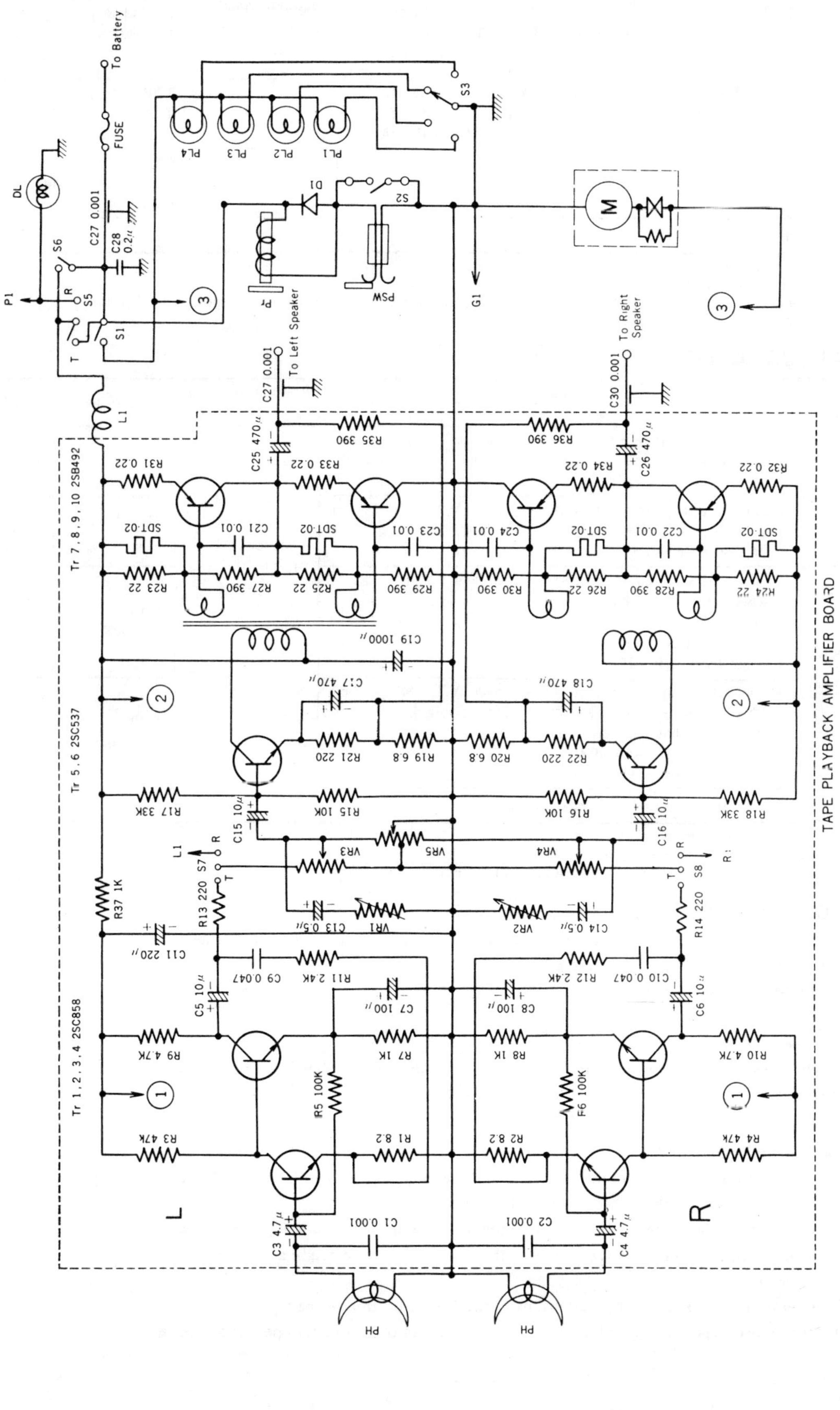

FM ADJUSTMENT

Step	Adjusting circuit	Connections		SG frequency	Position of tuning dial	Adjustment	V.T.V.M Oscilloscope
		Input	Output				
1	IF	Connect sweep generator to FM ANT Terminal.	Connect oscilloscope to test point TP1.	10.7 MHz (0% modulation)	Near max. capacity of VC. at position with no unrequired signal.	T203, T204 T101~T107	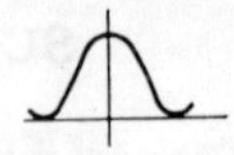
2	Ratio Det.		Connect oscilloscope to test point TP2 (IFT T108)	88 MHz (400 Hz 30% modulation)		IFT T108 (secondary)	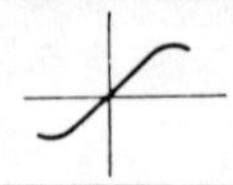
3	RF	Connect FM SG. to FM ANT. terminals.	Connect VTVM to Speaker Terminal	108 MHz (400 Hz 30% modulation)		T201, T202 T205	Max.
4						CT201 CT202 CT203	Max.
5	Repeat adjustments.						

PREPARE
1. Variable capacitor completely closed.
2. Set the dial pointer to very left line of dial scale.
3. Connect sweep generator, FM SG, V.T.V.M. and oscilloscope. FM antenna input impedance is 75 ohm.
4. Use a screwdriver with plastic grip for all adjustments.

FM MPX ADJUSTMENT

Step	Adjusting circuit	Connections		Position of tuning dial	Adjustment	V.T.V.M
		Input	Output			
1	SCA signal filter	Add SCA signal to FM stereo SG. to FM ANT. terminals. FM stereo signal OFF.	Connect V.T.V.M. to Test point TP3	Near max. capacity of VC. at position with no unrequired signal.	T109	Min.
2	19 KHz pilot signal	Connect FM stereo SG to FM ANT. terminals. 19 KHz signal ON. SCA signal OFF.	Connect V.T.V.M. to test point TP4		T110, T111 T112	Max.
3	FM stereo signal separation	Connect FM stereo SG to FM ANT. terminals. 19 KHz signal ON. Main channel, sub channel signal ON. Add 1000 Hz signal from Left CH.	Connect V.T.V.M. to Right speaker terminal.		T110 VR (10KΩ)	Min.
4		Connect FM stereo SG to FM ANT. terminals. 19 KHz signal ON. Main channel, sub channel signal ON. Add 1000 Hz signal from Right CH.	Connect V.T.V.M. to Left speaker terminal.		T110 VR (10KΩ)	Min.
5	Repeat steps 3 thru 6. Set at position with max. channel separaticn.					

PREPARE
1. Variable capacitor completely closed.
2. Connect FM stereo SG. and V.T.V.M.
3. Use a screwdriver with plastic grip for all adjustments.

HEAD ADJUSTMENT

Before starting this adjustment, the head and all tools to be used for alignment must be demagnetized.

The following equipments are required for head adjustment;

1) VTVM or VOM (precision circuit tester).
2) Load resistor (8 ohms, 10 watts).
3) DC power source or auto battery (12 volts).
4) Head alignment test tape cartridge (containing 1,000 Hz, −10 dB recorded signals on all channels).
5) Azimuth adjusting tape cartridge with 8,000 Hz, −10 dB recorded signal.
6) Small screwdriver.

HEAD ADJUSTMENT SEQUENCE

1) When the head is in the high position, the pole piece and the tape guide should be in parallel at the top edge, as shown in FIG. 1. It may be adjusted with the head height adjusting screw.
2) Set for maximum output with the 8 KHz portion of the test tape, using azimuth adjusting screw.
3) Set for maximum output with the 1 KHz portion of the tape. Using head height adjusting screw.
4) Repeat above steps 2 & 3.
5) Check for uniform output on all four programs, using the 1 KHz portion of the tape.
6) The adjustment must be performed several times over all tracks in order to attain perfect alignment.

REVISED PARTS/PRICE LIST FOR MODEL NO.

EMX-6810

DIA. NO.	PART NUMBER	DESCRIPTION	LIST PRICE
I.P.T	75A12787	Input Transformer	2.90
L 1	75A12256	Choke Coil	1.38
T 1	14A12878	I.F.T.	2.00
T 2	14A12879	I.F.T.	2.00
T3,5	14A12880	I.F.T.	2.00
T4,6	14A12881	I.F.T.	2.00
T 7	14A12882	I.F.T.	2.00
T 8	14A12883	I.F.T.	2.00
T 9	14A12884	Choke Coil	1.16
T 10	14A12885	MPX Transformer	2.00
T 11	14A12886	MPX Transformer	3.20
T 12	14A12887	MPX Transformer	2.00
T 13	14A12888	Choke Coil	1.16
TR-1, 2,3,4	6A12788	Transistor	1.60
TR-5, 6,111 108,109 110	8A12789	Transistor	1.70
TR-7 8,9,10	8A11667	Transistor	2.80
TR-101,102 103,104	6A12889	Transistor	2.20
TR-105,106 107	6A11301	Transistor	2.20
D101 112	1A11306	Diode	1.44
D113 114	1A12890	Diode	1.44
	5A12790	Thermistor	1.50
D 1	1A12214	Silicon Rectifier, SD-1	.60
VR	62A12501	Variable Resistor, 10 K Ohm	.60
VR 1 2	80A12891	Tone Control	2.76
VR 3 4,5	80A12892	Volume Control W/Push Switch	2.76
	71A12259	Antenna Socket	.50
	130A12260	Fuse Holder	.50
	108A12264	Fuse	.24
PL1, 2,3,4,	130A12793	Lamp, Channel Indicator and MPX	.24

DIA. NO.	PART NUMBER	DISCRIPTION	LIST PRICE
DL			
S.B.L.			
	130A12894	Lamp, Dial Scale	.24
	130A12795	Socket, Set Connection	.50
	130A12794	Plug, Set Connection (Cable Assy.)	3.00
	78A12896	FM Tuner Block	26.50
	12A12797	Speaker Lead (Gray)	.25
	12A12797	Speaker Lead (Green)	.25
	71A12268	Set Ground Lead	.25
	52A12898	Lamp Holder, Channel Indicator	1.00
	130A12796	Socket Insulating Tube	.24
	45A12800	Plate Spring, Antenna Mtg.	.50
	52A12801	Insulation Plate, Antenna Mtg.	.10
	52A12899	Dial Lamp Holder	.30
	45A12900	Dial Lamp Terminal	.05
S 6	94A12901	Power Push Switch	2.00
S 7,8	94A12902	Micro Switch Radio/Tape Change	2.00
C 3,4	20A12804	Electrolytic Cap. 4.7MFD, 6.3 WV	1.04
C 5,6, 15,16 121	20A12805	Electrolytic Cap. 10MFD, 6.3 WV	1.04
C 7,8	20A12806	Electrolytic Cap. 100MFD, 6.3 WV	1.50
C 11	20A12808	Electrolytic Cap. 220MFD, 16 WV	1.80
C 13, 14	20A12809	Electrolytic Cap. 0.5MFD, 10 WV	1.04
C 17, 18	20A12810	Electrolytic Cap. 470MFD, 6.3 WV	2.00
C 19,	20A12811	Electrolytic Cap. 1000MFD. 16 WV	2.40
C-25, 26	20A12813	Electrolytic Cap. 470MFD, 10WV	2.00
C 27, 29,30	23A12255	Ceramic Cap. 0.001MFD	.24
C 123, 130,142, 143,144, 145	20A12903	Electrolytic Cap. 10MFD, 16 WV	1.04
C128, 147	20A12526	Electrolytic Cap. 1MFD, 10 WV	1.04
C149, 150	20A12904	Electrolytic Cap. 100MFD 10 WV	1.50
C153	20A12905	Electrolytic Cap. 33MFD 10 WV	1.04
1	135A12906	Mechanism Chassis	5.00
2	52A12815	Bracket, Cartridge Slot Reinforcing	.50

REVISED PARTS/PRICE LIST FOR MODEL NO.

EMX-6810

DIA. NO.	PART NUMBER	DESCRIPTION	LIST PRICE
3	54A12816	Cartridge Pressure W/Roller	1.30
4	54A12817	Roller, Cartridge Pressure, Side	.25
5	45A12200	Roller Shaft For Roller (4)	.10
6	55A12818	Roller, Cartridge Pressure, Upper	.24
7	69A12819	Roller Shaft For Roller	.30
8	45A12820	Spring, Cartridge Pressure	.30
9	94A12907	Spring Switch, AC Power (S1,5)	2.00
10	130A12822	DC Motor	16.50
11	69A12174	Motor Pulley	.30
12	67A12175	Belt, Flywheel Drive	1.10
13	58A12823	Board, Head and Cam	1.60
14	45A12824	Flywheel Bearing	2.00
15	52A12825	Stopper, Tape Twining	.20
16	54A12826	Flywheel W/Ball	6.50
17	52A12827	Solenoid Contact	1.60
18	45A12828	Bearing, Head Trip	.50
19	52A12829	Tape Guide	1.00
20	52A12830	Board, Head Mounting	.30
21	52A12831	Cam Coupler Plate	.20
22	45A12832	Spring, Head Height Adjusting	.30
23	45A12833	Head Holder	.70
24	45A12832	Azimuth Adjustment Spring	.30
25	45A12835	Spring, Head Pushing Up, Large	.30
26	45A12836	Spring, Head Pushing Up, Small	.30
27	52A12837	Bracket, Cam Holding	.80
28	135A12838	Cam	3.00
29	69A12839	Cam Shaft	.50
30	135A12840	4-Channel Selector Cam	3.00
31	58A12841	Printed Circuit Board	1.25
32	130A12842	Rotary Brush (Consist of 33 and 34)	.80
33	130A12843	Brush	.40
34	130A12844	Boss	.40
35	135A12845	Pawl Assembly W/Pawl 36	1.10
36	135A12846	Pawl	.50
37	130A12847	Solenoid	2.50
38	45A12848	Solenoid Spring	.30
39	130A10698	Play Back Head	14.30

LUBRICATION

Lubricate those parts shown in Fig. 9

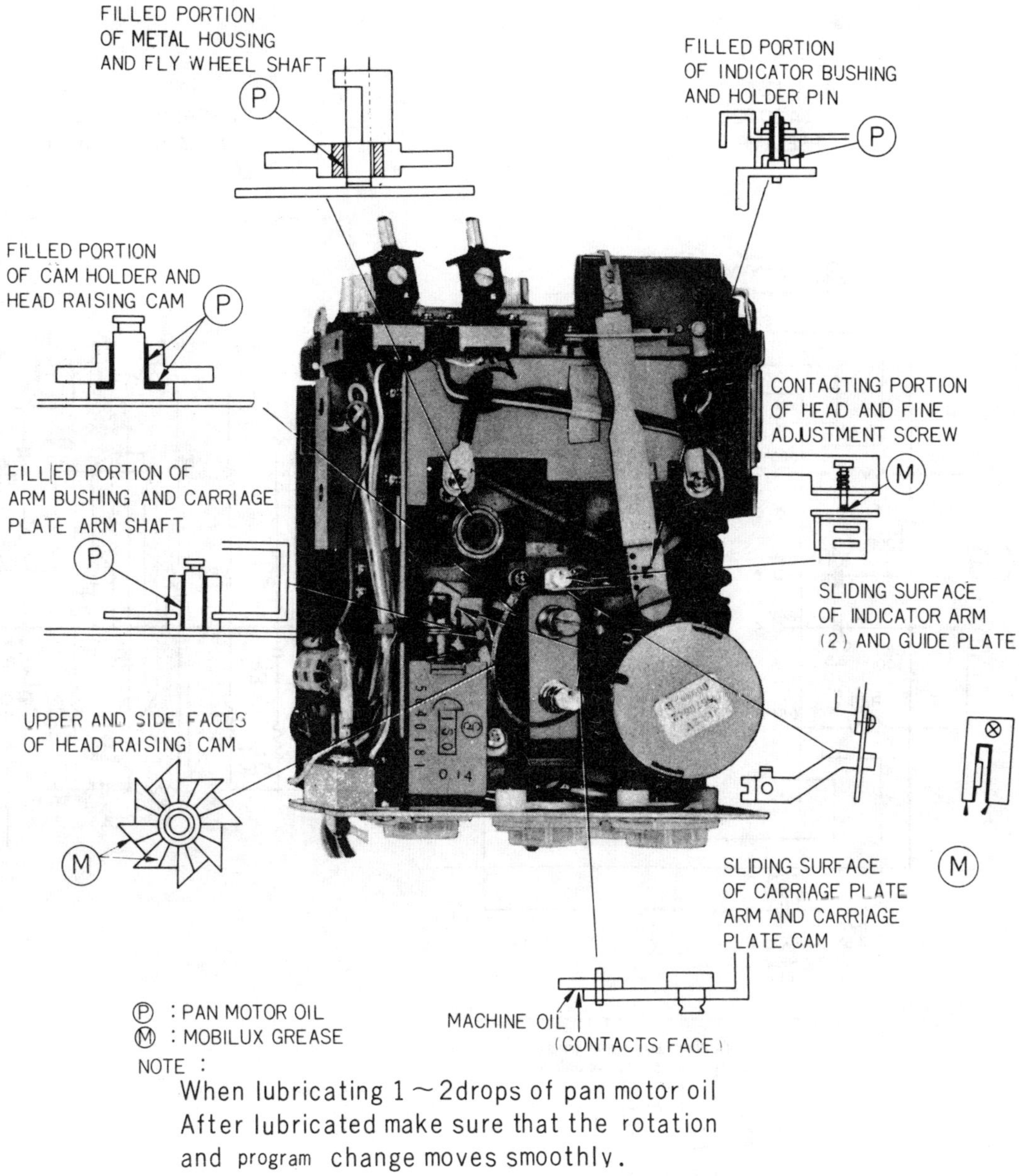

Fig. 9

ADJUSTMENT

An adjustment hole is provided on the upper cover so that adjustments can be carried out without removing the upper cover, as shown in Fig. 10.

This hole is plugged by a rubber bushing to protect dust from getting in.

CIRCUIT DIAGRAM

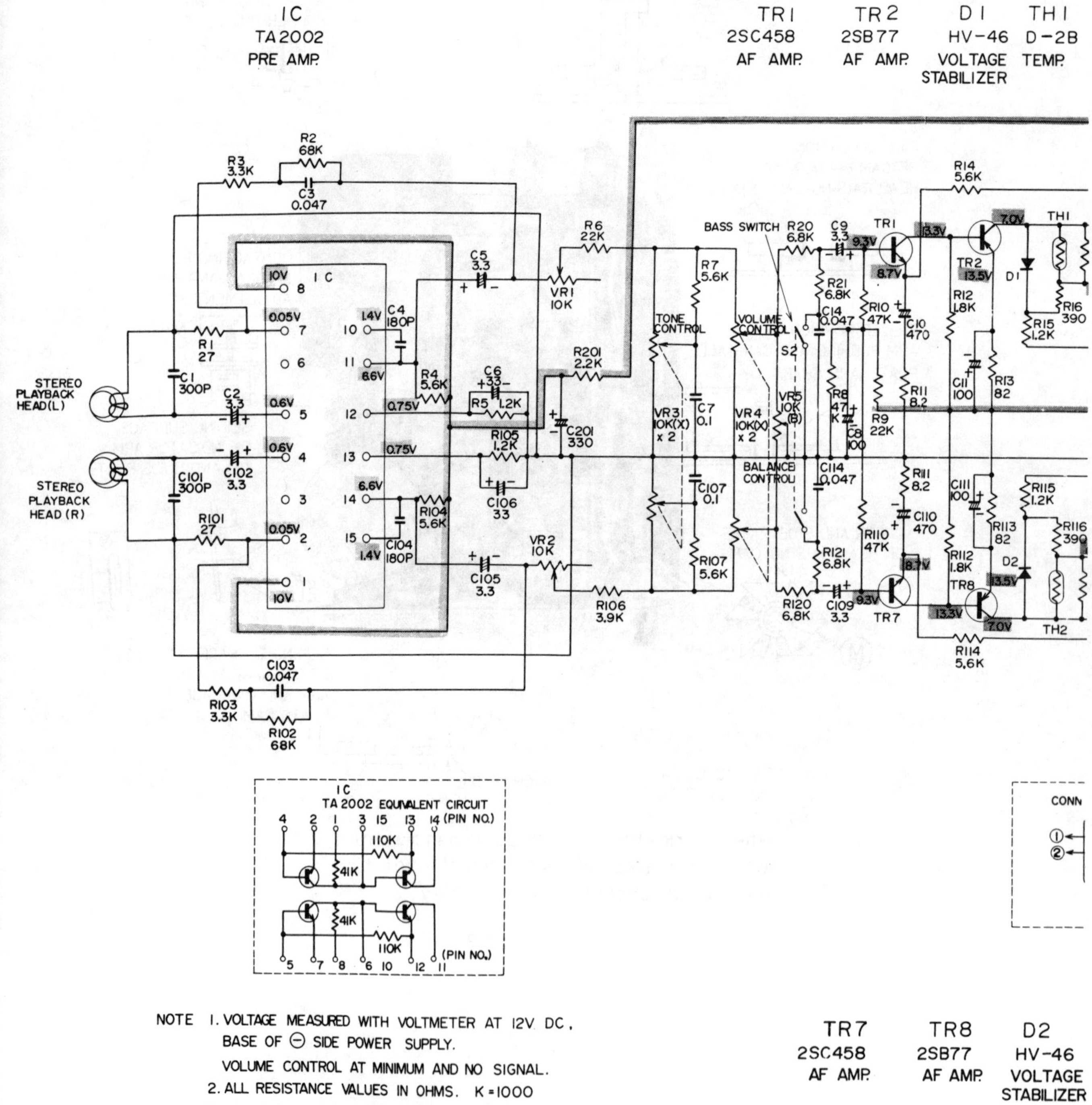

NOTE 1. VOLTAGE MEASURED WITH VOLTMETER AT 12V DC, BASE OF ⊖ SIDE POWER SUPPLY. VOLUME CONTROL AT MINIMUM AND NO SIGNAL.

2. ALL RESISTANCE VALUES IN OHMS. K = 1000

3. ALL CAPACITANCE VALUES IN μF EXCEPT "P" NOTED. P = $\mu\mu$F = PF

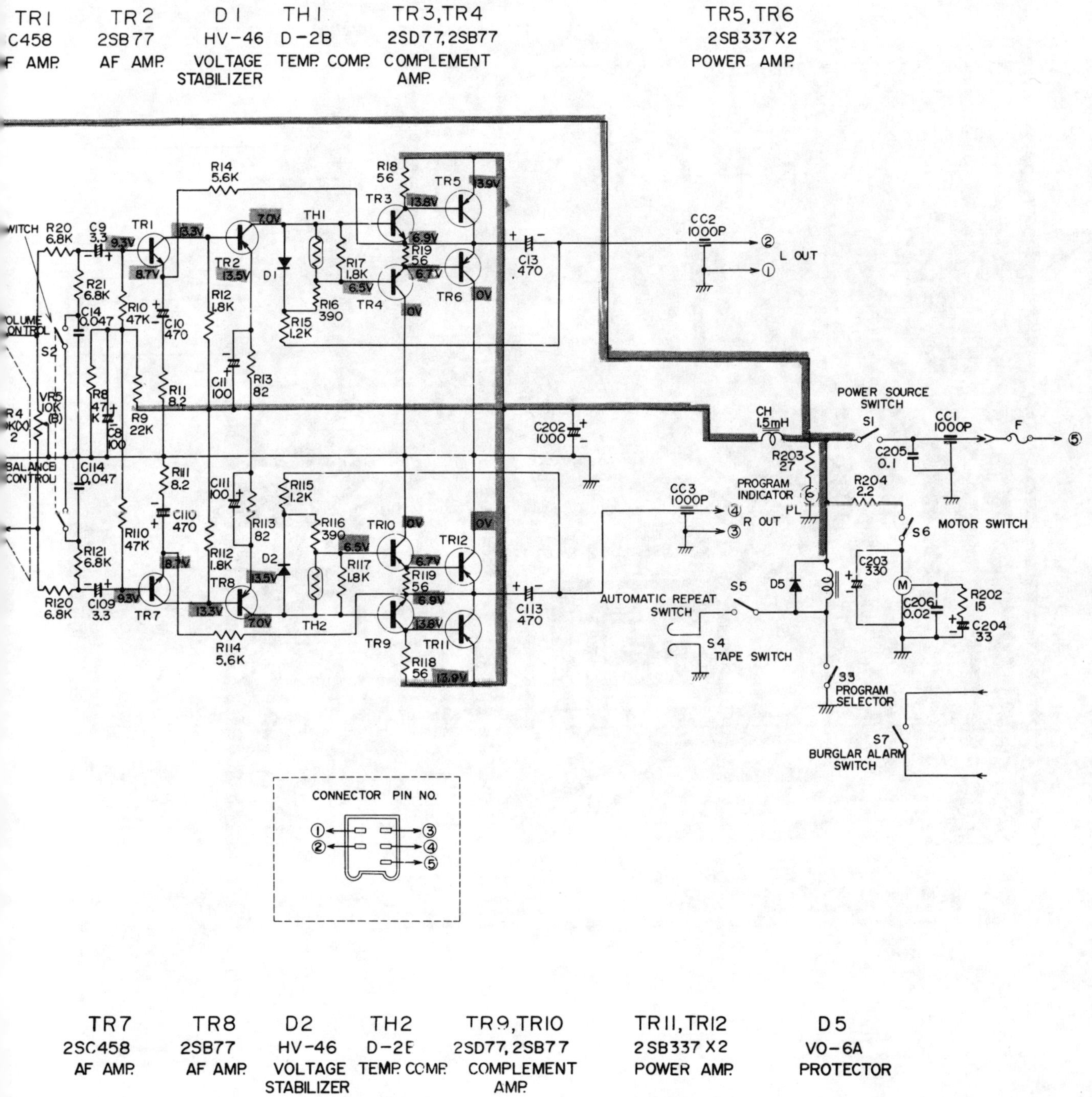
TR1
C458
F AMP.
TR2
2SB77
AF AMP.
D1
HV-46
VOLTAGE STABILIZER
TH1
D-2B
TEMP. COMP.
TR3,TR4
2SD77,2SB77
COMPLEMENT AMP.
TR5,TR6
2SB337 X2
POWER AMP.
L OUT
R OUT
POWER SOURCE SWITCH
PROGRAM INDICATOR
MOTOR SWITCH
AUTOMATIC REPEAT SWITCH
TAPE SWITCH
PROGRAM SELECTOR
BURGLAR ALARM SWITCH
CONNECTOR PIN NO.
TR7
2SC458
AF AMP.
TR8
2SB77
AF AMP.
D2
HV-46
VOLTAGE STABILIZER
TH2
D-2B
TEMP. COMP.
TR9,TR10
2SD77, 2SB77
COMPLEMENT AMP.
TR11,TR12
2SB337 X2
POWER AMP.
D5
VO-6A
PROTECTOR

CIRCUIT BOARD DIAGRAM

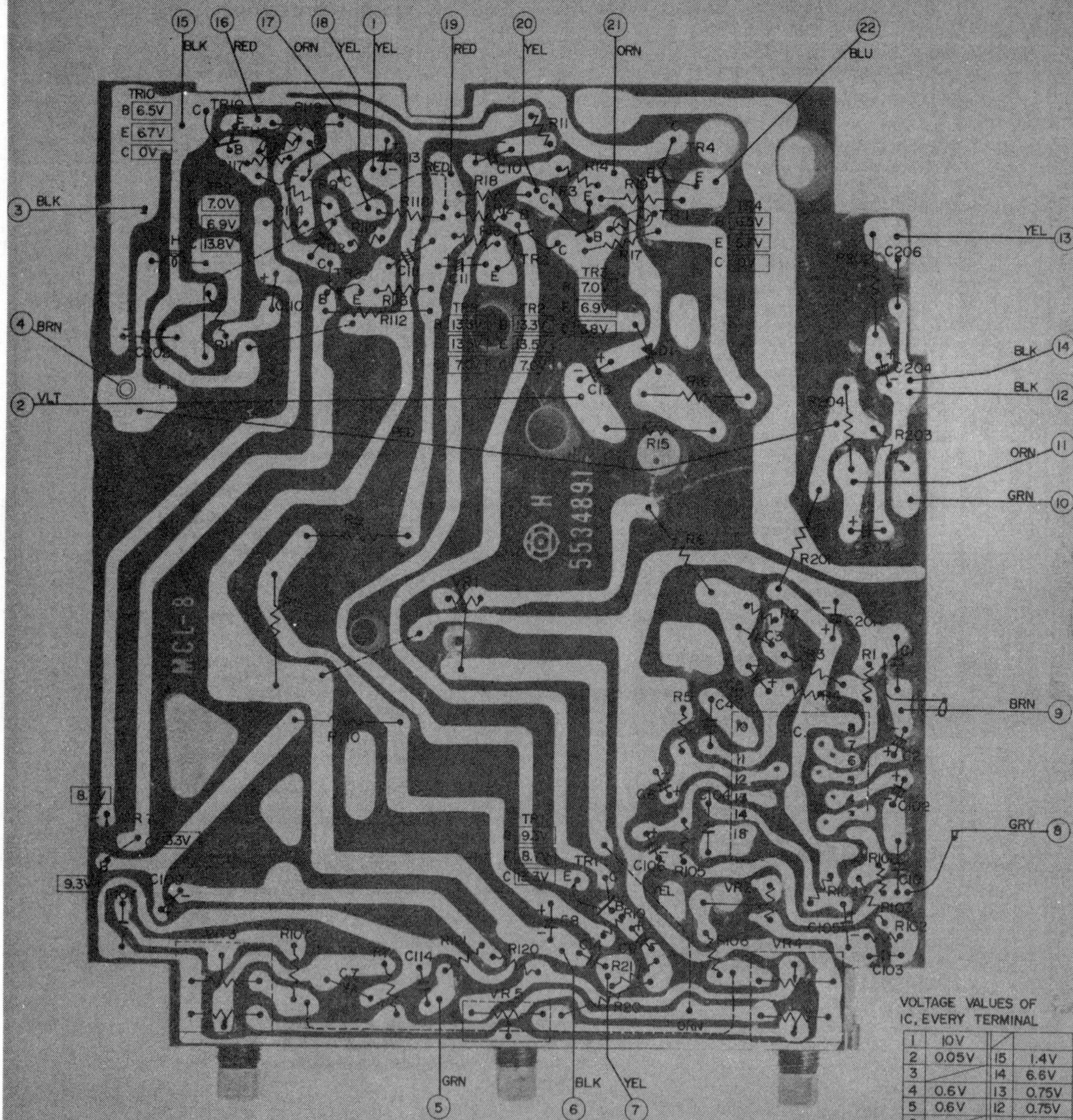

VOLTAGE VALUES OF IC, EVERY TERMINAL

1	10V		
2	0.05V	15	1.4V
3		14	6.6V
4	0.6V	13	0.75V
5	0.6V	12	0.75V
6		11	6.6V
7	0.05V	10	1.4V
8	10V		

REPLACEMENT PARTS

Symbol No.		Stock No.	Description		
CAPACITORS:					
C1	101	0248695	Ceramic, discal	300pF ±5%	50V
C2	102	0252613	Electrolytic	3.3μF	25V
C3	103	0275115	Mylar	0.047μF	50V
C4	104	0248730	Ceramic, discal	180pF ±10%	50V
C5	105	0252613	Same as C2 102		
C6	106	0252223	Electrolytic	33μF	6.3V
C7	107	0276111	Mylar	0.1μF	50V
C8		0252531	Electrolytic	100μF	15V
C9	109	0252613	Electrolytic	3.3μF	25V
C10	110	0252335	Electrolytic	470μF	10V
C11	111	0252231	Electrolytic	100μF	6.3V
C13	113	0252335	Same as C10 110		
C14	114	0275115	Same as C3, C103		
C 201		0252533	Electrolytic	330μF	16V
C 202		0256010	Electrolytic	1000μF	16V
C 203		0252533	Same as C201		
C 204		0252623	Electrolytic	33μF	25V
C 205		0276111	Same as C7 107		
C 206		0245018	Ceramic, discal	0.02μF	15V
CC 1, CC 2, CC 3		0234413	Cylindric ceramic	1000pF	500V
RESISTORS:					
R1	101	0137764	Carbon film	27Ω ±10%	¼SD
R2	102	0137911	Carbon film	68KΩ ±10%	¼SD
R3	103	0137857	Carbon film	3.3KΩ ±10%	¼SD
R4	104	0137860	Carbon film	5.6KΩ ±10%	¼SD
R5	105	0137852	Carbon film	1.2KΩ ±10%	¼SD
R6		0114468	Carbon film	22KΩ ±10%	¼P
R 106		0137858	Carbon film	3.9KΩ ±10%	¼SD
R7	107	0137860	Carbon film	5.6KΩ ±10%	¼SD
R8		0114529	Carbon film	47KΩ ±10%	¼P
R9		0114525	Carbon film	22KΩ ±10%	¼P
R10		0137909	Carbon film	47KΩ ±10%	¼SD
R 110		0114529	Same as R8		
R11	111	0137758	Carbon film	8.2Ω ±10%	¼SD
R12		0137854	Carbon film	1.8KΩ ±10%	¼SD
R 112		0114464	Carbon film	1.8KΩ ±10%	¼P
R13	113	0137770	Carbon film	82Ω ±10%	¼SD
R14	114	0137860	Carbon film	5.6KΩ ±10%	¼SD
R15		0114462	Carbon film	1.2KΩ ±10%	¼P
R 115		0137852	Same as R5 105		
R16		0114448	Carbon film	390Ω ±10%	¼P
R 116		0137808	Carbon film	390Ω ±10%	¼SD
R17	117	0137854	Same as R12		
R18	118	0137768	Carbon film	56Ω ±10%	¼SD
R19		0114380	Carbon film	56Ω ±10%	¼P
R 119		0137768	Same as R18 118		
R20	120	0137861	Carbon film	6.8KΩ ±10%	¼SD
R21	121	0137861	Same as R20 120		
R 201		0114465	Carbon film	2.2KΩ ±10%	¼P
R 202		0134291	Composition	15Ω ±10%	½GF
R 203		0134294	Composition	27Ω ±10%	½GF
R 204		0134281	Composition	2.2Ω ±10%	½GF
VR 1		0151808	Semi variable	10KΩ	
VR 2		0151808	Same as VR1		
VR 3		0151399	Variable resistor	10KΩ (X) ×2	
VR 4		0151399	Same as VR3		
VR 5		0151401	Variable resistor	10KΩ (B)	
TRANSISTORS:					
TR 1		0573481		2SC458	
TR 2		0573005		2SB77	
TR 3		5320306		2SD77	
TR 4		5320296		2SB77	
TR 5		0573040		2SB337	
TR 6		0573040	Same as TR5		

Symbol No.	Stock No.	Description	
TR 7	0573481		2SC458
TR 8	0573005	Same as TR2	
TR 9	5320306	Same as TR3	
TR 10	5320296	Same as TR4	
TR 11	0573040	Same as TR5	
TR 12	0573040	Same as TR5	
IC	5356001	Hybrid IC	TA2002
D 1	5340021		HV-46
D 2	5340021	Same as D1	
D 5	5330104	Diode	VO-6A
TH 1	0576038	Thermistor	D-2B
TH 2	0576038	Same as TH1	
COIL:			
CH	5220001	Choke	
MISCELLANEOUS:			
		for Final assembly	
F	0591138	Fuse	
	5741903	Connector H	
	8821117	Nut-5mmφ (4 req'd) } for body mounting	
	8813127	Spring washer-5mmφ (4 req'd) } for body mounting	
	8811117	Washer-5mmφ (4 req'd) } for body mounting	
	8832116	Bolt-5mmφ×16mm (3 req'd) } for body mounting	
	8785720	Tapping screw-5mmφ×20mm	
	7206831	Support strap	
	7772271	Wing nut	
	7716481	Head cleaning stick	
		for Case assembly	
	6214714	Escutcheon ass.	
	8744406	Binding screw-3mmφ×6mm (4 req'd) for escutcheon mounting	
	7207443	Upper cover	
	7664741	Cap	
	8744406	Binding screw 3mmφ×6mm (3 req'd) } for upper cover mounting	
	8744404	Binding screw-3mmφ×4mm } for upper cover mounting	
	7209281	Bottom cover ass.	
	8744406	Binding screw-3mmφ×6mm } for bottom cover mounting	
	7771883	Screw with locking washer } for bottom cover mounting	
	5740661	Earth cord	
	6267281	Knob ass.	
	6267301	Balance knob ass.	
	0711312	Pan head screw-2.6mmφ×12mm } (3 req'd) for knob mounting	
	8813123	Spring washer-2.6mmφ } (3 req'd) for knob mounting	
	8812113	Washer-2.6mmφ } (3 req'd) for knob mounting	
	6267261	Swtich button	
	5630651	Crose switch	
	0744304	Binding screw-2.6mmφ×4mm (2 req'd) for crose switch mounting	
	8821114	Nut-3mmφ	
	0638651	Staple	
	8744406	Binding screw-3mmφ×6mm	
	7623918	Name plate	
	7219011	Hanger	
	8813127	Spring washer 5mmφ } (4 req'd) for hanger mounting	
	8811117	Washer-5mmφ } (4 req'd) for hanger mounting	
	8832110	Bolt-5mmφ×10mm } (4 req'd) for hanger mounting	
		for Chassis assembly	
①	7224781	Cartridge guide ass.	
②	0532175	Slide switch	
	0923972	Staple	
	8741405	Binding screw-3mmφ×5mm	
	7771833	Screw with spring washer-2mmφ	

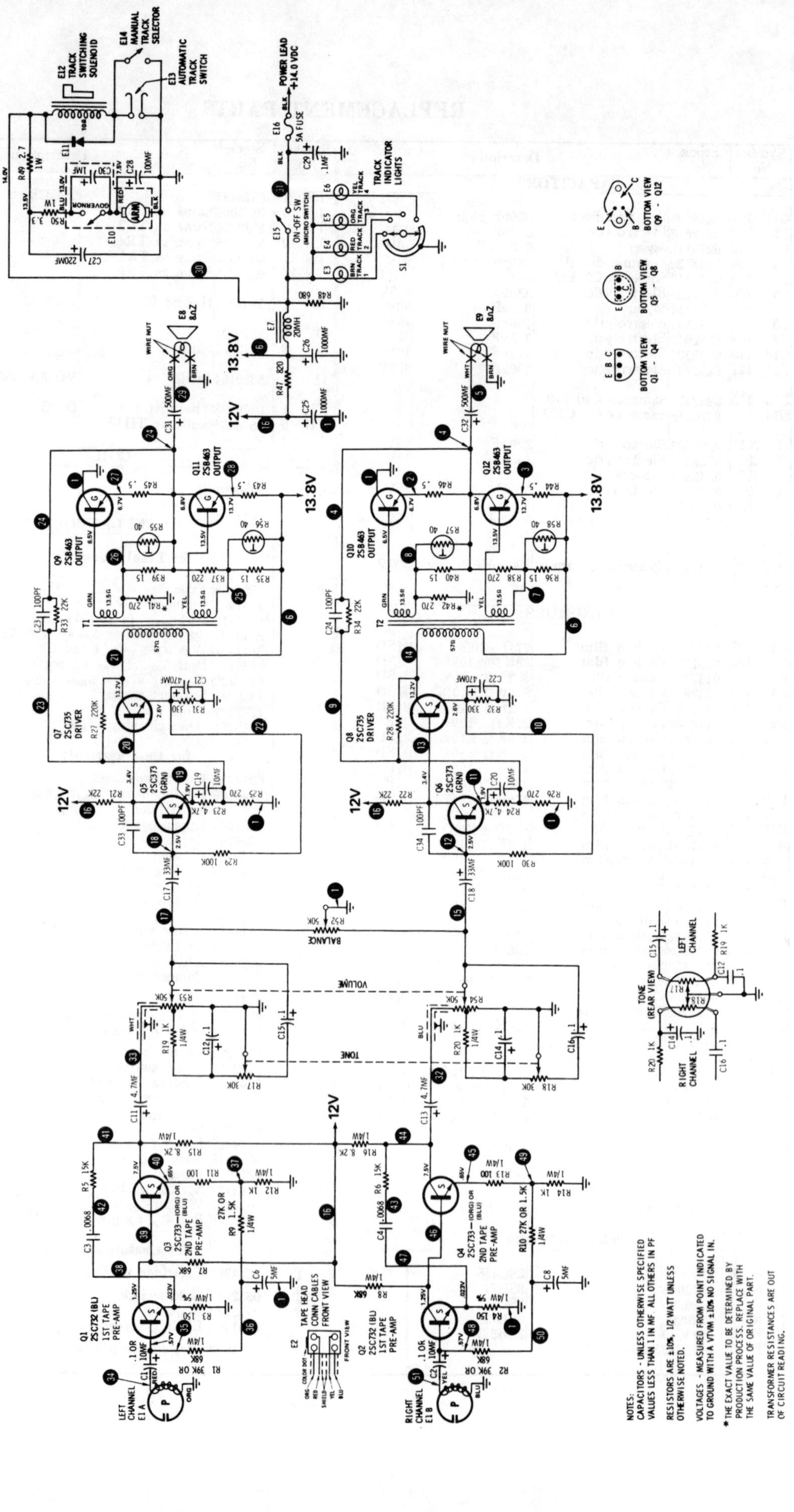
NOTES:
CAPACITORS - UNLESS OTHERWISE SPECIFIED VALUES LESS THAN 1 IN MF. ALL OTHERS IN PF
RESISTORS ARE ±10%, 1/2 WATT UNLESS OTHERWISE NOTED.
VOLTAGES - MEASURED FROM POINT INDICATED TO GROUND WITH A VTVM ±10% NO SIGNAL IN.
* THE EXACT VALUE TO BE DETERMINED BY PRODUCTION PROCESS. REPLACE WITH THE SAME VALUE OF ORIGINAL PART.
TRANSFORMER RESISTANCES ARE OUT OF CIRCUIT READING.

HEAD ADJUSTMENTS

NOTE: DO NOT operate tape player upside down with a loaded tape cartridge as cartridge may jam. Head adjustments are normally required for cases of crosstalk, poor high frequency response, or if the head has been replaced. A suitable test tape should be used for the height and azimuth adjustments.

1. HEAD HEIGHT ADJUSTMENT (REFER TO FIG. 3) - This is an adjustment to move the head up or down so that it can be positioned exactly in line with the pre-recorded information on the tape. Misadjustment of this screw would cause cross-talk and/or poor separation.

For convenience in making this adjustment, cut off and hollow out the rear end of a 1/4" nut driver so that a thin bladed screwdriver can be inserted through the center of the driver shank. (See Fig. 3).

A. Player should be right side up, even though adjustments are made from the bottom.
B. Connect output loads across each channel.
C. Insert test tape cartridge into tape slot and index head to tracks 2 and 6 (program number 2.)
D. Set balance control for maximum output from left channel and tone control to treble (maximum clockwise rotation.)
E. Connect scope (or VTVM-AC scale) across left channel output (track 2 information - 1Khz will be presented.)
F. Pull player forward halfway off bench. Lift front of player 30° - 45° so that adjustment screw hole at bottom is accessible.
G. Break cement seal around screw and nut and with a 1/4" spintight loosen nut.
H. Adjust height adjustment screw for a null; make sure there is a peak on either side of the null. The reason for the null is that the 1Khz information is recorded on the guard bands adjacent to track 2 on the test tape. This adjustment will minimize crosstalk.
I. After adjusting for null, lock the adjustment by tightening the locknut without moving the screw.

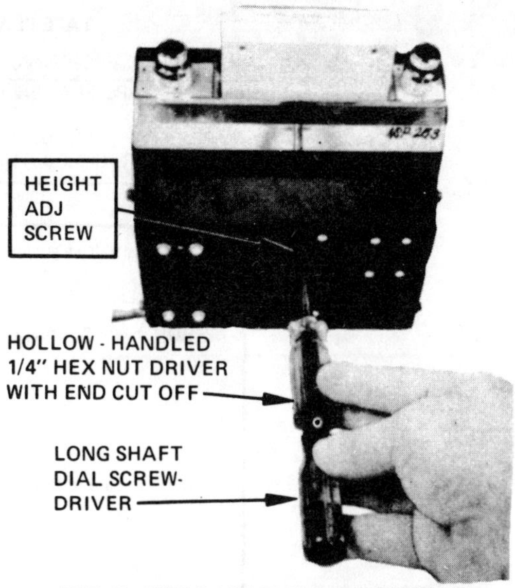

FIG. 3 HEAD HEIGHT ADJUSTMENT

2. AZIMUTH ADJUSTMENT (REFER TO FIG. 4) - This is an adjustment for maximum high frequency response. The adjustment is made from the top side of the player.

A. Set balance control for maximum right channel output.
B. Connect scope (or VTVM-AC scale) across the output of the right channel (track 6 information - 8Khz will be presented.)
C. Use a 1/16" Allen Head wrench and adjust for maximum output.
D. Repeat Steps 1 and 2 to optimize the adjustments.
E. Lock height and azimuth screws with glyptal (Part No. 11-40059A53) or other non-hardening cement.

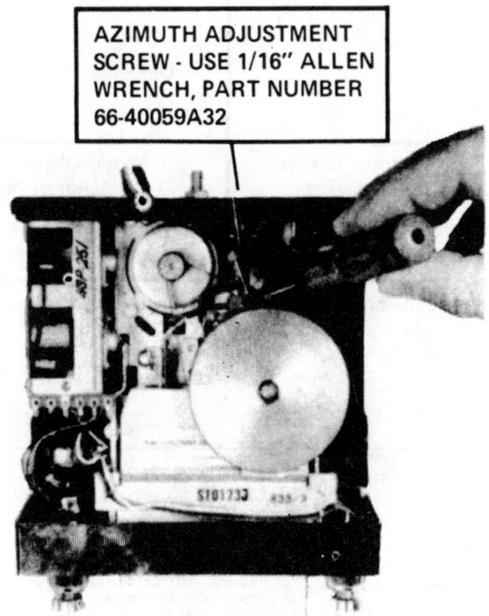

FIG. 4 AZIMUTH ADJUSTMENT

OTHER ADJUSTMENTS

1. INDEXING SOLENOID ADJUSTMENT (REFER TO FIG. 5). This adjustment should be made if the indexing solenoid (40) is replaced or does not index down to 11 volts. Adjust as follows:

A. Insert tape cartridge into tape player and adjust power supply to 11 volts.
B. Insert a screwdriver blade into slot on the solenoid adjustment screw. Turn the adjustment screw into the coil, and at the same time index the player. Continue to index while turning the screw until the solenoid just fails to index the tape head.
C. Back the screw out of the coil to where the head will properly index. Lock the retaining nut (43) with glyptal or other non-hardening cement.

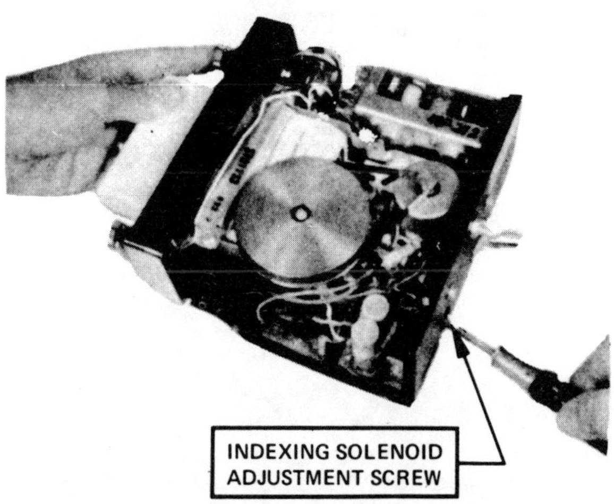

FIG. 5 INDEXING SOLENOID ADJUSTMENT

TAPE PLAYER TROUBLESHOOTING CHART

NOTE NO. 1: DO NOT OPERATE TAPE PLAYER UPSIDE DOWN WITH A LOADED TAPE CARTRIDGE AS CARTRIDGE MAY JAM – IF TAPE PLAYER IS OPERATED UPSIDE DOWN, USE A TAPE CARTRIDGE THAT HAS THE TAPE REMOVED.

NOTE NO. 2: MANY TAPE PLAYER PROBLEMS CAN BE CAUSED BY DEFECTIVE CARTRIDGES; BEFORE SERVICING THE TAPE PLAYER, CHECK THE CARTRIDGE BY SUBSTITUTING WITH A KNOWN GOOD CARTRIDGE.

SYMPTOM	CAUSE	REMEDY
Slow speed, wow and flutter.	1. Grease on drive surfaces.	Use isopropyl (rubbing) alcohol sparingly and wipe dry the following: outer edge of flywheel (110), motor pulley (part of 2), capstan shaft (109), and tape head (96.)
	2. Bent motor housing (8).	With cartridge inserted and operating, make certain drive belt does not ride too high or low on brass motor pulley; bend the motor housing using wide jawed pliers. Bend the housing away from the capstan shaft if belt rides along edge of flywheel; bend the motor housing toward the shaft if belt rubs against motor housing. CAUTION: Do not apply any pressure to the brass motor pulley.
	3. Mechanical binding.	Check for binding by first removing tape cartridge, turn on motor with actuating arm (25) and check flywheel speed with a strobe disc (Part No. 68-43109A08 or equivalent.) If speed is excessively slow, shut off the unit and remove drive belt (112); flywheel should rotate freely when turned by hand. If it does not, check for scored capstan shaft (109.) NOTE: If shaft is severely scored, replace capstan housing (27) also. If flywheel rotates freely when turned by hand, check for stretched drive belt (112), dirty or greasy flywheel or electrical problem.
	4. Defective motor (2).	Check motor by shorting out governor for an instant; motor should speed up, if motor does not speed up, it is defective. If motor does speed up, defect is probably in regulator circuitry.
Improper or erratic head indexing.	1. Indexing solenoid screw (42) out of adjustment.	Adjust per "Adjustment" section.
Crosstalk	1. Tape head (96) out of adjustment.	Check head height and azimuth adjustments with a test tape (Part No. 99-43309A01) as outlined in "Head Adjustments" section.
Tape breaks or winds up around capstan shaft.	1. Bent or burred edge on micro-switch actuating arm (25).	Replace actuating arm.

PARTS LIST

Ref No.	Part No.	Description
		ELECTRICAL PARTS
		CAPACITORS
C1,2		10 MF 16V lytic (in some sets)
C1,2		.1 MF 20% 100V mylar (in some sets)
C3,4		.0068 MF 10% 100V mylar
C6,8	D11623	5 MF 6V lytic
C11,13	D11624	4.7 MF 25V lytic
C12,14		.1 MF film cap
C15,16		.1 MF film cap
C17	D11625	33 MF 10V lytic
C18		33 MF 10V lytic
C19	D11626	10 MF 10V lytic
C20		10MF 10V lytic
C21	D11627	470 MF 6.3V lytic
C22		470 MF 6.3V lytic
C23		100 PF 50V disc
C24		100 PF 50V disc
C25	D11628	1000 MF 16V lytic
C26		1000 MF 16V lytic
C27	D11629	220 MF 16V lytic
C28	D11630	100 MF 16V lytic
C29		.1 MF film cap
C31	D11631	500 MF 15V lytic
C32		500 MF 15V lytic
C33		100 PF 50V disc
C34		100 PF 50V disc

Ref. No.	Part No.	Description
		TRANSISTORS
Q1,2	2SC732	2SC732 blu (1st tape preamp)
Q3,4	2SC733	2SC733 org (2nd tape preamp-in some sets)
Q3,4	2SC733	2SC733 blu (2nd tape preamp-in some sets)
Q5,6	2SC733	2SC733 grn (pre-driver)
Q7,8	2SC735	2SC735 yel (driver)
Q9,10	2SB463	2SB463 (output)
Q11,12	2SB463	2SB463 (output)
		RESISTORS & CONTROLS
R3,4	6-131276	150 5% 1/4W
R17,18		
R53,54	D9391	Control,multiple(vol-tone track selector)
R43	D16227	.5 20% 1W WW
R44	D16227	.5 20% 1W WW
R45	D16227	.5 20% 1W WW
R46	D16227	.5 20% 1W WW
R49	D16228	2.7 10% 1W WW
R50	D16229	3.3 10% 1W WW
R52	D9392	Control, 50K balance
R55	D16230	Thermistor,14D46S (40 ohm)
R56		Thermistor, 14D46S (40 ohm)
R57		Thermistor, 14D46S (40 ohm)
R58		Thermistor, 14D46S (40 ohm)
		TRANSFORMERS
T1	D3217	Driver
T2	D3217	Driver
		MISCELLANEOUS ELECTRICAL PARTS
E1	D-19788	Head, stereo playback (also ref #96)
E2	D-19720	Cable, tape head (also ref #93)
E3		
E6	D-19789	Program Indicator, lights
E7	D-21488	Choke, filter
E8,9	D 1259	Speaker, 5-1/4" VC: 8 ohm
E10	D-6121	Motor, regulated DC (also ref #2)
E11	D21489	10D1, rectifier (solenoid)
E12	D10179	Coil, solenoid (complete)
E13	D4272	Switch, contact (also ref #100)
E14	D4273	Switch, track selector: man (part of R17)
E15	D4274	Switch, micro (also ref #24)
E16	65-40348G01	Fuse

SCHEMATIC DIAGRAM

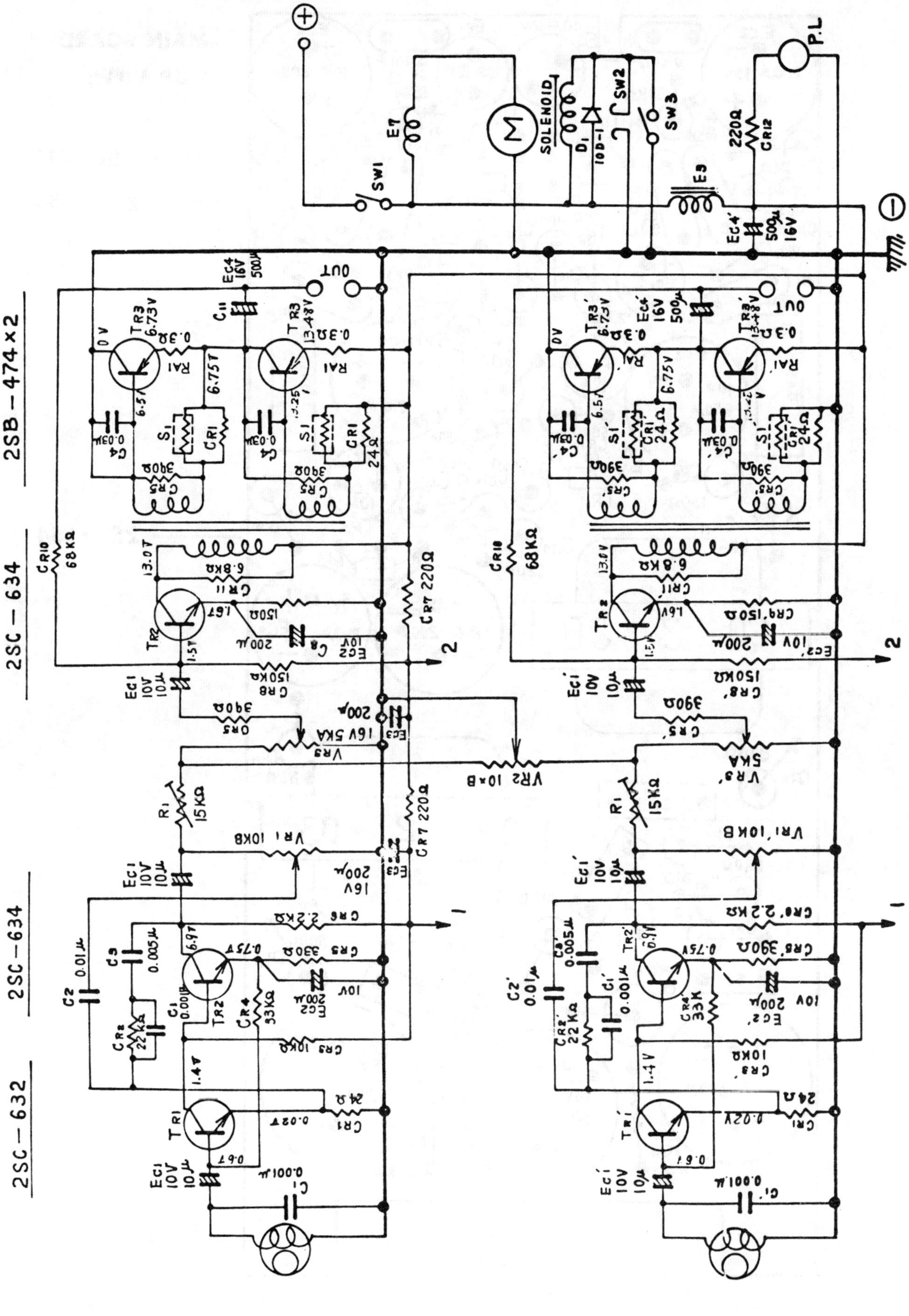

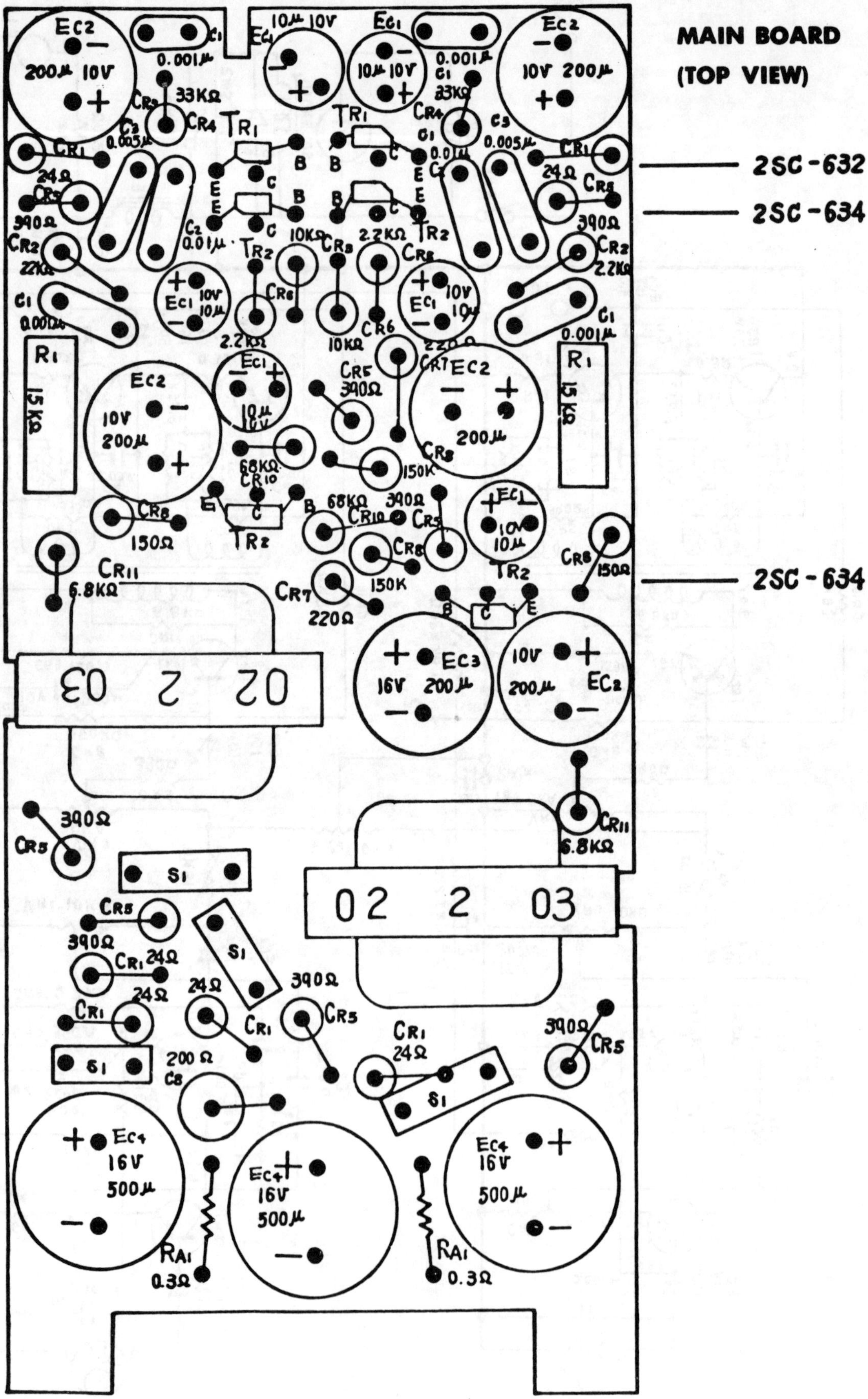
MAIN BOARD
(TOP VIEW)
2SC-632
2SC-634
2SC-634

REF. NO.	DESCRIPTION	PART NO.	LIST PRICE
	TRANSISTORS (cont'd)		
TR2	2SC634	09-302134	$ 2.10
TR3	2SB474	09-301030	2.95
D1	Diode, 10D-1	09-306054	.96
S1	Thermistor, SDT-04 (4 used)	09-307014	.46
	RESISTORS		
	ALL RESISTORS NOT SHOWN ON THIS PARTS LIST ARE CARBON, 1/8 WATT. FOR SPECIFIC VALUES CONSULT SCHEMATIC DIAGRAM.		
R1,1'	Control, Sensitivity 15K ohm	65-164008	.46
	CAPACITORS		
	ALL CAPACITORS NOT SHOWN ON THIS PARTS ARE MYLAR, 50 WV. FOR SPECIFIC VALUES CONSULT SCHEMATIC DIAGRAM.		
EC1,1'	Electrolytic, 10MF/10V	77-336106	.74
EC2,2'	Electrolytic, 200MF/10V	77-336207	.74
EC3	Electrolytic, 220MF/16V	77-337227	.96
EC4,4'	Electrolytic, 500MF/16V	77-337507	.96

With theft alarm—Montgomery Ward FK1-16779A

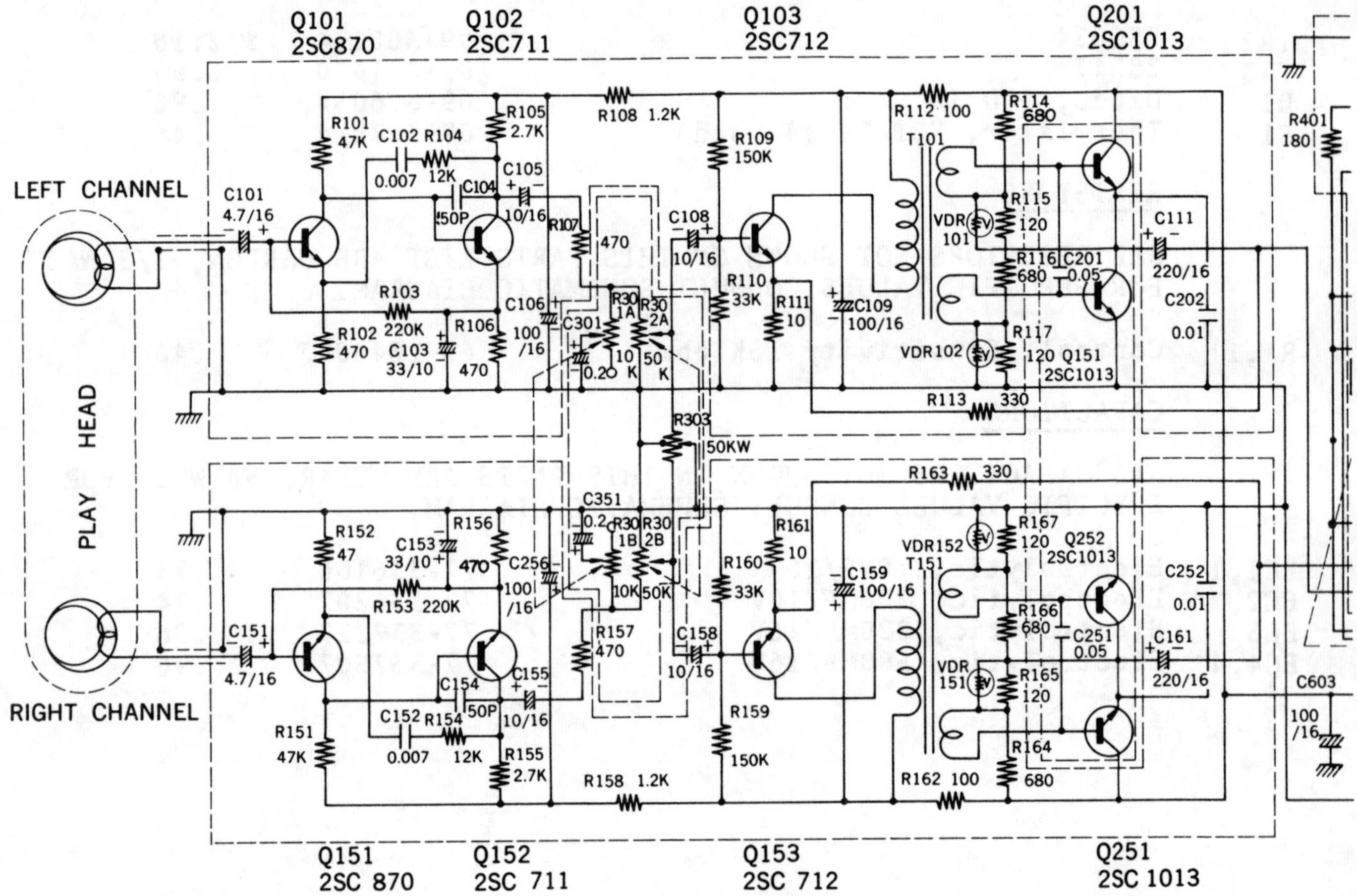

VOLTAGE CHART

REF SYMBOL	TRANSISTOR TYPE	EMITTER	BASE	COLLETOR
Q101,151	2SC 870	0.01	0.6	1.3
Q102,152	2SC 711	0.7	1.3	6.5
Q103,153	2SC 712	0.1	O 7	11.0
Q201,251	2SC 1013	7.0	7.6	14.0
Q202,252	2SC 1013	0	0.6	7.0

1 DC SUPPLY VOLTAGE +14VOLTS.
2 ALL MEASUREMENTS TO POSITIVE LINE ON PCB USING VTVM.
3 READINGS ARE ± 20%
4 VOLUME CONTROL IN MINIMUM OUTPUT POSITION.

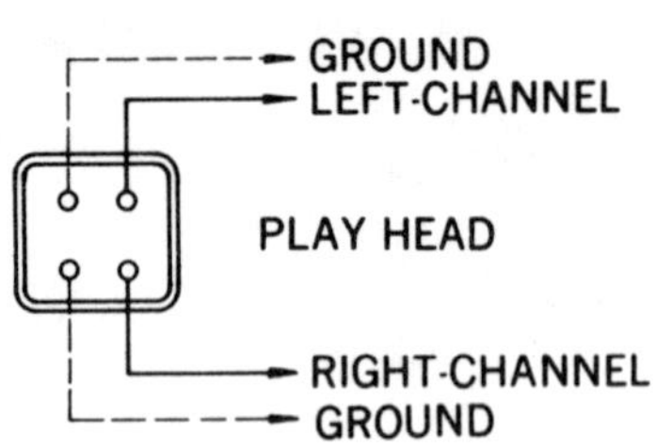

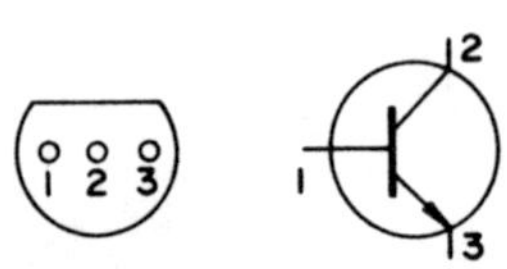

1 BASE
2 COLLECTOR
3 EMITTER
SILICON TRANSISTOR
Q101 Q151
Q102 Q152
Q103 Q153

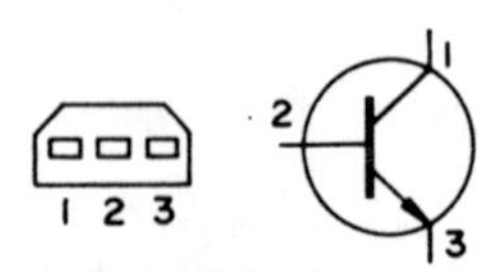

1 COLLECTOR
2 BASE
3 EMITTER
SILICON TRANSISTOR
Q201 Q251
Q202 Q252

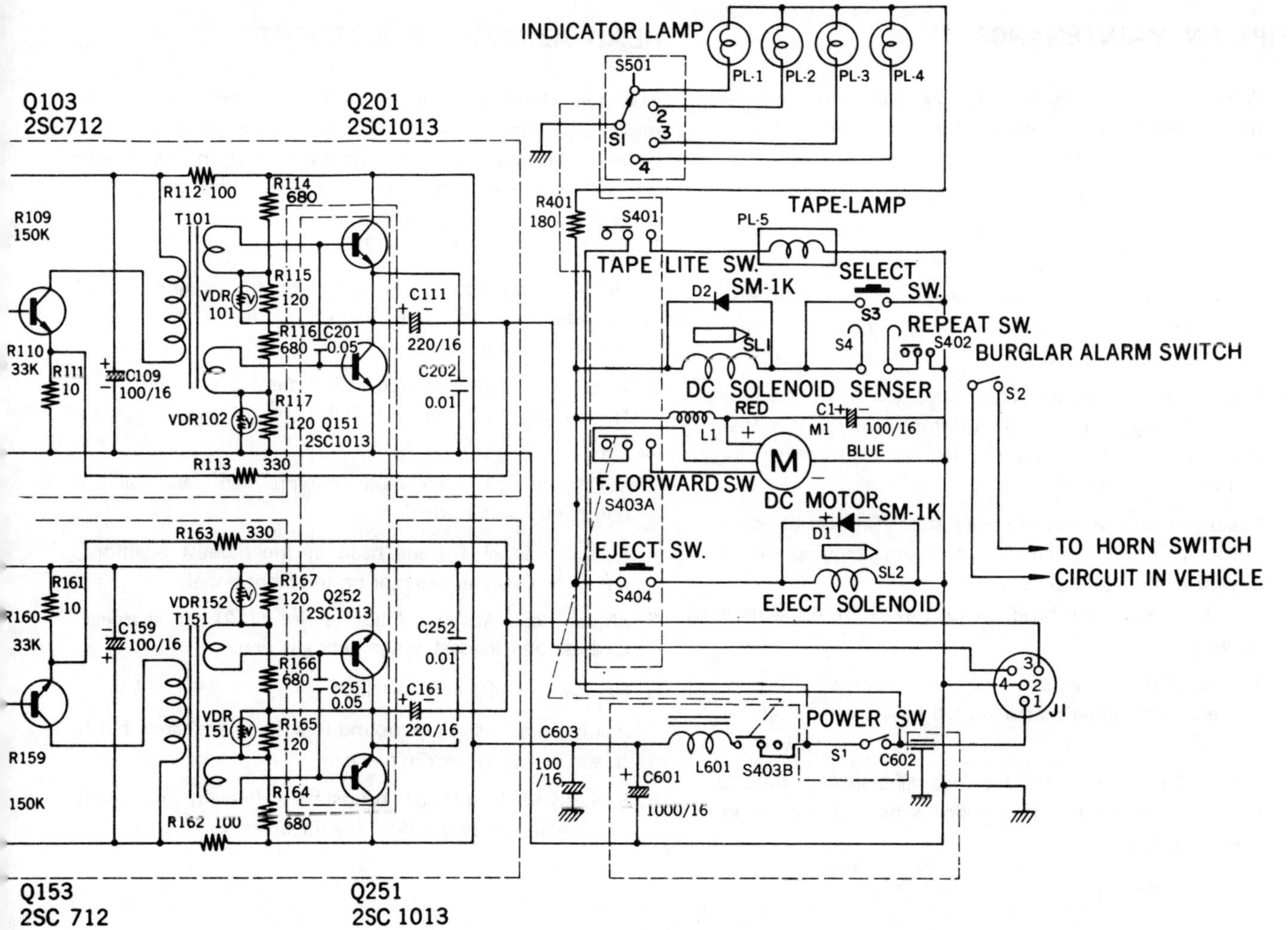

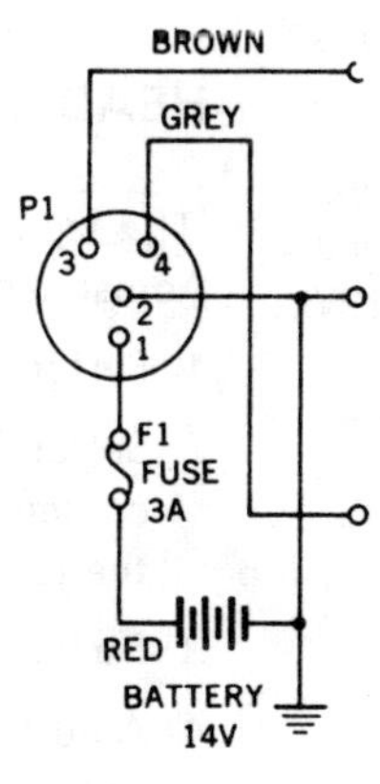

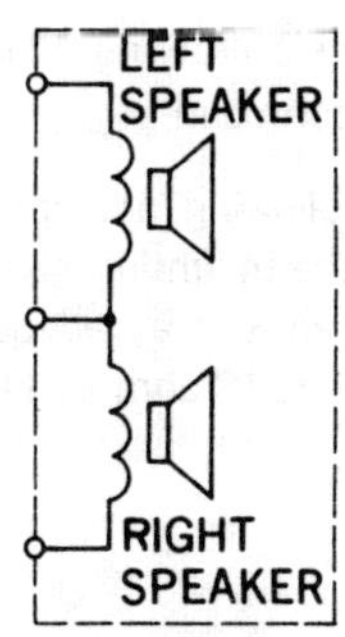

2 SPEAKER
CONNECTION
ARTICLE NO 61-6601

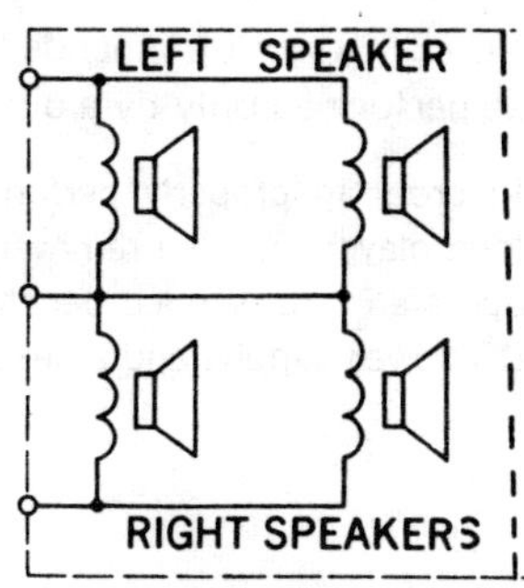

4 SPEAKER
CONNECTION
ARTICLE NO 61-6602

NOTES
1 ALL RESISTANCE VALUES SHOWN IN OHMS.
2 ALL CAPACITOR VALUES SHOWN IN MFD UNLESS NOTED.
3 ELECTROLYTIC CAPACITOR (CAPACITANCE/VOLTAGE).
4 FILM AND CERAMIC CAPACITORS.

TIPS ON MAINTENANCE

1. When not in use, remove cartridge from the player, or pull it out of its seated position so that the channel indicator is not lighted.
2. For maximum performance and enjoyment of your tapes, store them in their original case and protect them from direct sunlight when not in use.
3. Extreme heat or cold may cause erratic sound reproduction. Temperature inside the vehicle should be maintained between 20 and 100 degrees for proper operation.
4. Magnetism will damage your tapes. Do not place your tape cartridges near motors or transformers other than those of the tape player. Also, avoid touching the surface of the tape.
5. Clean the capstan and the head after 20 hours of use or twice a month using a soft cloth or a piece of isopropyl-alcohol-soaked cotton.
   Do not damage the head. Keep it away from magnetic materials.
6. The interior mechanism is built and adjusted precisely. All servicing should be performed by a qualified serviceman.
7. If the channel indicator does not light up or if there is no sound when the tape cartridge is inserted in position, check the fuse.

HEAD AZIMUTH ADJUSTMENT

The tape head azimuth, or tilt, must be adjusted for maximum high frequency response. Misalignment due to excessive vibration or mishandling will result in loss of both volume and high frequency response. Realignment is accomplished in the following manner.
(Refer to Figures 7, 16 and 17).

1. Remove cabinet top (17). (Refer to paragraph B of "Disassembly for Maintenance"), and make proper connections for power and speakers.
2. Insert test tape cartridge (Standard Equipment Item No. 21540) into the unit. (Unit should now be operating.)
3. Set Tone control to treble (maximum hi) and Balance control to center position.
4. Select channel 1 (tape head in the highest position), using the Channel Selector on the front panel.
5. Adjust Head Azimuth Adjust screw (112) for sharpest, clearest and loudest sound reproduction.
6. Remove the test tape cartridge.
7. Apply an adhesive compound to adjustment screw (112) to secure adjustment in place.
8. Proceed to Head Height Adjustment. (Height adjustment may be affected by Azimuth adjustment.)

ADJUSTMENTS

NOTE: It is recommended that the following adjustments be performed only by a qualified technican.

In order to properly perform the following adjustments the tape player must be removed from the mounting hangers and operated on a service bench equipped with an adequate 12V DC power supply, and a set of speakers (8 ohm VCI).

LUBRICATION

Wipe all visible grease from sliding surfaces. If the capstan shaft appears dirty at the top of the shaft bearing, remove the flywheel bearing bracket and complete flywheel and shaft after loosening the retaining screw (111).
Lubrication requirements are as follows:

Motor shaft: none
Capstan shaft: one drop light oil at slot in shaft bearing.
(CAUTION: Be sure no oil remains on capstan surface.)
Flywheel steel bearing: light silicone grease.
Head-shifting cam: light silicone grease.

HEAD HEIGHT ADJUSTMENT

The tape head may require height adjustment periodically. Adjustment can be made as follows:

1. Perform step 1 of head azimuth adjustment procedure.
2. Select a good 8 track tape cartridge having a musical content with which you are familiar and insert it into the tape player.
3. Perform steps 3 and 4 of head azimuth adjustment.
4. Adjust head height adjust screw (113) for maximum sound output from both left and right channel speaker systems. (Refer to Figures 7, 16 and 17.)
5. Select each of the other three channels and check that sound output is also at maximum level. Minor touch–up adjustment may be required to obtain optimum performance on all channels.
6. Remove the test tape cartridge.
7. Apply an adhesive compound to adjustment screw (113) to secure adjustment in place.

ELECTRICAL PARTS LIST

REF. No.	PART No.	DESCRIPTION
CAPACITORS		
C1.C603	CEUZ16V100M	100 MFD @16V, Electrolytic
C101, C151	CEUZ16V04.7M	4.7 MFD @16V, Electrolytic
C102, C152	CM50YM0.007MF	0.007MFD, 50V, Mylar
C103, C153	CEUZ10V33M	33 MFD @10V, Electrolytic
C104, C154	CC50YM50PF	50 PF, 50V, Discap
C105, C108 C155, C158	CEUZ16V10M	10 MFD @16V, Electrolytic
C106, C109 C156, C159	CEUZ16V100M	100 MFD @16V, Electrolytic
C111, C161	CEUZ16V220M	220 MFD @16V, Electrolytic
C201, C251	CM50YM0.05MF	0.05 MFD, 50V, Discap
C202, C252	CM50YM0.01MF	0.01 MFD, 50V, Mylar
C301, C351	CEUZ6V0.22M	0.22 MFD @6.3V, Electrolytic
C601	CEUZ16V1000M	1000 MFD @16V, Electrolytic
C602	UBL-306R1000P	1000PF, 15V, Feed-thru
C701	B10F2	500MFD @ 15V, Electrolytic (Noise Suppression, Figure 6)
RESISTORS		
R101, R151	CA5347	47K ohm, ¼W, 5%, Carbon
R102, R152	CA5047	47 ohm, ¼W, 5%, Carbon
R103, R153	CA5422	220K ohm, ¼W, 5%, Carbon
R104, R154	CA5312	12K ohm, ¼W, 5%, Carbon
R105, R155	CA5227	2.7K ohm, ¼W, 5%, Carbon
R106, R107 R156, R157	CA5147	470 ohm, ¼W, 5%, Carbon
R108, R158	CA5212	1.2K ohm, ¼W, 5%, Carbon
R109, R159	CA5415	150K ohm, ¼W, 5%, Carbon
R110, R160	CA5333	33K ohm, ¼W, 5%, Carbon
R111, R161	CA5010	10 ohm, ¼W, 5%, Carbon
R112, R162	CA5110	100 ohm, ¼W, 5%, Carbon
R113, R163	CA5133	330 ohm, ¼W, 5%, Carbon
R114, R116 R164, R166	CA5168	680 ohm, ¼W, 5%, Carbon
R115, R117 R165, R167	CA5112	120 ohm, ¼W, 5%, Carbon

REF. No.	PART No.	DESCRIPTION
VDR101 VDR102 VDR151 VDR152	MV-1	Varistor
R401	CB5118	180 ohm, ½W, 5%, Carbon
CONTROLS		
R301A, B	Not Available	10Kohm, Dual Pot., Tone — Refer to Mechanical Parts List (Part of 1A)
R302A, B	Not Available	50Kohm, Dual Pot., Volume — Refer to Mechanical Parts List (Part of 1A)
R303	Not Available	50Kohm, Pot., Balance — Refer to Mechanical Parts List (Part of 1A)
COILS AND TRANSFORMERS		
L1	2820500008	Coil, Filter
LL01	ETQD-4MH-1	Coil, Choke
T101, T151	TRF500T30	Transformer, Driver
SEMICONDUCTORS		
D1	SR1-FM2	Diode, Eject Solenoid
D2	SR1-FM2	Diode, Channel Select Solenoid
Q101, Q151	2SC870	Transistor, 2SC870, 1st Pre Amp
Q102, Q152	2SC711	Transistor, 2SC711, 2nd Pre Amp
Q103, Q153	2SC712	Transistor, 2SC712, Driver
Q201, Q251	2SC1013	Transistor, 2SC1013, Output
Q202, Q252	2SC1013	Transistor, 2SC1013, Output
MISCELLANEOUS		
F1	312003	Fuse, 3A

With FM—Panasonic CQ-909EU (Matsushita Electric Corp.)

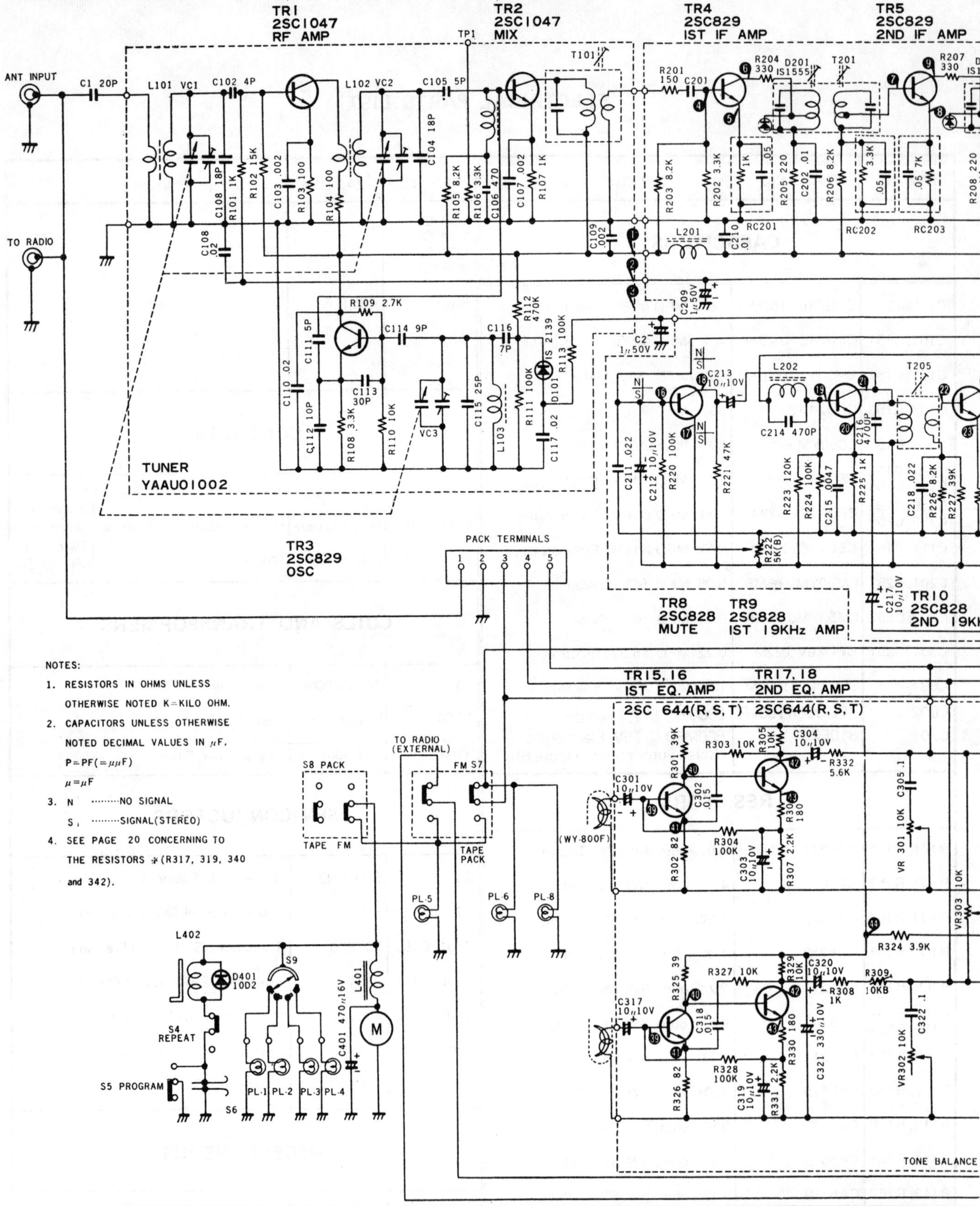

NOTES:

1. RESISTORS IN OHMS UNLESS OTHERWISE NOTED K=KILO OHM.
2. CAPACITORS UNLESS OTHERWISE NOTED DECIMAL VALUES IN μF, P=PF(=μμF) μ=μF
3. N ....NO SIGNAL
   S ....SIGNAL(STEREO)
4. SEE PAGE 20 CONCERNING TO THE RESISTORS ※(R317, 319, 340 and 342).

NO.	1	2	3	4	5	6	7	8	9	10	11	12	13	14	15
CIRCUIT TESTER >20KΩ/VOLT	8,0	0,77	−0,02	2.16	1,59	7,1	2,16	1,59	7,1	2,16	1,59	7,1	2,16	1,58	6,32
V.T.V.M	8,1	0,80	−0,04	2,26	1,58	7,67	2,26	1,58	7.67	2,26	1,58	7,67	2,29	1,61	6,3

NO.	16	17	18	19	20	21	22	23	24	25	26	27	28	29
CIRCUIT TESTER >20KΩ/VOLT	N 5,8 / S 3,8~5,8	6,1 / 6,1	6,2 / 6,4	2,5	2,5	7,9	1,3	0,79	7,9	N 4,3 / S 2,9	13,2 / 2,5	7,9 / 7,9	0,03 / 0,59	0 / 0
V.T.V.M	N 6,6 / S 3,8~6,6	6,1 / 6,1	6,3 / 7,9	3,0	2,5	7,9	1,4	0,80	7,9	N 4,1 / S 3,0	13,2 / 2,5	7,9 / 7,9	0,05 / 0,63	0 / 0

Panasonic CQ-909EU (Matsushita Electric Corp.)

PANASONIC CAR STEREO WITH RADIO

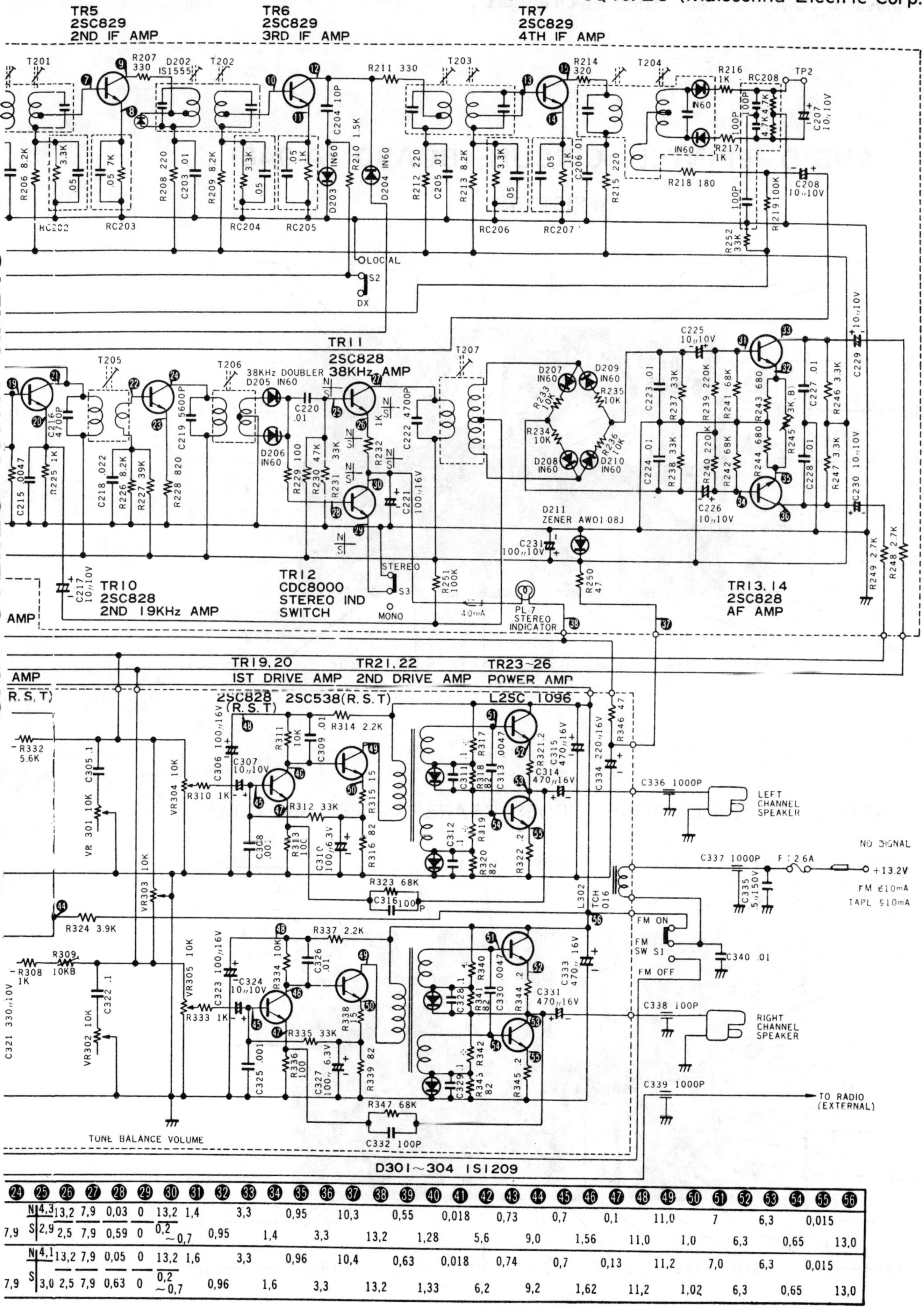

	24	25	26	27	28	29	30	31	32	33	34	35	36	37	38	39	40
N		4,3	13,2	7,9	0,03	0	13,2	1,4		3,3		0,95		10,3		0,55	
S	7,9	2,9	2,5	7,9	0,59	0	0,2 ~0,7		0,95		1,4		3,3		13,2		1,28
N		4,1	13,2	7,9	0,05	0	13,2	1,6		3,3		0,96		10,4		0,63	
S	7,9	3,0	2,5	7,9	0,63	0	0,2 ~0,7		0,96		1,6		3,3		13,2		1,33

	41	42	43	44	45	46	47	48	49	50	51	52	53	54	55	56
N	0,018		0,73		0,7		0,1		11,0		7		6,3		0,015	
S		5,6		9,0		1,56		11,0		1,0		6,3		0,65		13,0
N	0,018		0,74		0,7		0,13		11,2		7,0		6,3		0,015	
S		6,2		9,2		1,62		11,2		1,02		6,3		0,65		13,0

AUDIO PRINTED CIRCUIT BOARD (A-048)

RESISTORS

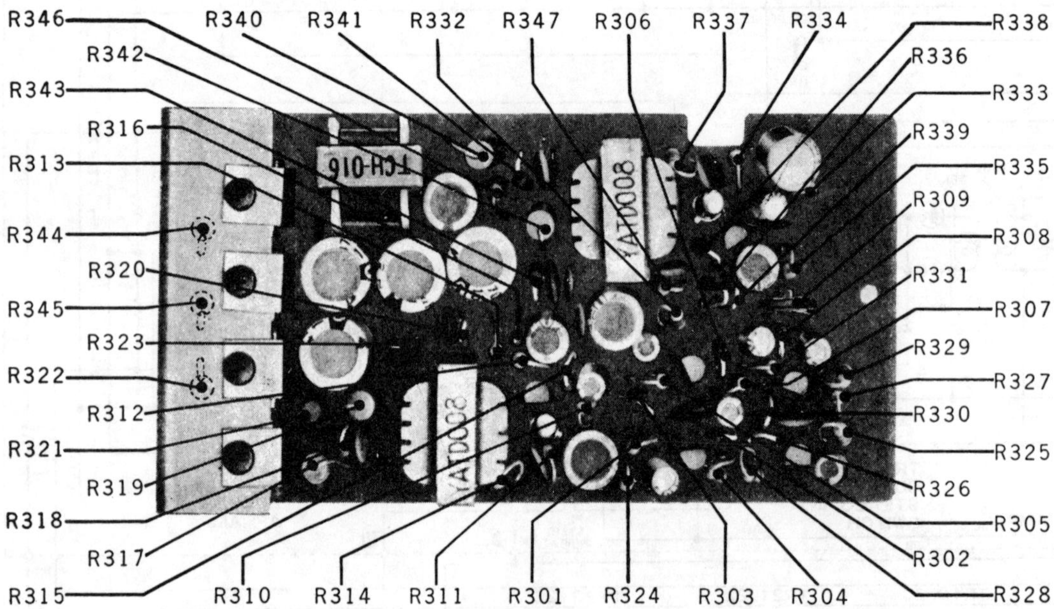

Fig. 8

CAPACITORS, TRANSISTORS, DIODES, TRANSFORMERS and COILS

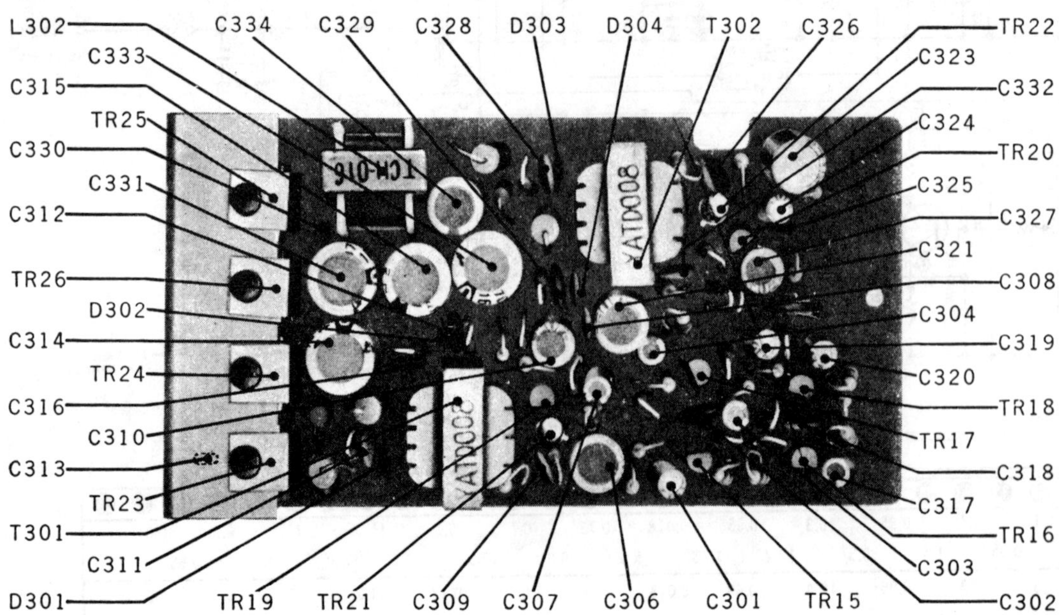

Fig. 9

TUNER ASSEMBLY YAAU01002

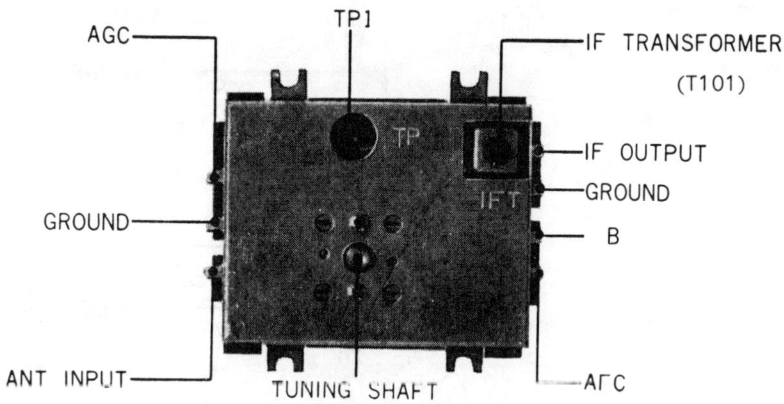

Fig. 10 TOP VIEW

Fig. 11 BOTTOM VIEW

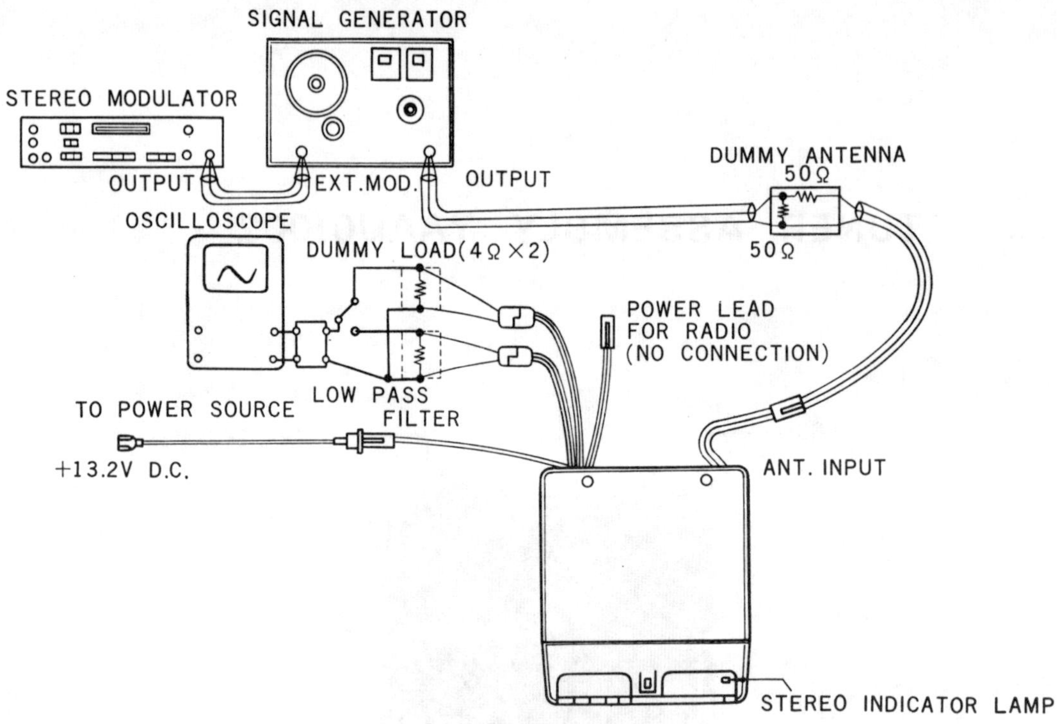

Fig. 20

5. BALANCE CONTROL ADJUSTMENT (R309)

Balance adjustment. (See Fig. 21)

(1) When test tape is available.

insert the test cartridge. (RCA #325), set the balance lever to the middle and adjust the semifixed resistor © which is located on the audio printed circuit board (A-048) with a nonmetallic screwdriver for equal output from right and left speaker.

(2) When test tape is not available.

Insert the cartridge in which nearly equal-level music is recorded for right and left channels, set the balance lever to ''middle'', and adjust the semifixed resistor © so that the sound source may be heard from the center.

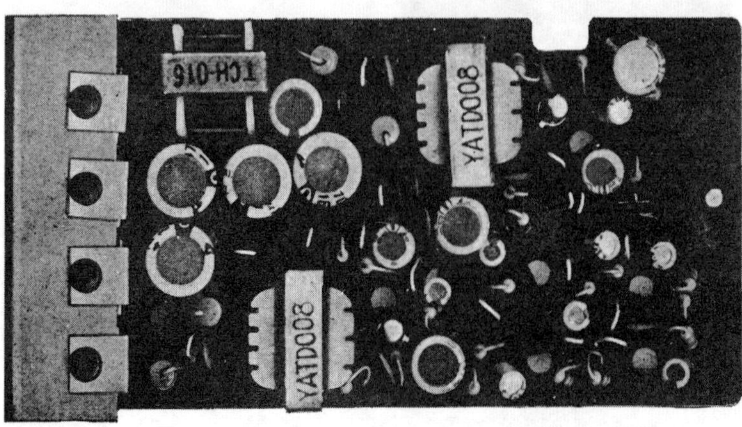

Fig. 21

6. HEAD HEIGHT AND AZIMUTH ADJUSTMENT

1) Head azimuth adjustment. (See Fig. 22)

Take off top cover before adjusting.

(1) When test tape is available.

Insert the test cartridge, set the program to proper track, and adjust the azimuth adjusting screw Ⓐ for maximum output from the speaker.

(2) When test tape is not available.

Insert the cartridge in which music containing treble is recorded, and rotate the azimuth adlusting screw Ⓐ for maximum treble.

2) Head height adjustment. (See Fig. 23)

Remove the rubber grommet on bottom cover.

You can adjust head height without taking the bottom covor off.

(1) When test tape is available.

Insert the test cartridge, set the channel to proper track and adjust the height adjusting screw Ⓑ for minimum output from the speaker.

(2) When test tape is not available.

Insert 8-track music cartridge, and adjust the height adjusting screw Ⓑ for maximum sound volume.

■ When the program is switched, ensure that tone from other channels is not mixed.

■ For head adjustment, be certain to perform azimuth adjustment first, then the height adjustment.

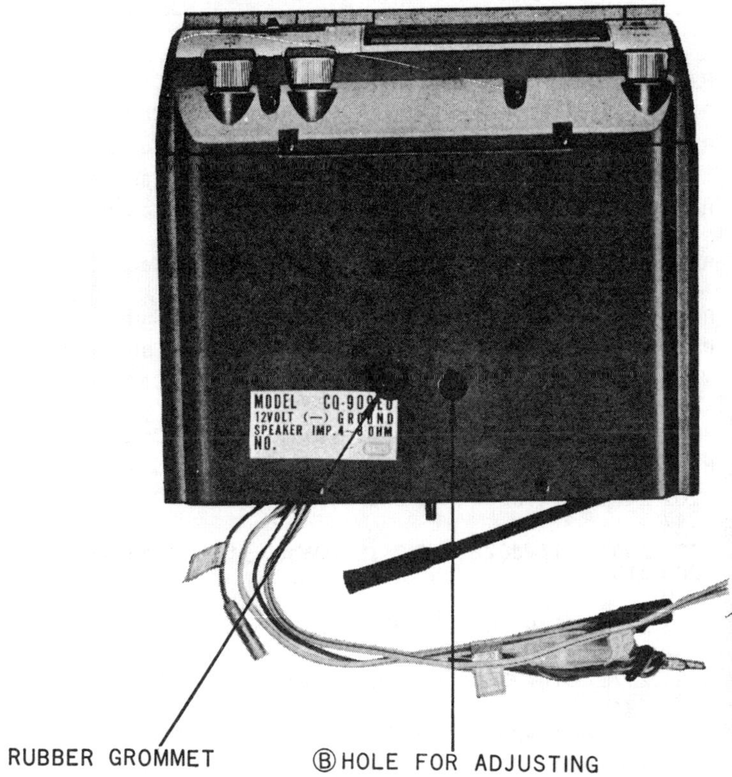

Fig. 22

Fig. 23

REPLACEMENT PARTS LIST

MODEL CQ-909EU

REF. NO.	PART NO.	DESCRIPTION
		TRANSISTORS
TR4,5,6,7	2SC829(B.C)	1st~4th IF Amp.
TR8	2SC828(R.S.T)	Mute
TR9	2SC828(R.S.T)	1st 19KHz Amp.
TR10	2SC828(R.S.T)	2nd 19KHz Amp.
TR11	2SC828(R.S.T)	38 KHz Amp.
TR12	YEANCDC80001	Stereo Indicator Switch
TR13, 14	2SC828(R.S.T)	AF Amp.
TR15, 16	2SC644(R.S.T)	1st EQ. Amp.
TR17, 18	2SC644(R.S.T)	2nd EQ. Amp.
TR19, 20	2SC828(R.S.T)	1st Drive Amp.
TR21, 22	2SC538(R.S.T)	2nd Drive Amp.
TR23, 24 25, 26	YAANL2SC1096	Power Amp.
		DIODES
D201,202	YEAD011	Limiter IS1555
D203 204 205,206 207,208 209,210	YEAD009	1N60
D211	YEAD010	Zener AW01-08J
D301,302 303,304	YAAD007	Temperature Compensator IS1209
D401	YAAD004	Spark Suppression 10D2
		CAPACITORS
C201,202 203,205 206,210	ECKD05103PJ	0.01μF 50WV, ±5% Ceramic
C204	YECCD05100F	10pF 50WV, Titanium
C216,222	ECQS1472JZ	4700pF 50WV, ±5% Styrol
C219	ECQS1562JZ	5600pF 50WV, ±5% Styrol
C214	ECQS1471JZ	470pF 50WV, ±5% Styrol
C215	YECQN05502K	0.0047μF 50WV, ±10% Polyester film
C220,223 224,227 228	YECQN05103K	0.01μF 50WV, ±10% Polyester film
C211,218	YECQN05203K	0.022μF 50WV, ±10% Polyester film
C207	ECEB10V10N	10μF 10WV, Electrolytic
C208,212 213,217 225,226 229,230	ECEA10V10N	10μF 10WV, Electrolytic
C231	ECEA10V100N	100μF 10WV, Electrolytic
C221	ECEA16V100N	100μF 16WV, Electrolytic
C209	ECEA50V1N	1μF 50WV, Electrolytic
C316,332	YECCD05101K	100pF 50WV, ±10% Titanium
C308,325	YACQN05102 M	0.002μF 50WV, ±20% Polyester film
C313,330	YACQN05472M	0.0047μF 50WV, ±20% Polyester film
C309,326 340	YACQN05103M	0.01μF 50WV, ±20% Polyester film
C302,318	YACQN05153K	0.015μF 50WV, ±10% Poryester film
C311,312 328,329	YECCD12104M	0.1μF 12WV, ±20% Ceramic
C310,327	ECEA6V100N	100μF 6WV, Electrolytic
C301,303 304,307 317,319 320,324	ECEA10V10N	10μF 10WV, Electrolytic
C321	ECEA10V330N	330μF 10WV, Electrolytic
C306,323	ECEA16V100N	100μF 16WV, Electrolytic
C334	ECEA16V220N	220μF 16WV, Electrolytic
C314.315 331,333	ECEA16V470Z	470μF 16WV, Electrolytic
C1	YECCD05200R	20pF 50WV, Titanium
C305,322	YECCD12104M	0.1μF 12WV, Ceramic
C335	ECHR1504M	0.5μF 150WV, Metallized
C2	ECEA50V1N	1μF 50WV, Electrolytic
C336,337 338,339	YACCL510205	1000pF×4 500WV, Feed Through
C401	ECEA16V470N	470μF 16WV, Electrolytic
		RESISTORS
R250	ERD12TJ470	47Ω ½Watt, ±5% Carbon
R229	ERD14VJ101	100Ω ¼Watt, ±5% Carbon
R201	ERD14VJ151	150Ω ¼Watt, ±5% Carbon
R218	ERD14VJ181	180Ω ¼Watt, ±5% Carbon
R205,208 212,215	ERD14VJ221	220Ω ¼Watt, ±5% Carbon
R204,207 211	ERD14VJ331	330Ω ¼Watt, ±5% Carbon
R243,244	ERD14VJ681	680Ω ¼Watt, ±5% Carbon
R214,228	ERD14VJ821	820Ω ¼Watt, ±5% Carbon
R216,217 225,232	ERD14VJ102	1KΩ ¼Watt, ±5% Carbon
R210	ERD14VJ152	1.5KΩ ¼Watt, ±5% Carbon
R248,249	ERD14VJ272	2.7KΩ ¼Watt, ±5% Carbon
R202,246 247	ERD14VJ332	3.3KΩ ¼Watt, ±5% Carbon
R203,206 209,213 226	ERD14VJ822	8.2KΩ ¼Watt, ±5% Carbon
R233.234 235,236	ERD14VJ103	10KΩ ¼Watt, ±5% Carbon
R231.237 238,252	ERD14VJ333	33KΩ ¼Watt, ±5% Carbon
R227	ERD14VJ393	39KΩ ¼Watt, ±5% Carbon
R221,230	ERD14VJ473	47KΩ ¼Watt, ±5% Carbon
R241,242	ERD14VJ683	68KΩ ¼Watt, ±5% Carbon
R220,224 251	ERD14VJ104	100KΩ ¼Watt, ±5% Carbon
R223	ERD14VJ124	120KΩ ¼Watt, ±5% Carbon
R239,240	ERD14VJ224	220KΩ ¼Watt, ±5% Carbon
R219	ERD14VJ104	100KΩ ¼Watt, ±5% Carbon
R315,338	ERD14VJ150	15Ω ¼Watt, ±5% Carbon

MODEL CQ-909EU

REF. NO.	PART NO.	DESCRIPTION
R302,316 318,320 326,339 341,343	ERD14VJ820	82Ω ¼Watt, ±5% Carbon
R313,336	ERD14VJ101	100Ω ¼Watt, ±5% Carbon
R306,330	ERD14VJ181	180Ω ¼Watt, ±5% Carbon
R308,310 333	ERD14VJ102	1KΩ ¼Watt, ±5% Carbon
R324	ERD14VJ392	3.9KΩ ¼Watt, ±5% Carbon
R303,305 311,327 329,334	ERD14VJ103	10KΩ ¼Watt, ±5% Carbon
R312,335	ERD14VJ333	33KΩ ¼Watt, ±5% Carbon
R301,325	ERD14VJ393	39KΩ ¼Watt, ±5% Carbon
R323,347	ERD14VJ683	68KΩ ¼Watt, ±5% Carbon
R304,328	ERD14VJ104	100KΩ ¼Watt, ±5% Carbon
R307,314 331,337	ERD14VJ222	2.2KΩ ¼Watt, ±5% Carbon
R332	ERD14VJ562	5.6KΩ ¼Watt, ±5% Carbon
R317,319 340,342	ERG2PSK201	200Ω 2Watt, ±10% Metal Oxide Film
R317,319 340,342	ERG2PSK221	220Ω 2Watt, ±10% Metal Oxide Film
R317,319 340,342	ERG2PSK271	270Ω 2Watt, ±10% Metal Oxide Film
R317,319 340,342	ERG2PSK301	300Ω 2Watt, ±10% Metal Oxide Film
R317,319 340,342	ERG2PSK331	330Ω 2Watt, ±10% Metal Oxide Film
R321,322 344,345	ERM12VKR20	0.2Ω ½Watt, ±10% Wire Wound
R346	ERD12TJ470	47Ω ½Watt, ±5% Carbon
VARIABLE RESISTORS		
R245	EVLT0AA00B33	3KΩ B Stereo Separation, Semifixed
R222	EVLS0AA00B53	5KΩ B Muting, Semifixed
R309	EVLT0AA00B14	10KΩ B Semifixed
VR301 302,304 305	EVKP0AK20A14	10KΩ A Volume Control & Tone Control
VR303	EVAQ 0AA00B14	10KΩ B Balance Control
COILS AND TRANSFORMERS		
T201,202 203	YEIF14S1015B	FM IFT 10.7 MHz
T204	YEIF14S1016D	Discriminator Transformer 10.7 MHz
T205	YALM10S001	MPX Coil 19 KHz
T206	YALM10S002	MPX Coil 19 KHz
T207	YALM10S003	MPX Coil 38 KHz
L202	ELQ4CI(RLM-1×1)	SCA Trap Coil 67 KHz
L201	YELQ09C006	RFC
T301,302	YATD024F008	Input Transformer
L302	YBTQ025F016	Filter Choke Coil
L101,102 103 T101	YAAU01002	Tuner Ass'y
L401	YATQ014F003	Filter Choke Coil

REF. NO.	PART NO.	DESCRIPTION
COMPONENT COMBINATIONS		
RC201, 203,205 207	YEXA001	R : 1KΩ C : 0.01μF
RC202, 204,206	YEXA002	R : 3.3KΩ C : 0.01μF
RC208	EXA5DL01C	R : 4.7KΩ×2 C : 100pF×3
MISCELLANEOUS		
M100	YAFC02029	Escutcheon Ass'y
M101	YAFX047006	Pack Guide
M102	YAFX001034	Cartridge Door
M103	YAFW01103	Cartridge Door Installation shaft
M104	YAFX011131	Spring for Cartridge Door
M105	YAFA03013	Top Cover
M106	YAFA11032	Bottom Cover
M107	YAFE01025	Volume Control & Tone Control Knob
M108	YAFJ08010	Program Indicator Plate
M109	YAFJ08013	Dial Plate
M110	YAFV02005	Light Shielding Cover
M111	YAFX010024	Drum
M112	YAFX011129	Spring for Dial Cord
M113	YEFX007035	Cord Clamp 004
M114	YEFX007020	Cord Clamp 020
M115	YAFR05001	Rubber Grommet
M116	YEFR01005	Rubber Bushing
M117	YEFH07012	Dial Cord Guide E-012
M118	YAFA11025	Rear Panel
M119	YAFA10020	Side Plate (Right)
M120	YAFA10021	Side Plate (Left)
M121	YAFE01036	Tuning Knob w/Shaft
M122	YAFX030006	Balance lever
M123	YAFX036011	Balance Joint
M124	YAFK04016	Lamp socket for Program Selector
M125	XUC4FY	E-Washer 4φ
M126	XUC2FY	E-Washer 2φ
M127	YEJN01004FX	Hexagonal Nut 8φ
M128	XSB3+6FZS	Screw (M3×6); BLACK
M129	XSB3+6FBS	Screw (M3×6)
M130	XSB3+5FBS	Screw (M3×5)
M131	XSB26+5FB	Screw (M2.6×5)
M132	XSS26+6FZK	Screw (M2.6×6), BLACK
M133	XSB3+5Z	Screw (M3×5) Poly carbonate
M134	YAFH01005	Dial Pointer
S1,2,3,4, 5	YAAS12003	Switch Ass'y
PL7	YAAL01007	Pilot Lamp for Stereo Indicator
PL1,2,3,4	YAAL01010	Pilot Lamp for Program Selector
PL5,6	YEAL01007	Pilot Lamp, S7 12V 2W
PL8	YEAL01017	Pilot Lamp, S9 12V 3W

SCHEMATIC DIAGRAM

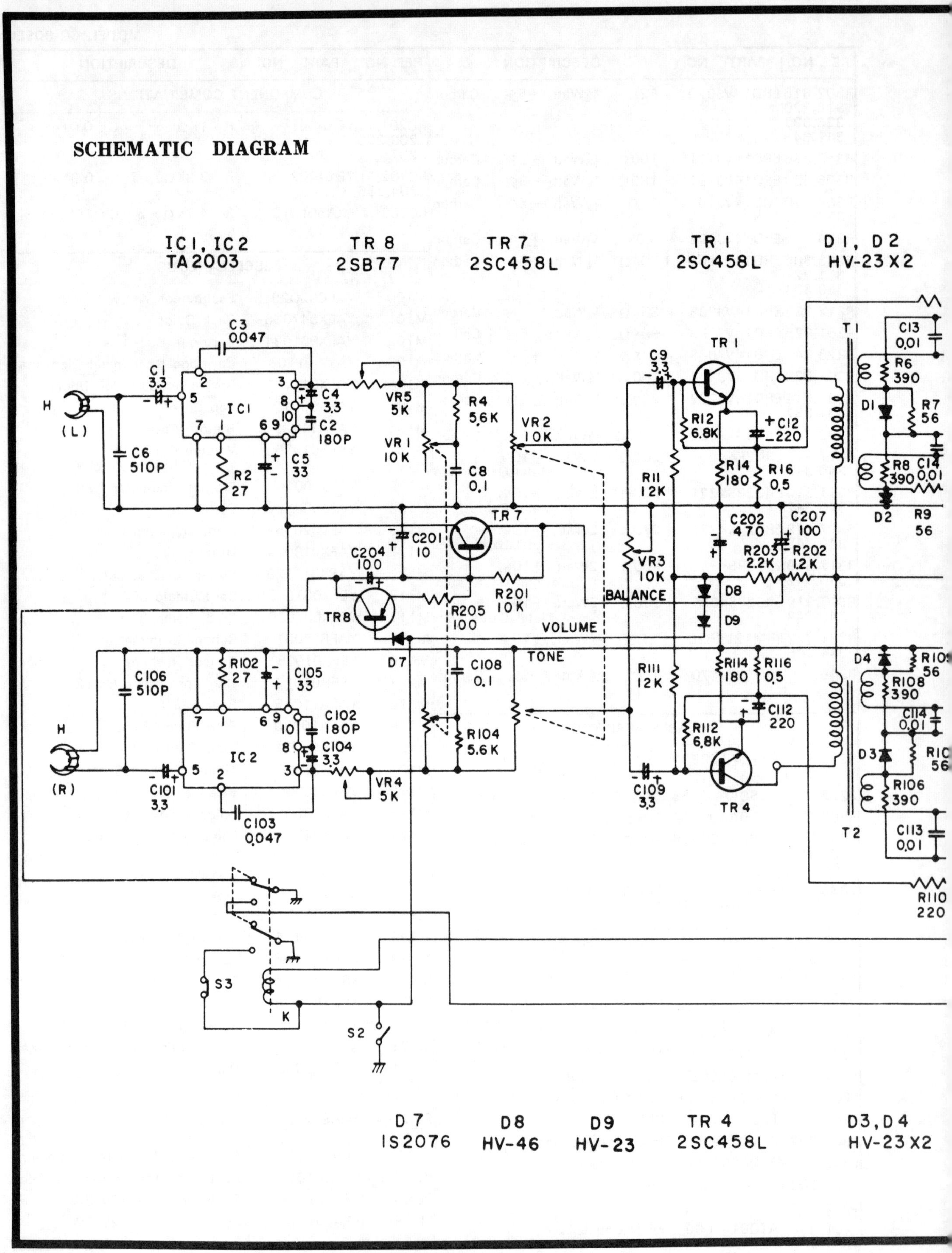

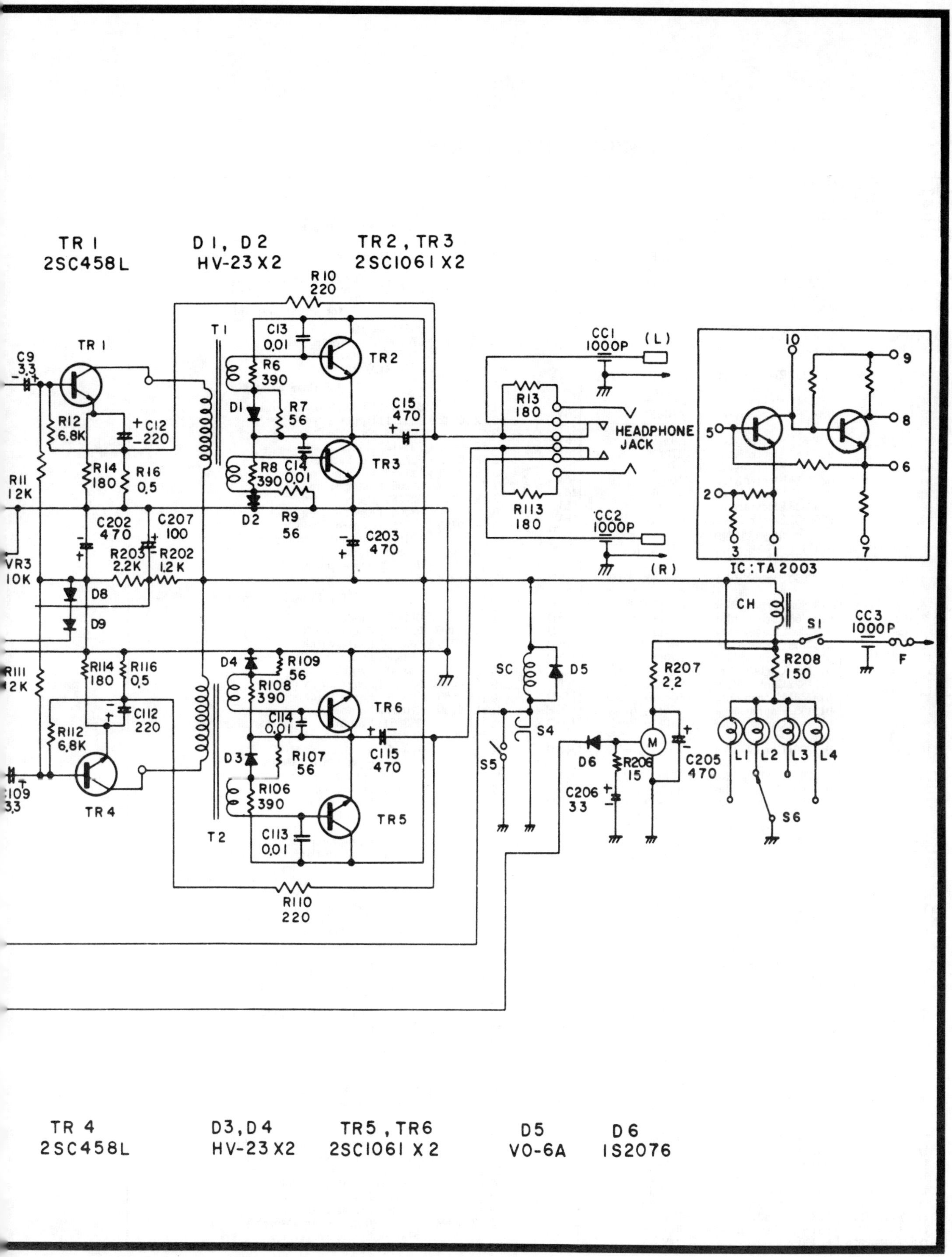
TR 1
2SC458L
D 1, D 2
HV-23 X 2
TR 2, TR 3
2SC1061 X 2
R 10
220
T 1
C13
0.01
TR 1
TR 2
C9
3.3
R6
390
D1
R7
56
C15
470
R12
6.8K
C12
220
R8
390
C14
0.01
TR 3
R14
180
R16
0.5
R11
12K
D2
R9
56
C202
470
C207
100
R203
2.2K
R202
1.2 K
C203
470
VR3
10K
D8
D9
CC1
1000P
(L)
R13
180
HEADPHONE
JACK
R113
180
CC2
1000P
(R)
IC : TA 2003
CH
S1
CC3
1000P
F
R114
180
R116
0.5
R111
12K
D4
R109
56
R108
390
SC
D5
R207
2.2
R208
150
C112
220
R112
6.8K
C114
0.01
TR 6
C115
470
S4
S5
D3
R107
56
D6
R206
15
M
C205
470
L1
L2
L3
L4
C109
3.3
TR 4
R106
390
C206
33
S6
TR 5
T 2
C113
0.01
R110
220
TR 4
2SC458L
D3, D4
HV-23 X 2
TR5, TR6
2SC1061 X 2
D5
VO-6A
D 6
1S2076

REPLACEMENT PARTS

Symbol No.	ARS PART No.	MFR'S PART No.	Description		
CAPACITORS:					
C 1,101		0252613	Electrolytic	3.3μF	25V
C 2,102		0248730	Ceramic	180pF	
C 3,103		0275115	Mylar	0.047μF	
C 4,104		0252613	Electrolytic	3.3μF	25V
C 5,105		0252223	Electrolytic	33μF	6.3V
C 6,106		0249515	Ceramic	510pF	
C 8,108		0276111	Mylar	0.1μF	
C 9,109		0252613	Electrolytic	3.3μF	25V
C 12,112		0252232	Electrolytic	220μF	6.3V
C 13,113		0275111	Mylar	0.01μF	
C 14,114		0275111	Mylar	0.01μF	
C 15,115		0256083	Electrolytic	470μF	16V
C 201		0252621	Electrolytic	10μF	25V
C 202		0256083	Electrolytic	470μF	16V
C 203		0256083	Electrolytic	470μF	16V
C 204		0252331	Electrolytic	100μF	16V
C 205		0256083	Electrolytic	470μF	16V
C 206		0252623	Electrolytic	33μF	25V
C 207		0252531	Electrolytic	100μF	16V
RESISTORS:					
R 2,102		0137764	Carbon tilm	27Ω±10%	¼SD
R 4,104		0137860	Carbon film	5.6kΩ±10%	¼SD
R 6,106		0137808	Carbon film	390Ω±10%	¼SD
R 7,107		0137768	Carbon film	56Ω±10%	¼SD
R 8,108		0137808	Carbon film	390Ω±10%	¼SD
R 9,109		0137768	Carbon film	56Ω±10%	¼SD
R 10,110		0137805	Carbon film	220Ω±10%	¼SD
R 11		0137902	Carbon film	12kΩ±10%	¼SD
R 111		0114522	Composition	12kΩ±10%	¼P
R 12,112		0137861	Carbon film	6.8kΩ±10%	¼SD
R 13,113		0131652	Composition	180Ω±10%	¼GF
R 14,114		0137804	Carbon film	180Ω±10%	¼SD
R 16,116		0172001	Carbon film	0.5Ω±10%	¼SD
R 201		0137901	Carbon film	10kΩ±10%	¼SD
R 202		0137852	Carbon film	1.2kΩ±10%	¼SD
R 203		0137855	Carbon film	2.2kΩ±10%	¼SD
R 205		0137801	Carbon film	100Ω±10%	¼SD
R 206		0134291	Composition	15Ω±10%	½GF
R 207		0134281	Composition	2.2Ω±10%	½GF
R 208		0134363	Carbon film	150Ω±10%	
VR 1	P-1052	0171510	Variable	10kΩ×(tone)	
※VR 2		0171508	Variable	10kΩA(VR)	
※VR 3		0171512	Variable	10kΩW(blance)	
VR 4	P-6016	0151360	Semi variable	5kΩ	
VR 5	P-6016	0151360	Semi variable	5kΩ	
TRANSISTORS:					
TR 1		5320064		2SC458D	
TR 2		5320671		2SC1061T(B)	
TR 3		5320671		2SC1061T(B)	
TR 4		5320064		2SC458D	
TR 5		5320671		2SC1061T(B)	
TR 6		5320671		2SC1061T(B)	
TR 7		5320064		2SC458D	
TR 8		0573119		2SB77C	
IC 1		5356011	Hybrid IC	TA2003	
IC 2		5356011	Hybrid IC	TA2003	
D 1		0576054	Varistor	HV-23	
D 2		0576054	Varistor	HV-23	
D 3		0576054	Varistor	HV-23	
D 4		0576054	Varistor	HV-23	
D 5		5330104	Diode	VO-6A	
D 6	R-7027	5330131	Diode	1S2076	
D 7	R-7027	5330131	Diode	1S2076	
D 8		5340021	Varistor	HV-46	
D 9	R-7026	0576054	Varistor	HV-23	

Symbol No.	ARS PART No.	MFR'S PART No.	Description	
TRANSFORMERS:				
T 1	TD-0014	0441114	Driver	2kΩ:200Ω
T 2	TD-0014	0441114	Driver	2kΩ:200Ω
COIL:				
CH	CB-2085	5220001	Choke	1.5mH
			for Final assembly	
		0591138	Fuse	
		5741005	Connector (E)	
		7206831	Support strap	
		7772271	Wing nut	
			for Case assembly	
1		6215651	Escutcheon assembly	
2	Z-0092	7234111	Upper cover	
3		7664741	Cap	
4	Z-0093	7234121	Under cover	
5	K-0752	6271811	FF button	
6	K-0753	6271821	Play button	
7		6320661	Selec spring	
8	K-0754	6271831	Eject button	
9		6320651	Eject spring	
10	K-0755	6271801	Control knob	
11		6216021	Button assembly	
12	MB-0016	7235311	Hanger	
		7625022	Name plate	
13	W-1238	5740661	Earth wire	
			for Chassis assembly	
14		6719631	Guide plate	
15	RA-4660	7229851	Roller shaft assembly	
16	S-7010	5632182	Leaf switch	
17		5535361	Lamp P.C.B assembly	
18		7234451	Switch holder (1)	
19	S-8024	5632131	Program switch	
20	H-0663	7234461	Lamp holder	
21	L-0250	5760582	Lamp	
22		5536541	Volume P.C.B assembly	
23	S-7012	5631393	Push switch	
24		7234471	Switch holder (2)	
25		7234411	Base plate assembly	
26		7234391	Spring holder	
27		7234392	Spring holder	
28		6342221	Lever return spring	
29		6320641	Pressure spring	
30		6311361	Pause arm spring	
31	J-0266	5670451	Headphone jack	
32		7206541	Head shifting plate assembly	
33		6710032	Tape guide piece for volume P.C.B strap	
34		7504741	Adjusting screw plate assembly	
35		6311641	Adjusting spring	
36		5440521	Head assembly	
37	RA-4662	6342032	Head shifting cam assembly	
38		7586781	Head shifting plate shat	
39		7230901	"E" ring	
40		6315461	Adjusting spring	
41	RA-4663	7206602	Return cam	
42		6315471	Push plate cam spring	
43		7230901	"E" ring	
44		6314811	Head shifting plate spring	
45		7206582	Push plate arm assembly	
46	S-9003	5640181	DC solenoid (E)	
47		7230901	"E" ring	
48		7501783	Push plate shaft	
49		6314921	Plunger spring	
50		7234201	Eject spring holder	
51		6316962	Return spring	
52		7234431	Tape guide	
53	S-8025	5790431	Track switch connector	
54		6720761	Spring holder assembly	
55	RA-3081	0970811	Flywheel assembly	
56		7234221	Thrust sheet assembly	
57		6719404	Metal assembly	
58		6350732	Belt	
59		5570971	Motor assembly	
60	Z-0094	7234131	Rear cover assembly	
		5741402	Connector (A)	

Chapter 6

Schematics, Board Layouts, Wiring, and Parts Lists for Representative Portable Tape Players

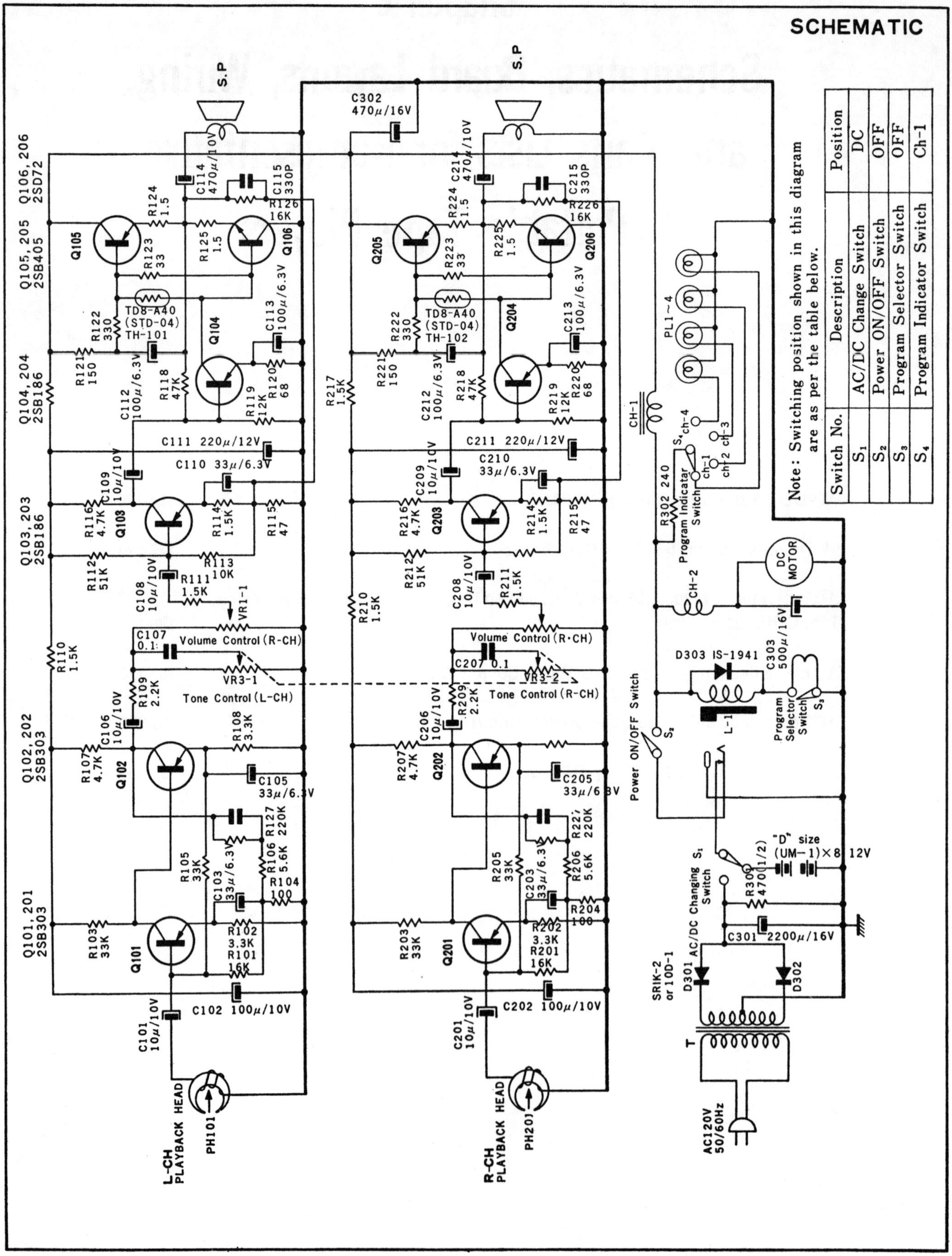
SCHEMATIC
Note: Switching position shown in this diagram are as per the table below.
Switch No. | Description | Position
S1 | AC/DC Change Switch | DC
S2 | Power ON/OFF Switch | OFF
S3 | Program Selector Switch | OFF
S4 | Program Indicator Switch | Ch-1
L-CH PLAYBACK HEAD
PH101
R-CH PLAYBACK HEAD
PH201
Volume Control (R-CH)
Tone Control (L-CH)
Volume Control (R·CH)
Tone Control (R-CH)
Power ON/OFF Switch
Program Selector Switch
Program Indicator Switch
AC/DC Changing Switch
DC MOTOR
AC120V 50/60Hz
S.P

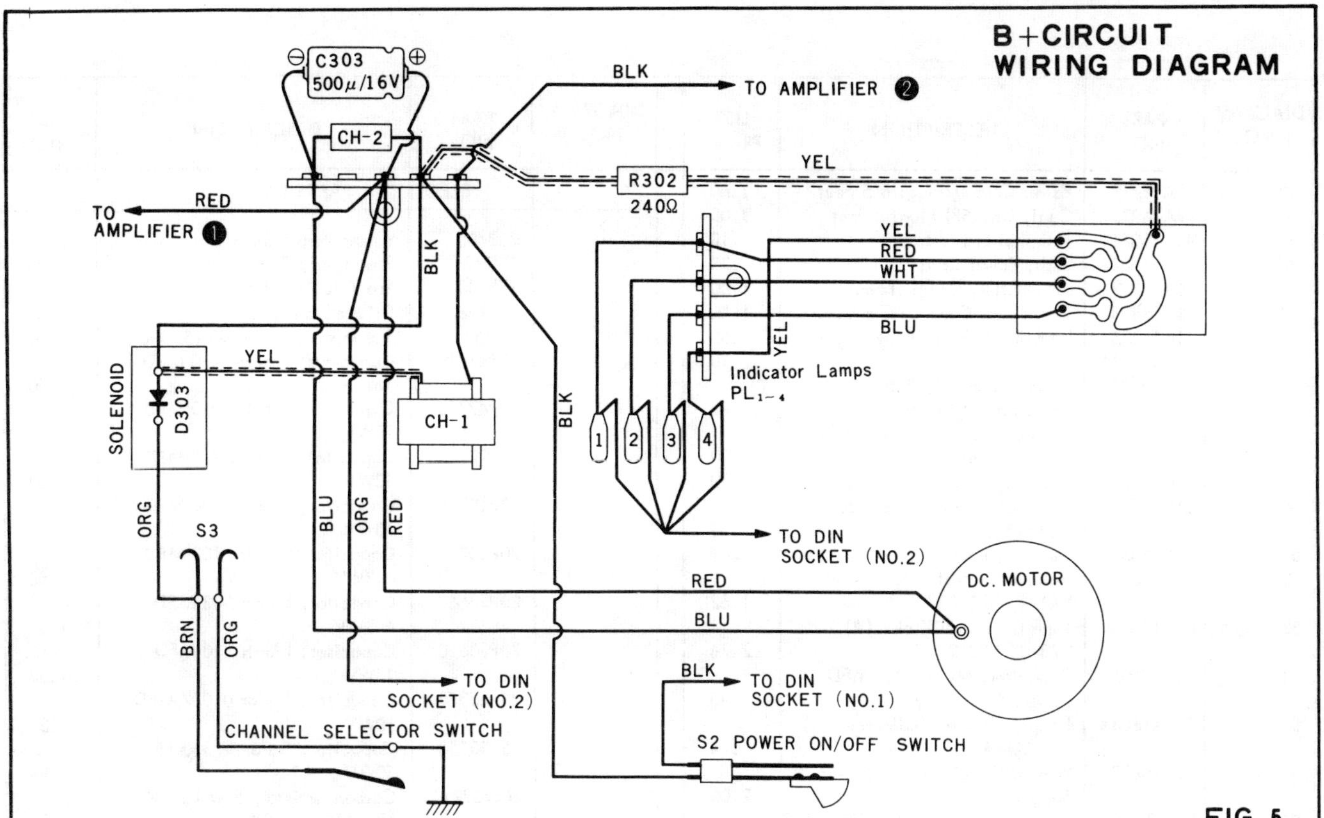
B+CIRCUIT WIRING DIAGRAM
C303 500μ/16V
CH-2
BLK
TO AMPLIFIER 2
YEL
R302 240Ω
TO AMPLIFIER 1
RED
YEL
RED
WHT
BLU
BLK
YEL
Indicator Lamps PL1-4
SOLENOID
D303
CH-1
BLK
ORG
S3
BLU
ORG
RED
TO DIN SOCKET (NO.2)
RED
BLU
DC. MOTOR
BRN
ORG
TO DIN SOCKET (NO.2)
BLK
TO DIN SOCKET (NO.1)
CHANNEL SELECTOR SWITCH
S2 POWER ON/OFF SWITCH

FIG. 5

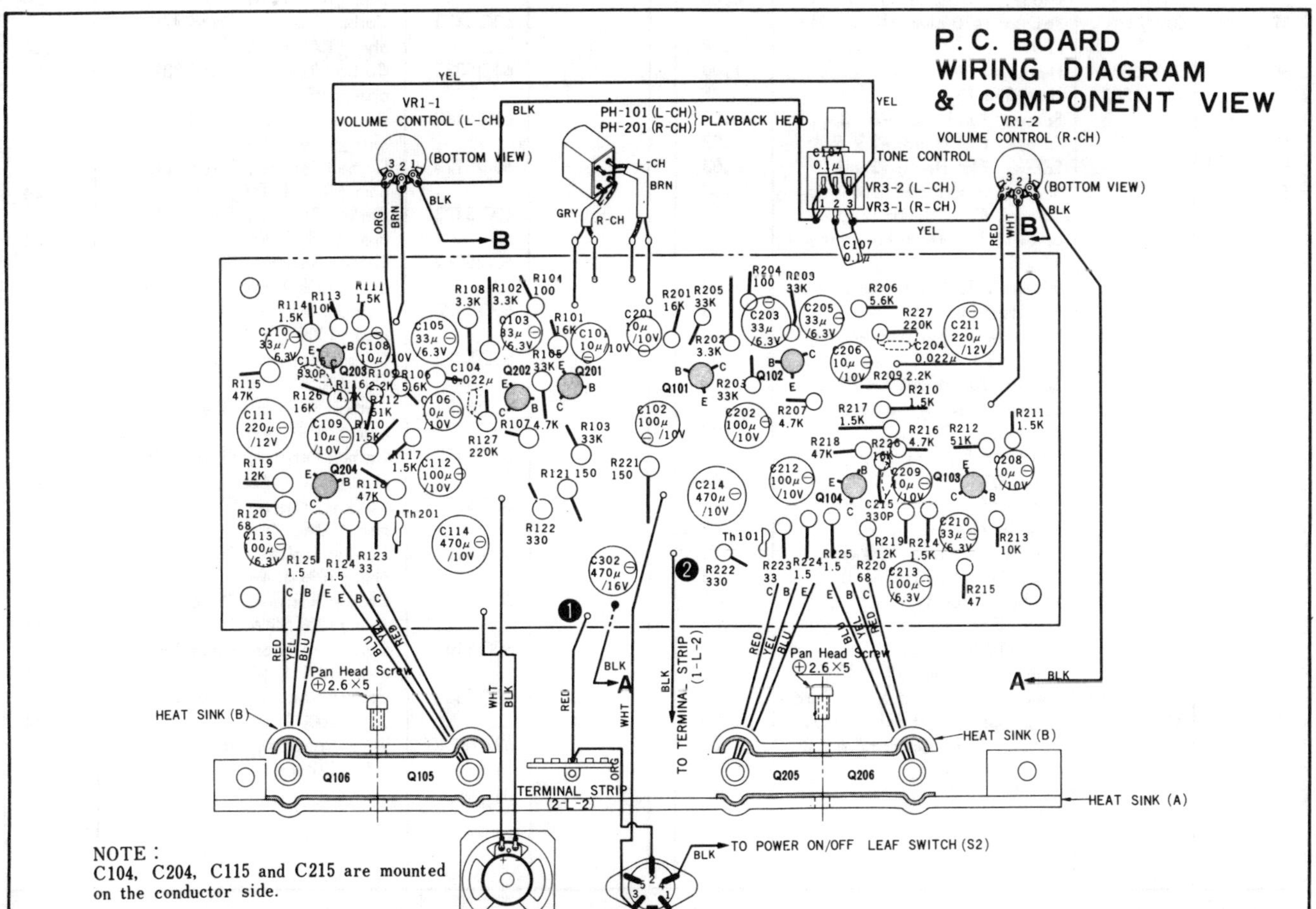
P. C. BOARD WIRING DIAGRAM & COMPONENT VIEW
YEL
VR1-1 VOLUME CONTROL (L-CH)
(BOTTOM VIEW)
BLK
PH-101 (L-CH) PH-201 (R-CH) PLAYBACK HEAD
VR1-2 VOLUME CONTROL (R-CH)
TONE CONTROL
VR3-2 (L-CH)
VR3-1 (R-CH)
(BOTTOM VIEW)
C107 0.1μ
HEAT SINK (B)
Pan Head Screw ⊕2.6×5
TERMINAL STRIP (2-L-2)
TO TERMINAL STRIP (1-L-2)
HEAT SINK (A)
Q106
Q105
Q205
Q206
TO POWER ON/OFF LEAF SWITCH (S2)
NOTE : C104, C204, C115 and C215 are mounted on the conductor side.
SPEAKER (L-CH)
DIN SOCKET (BOTTOM VIEW)

FIG. 6

DIAGRAM NO.	PART NO.	DESCRIPTION	LIST PRICE
39	12A16819	Power Cord w/Plug; 6.5 Feet	1.60
40	12A16820	Cord, Din (5P) Plug; 8 Feet	3.00
41	86A16821	Terminal Strip, 1-L-1	.10
42	135A16822	Bush, Power Cord	.20
43	55A16823	Lid, Battery Compartment	1.00
44	55A16824	Lid, Power Cord Compartment	1.50
45	43A16825	Net (B), Cabinet Body; Right	.50
46	135A16826	Caution Label	.10
47	67A16827	Foam Rubber (B), Battery Compartment Cover	.40
48	52A16828	Bracket, Variable Resistors & Indicator Lamps	.90
49	52A16829	Contact, Channel Selector	.10
50	43A16830	Insulator, Channel Selector Contact	.10
51	67A16831	Cushion, Rubber	.04
52	80A16832	Volume Control 10K ohm (B) VR1-1, VR1-2	1.62
53	80A16833	Tone Control 50K ohm (A) x 2, VR3-1, VR3-2	2.76
54	25-10350	Capacitor, Mylar; 0.1 MFD 50WV, C107, 207	.46
55	130A16834	Lamps, Channel Indicator; 6V, 30MA	.24
56	58A16835	Printed Circuit Board, Power Supply	2.00
57	20P22816	Capacitor Electrolytic; 2200 MFD 16WV C301	4.80
58	1A12690	Diode, SR1K-2 or 10D-1	2.88
59	60S471	Resistor, 470 ohm ±10% 1/2W, R301	.18
60	52A16836	Heat Sink (A)	1.00
61	52A16837	Heat Sink (B)	.25
62		Screws, Pan Head; -2 x 3	.03
63		Screws, Flat Head; -2 x 6	.03
64		Screws, Flat Head; -2 x 8	.03
65		Screws, Pan Head; +2.6 x 5	.03
66		Screws, Oval Head; +2.6 x 7	.03
67		Screws, Pan Head; Polycarbonate +3 x 5	.03
68		Screws w/Spring Washer, Pan Head; +3 x 6	.03
69		Screws w/Spring Washer, Pan Head; +3 x 8	.03
70		Screws, Pan Head; 3 x 20	.03
71		Screws w/Spring Washer, Pan Head; + 3 x 10	.03
72		Pan Head Tapping Screw, Type (B); +3 x 8	.03
73		Pan Head Tapping Screws, Type (B) Cutting; +3 x 8	.03
74		Pan Head Tapping Screws, Type (B) Cutting; +3 x 10	.03
75		Washers, Flat 3φ	.01
76		Nuts, 3 φ	.02
77		Lugs, Terminal; 3φ	.05
78		Lugs, Wire Binder; L = 40mm	.05
79	135A16838	Bands, Power Cord & Din Cord	.20
80	130A16839	Amplifier Circuit Board Assy.	22.00
81		Screws, Oval Head; +2.6 x 8	.03

DIAGRAM NO.	PART NO.	DESCRIPTION	LIST PRICE
		ELECTRICAL PARTS LIST	
	2SB405	Power Amp Transistor	2.80
	2SD72	Power Amp Transistor	2.80
	2SB303	Pre Amp Transistor	1.70
	2SB186A	Driver Transistor	2.80
	5A16841	Thermistor, TD8-A40 (STD-04)	1.50
	20P47716	Capacitor, Electr. 470 MFD 16WV	2.00
	20P47710	Capacitor, Electr. 470 MFD 10WV	2.00
	20P22712	Capacitor, Electr. 220 MFD 12WV	1.80
	20P10710	Capacitor, Electr. 100 MFD 10WV	1.50
	20P1076	Capacitor, Electr. 100 MFD 6.3WV	1.50
	20P3366	Capacitor, Electr. 33 MFD 6.3WV	1.04
	20P10610	Capacitor, Electr. 10 MFD 10WV	1.04
	25-22350	Capacitor, Mylar 0.022 MFD 50WV	.34
	15-33150	Capacitor, Ceramic 330 PF 50WV	.24
	60QS224	Carbon Resistor, Fixed 220K ohm ±10% 1/4W	.24
	60QS513	Carbon Resistor, Fixed 51K ohm ±10% 1/4W	.24
	60QS473	Carbon Resistor, Fixed 47K ohm ±10% 1/4W	.24
	60QS333	Carbon Resistor, Fixed 33K ohm ±10% 1/4W	.24
	60QS163	Carbon Resistor, Fixed 16K ohm ±10% 1/4W	.24
	60QS123	Carbon Resistor, Fixed 12K ohm ±10% 1/4W	.24
	60QS103	Carbon Resistor, Fixed 10K ohm ±10% 1/4W	.24
	60QS562	Carbon Resistor, Fixed 5.6K ohm ±10% 1/4W	.24
	60QS472	Carbon Resistor, Fixed 4.7K ohm ±10% 1/4W	.24
	60QS332	Carbon Resistor, Fixed 3.3K ohm ±10% 1/4W	.24
	60QS222	Carbon Resistor, Fixed 2.2K ohm ±10% 1/4W	.24
	60QS152	Carbon Resistor, Fixed 1.5K ohm ±10% 1/4W	.24
	60QS334	Carbon Resistor, Fixed 330K ohm ±10% 1/4W	.24
	60QS151	Carbon Resistor, Fixed 150 ohm ±10% 1/4W	.24
	60QS470	Carbon Resistor, Fixed 47 ohm ±10% 1/4W	.24
	60QS101	Carbon Resistor, Fixed 100 ohm ±10% 1/4W	.24
	60QS68	Carbon Resistor, Fixed 6.8 ohm ±10% 1/4W	.24
	63A16840	Wire Wound Resistor 1.5 ohm ± 5% 1/2W	.30

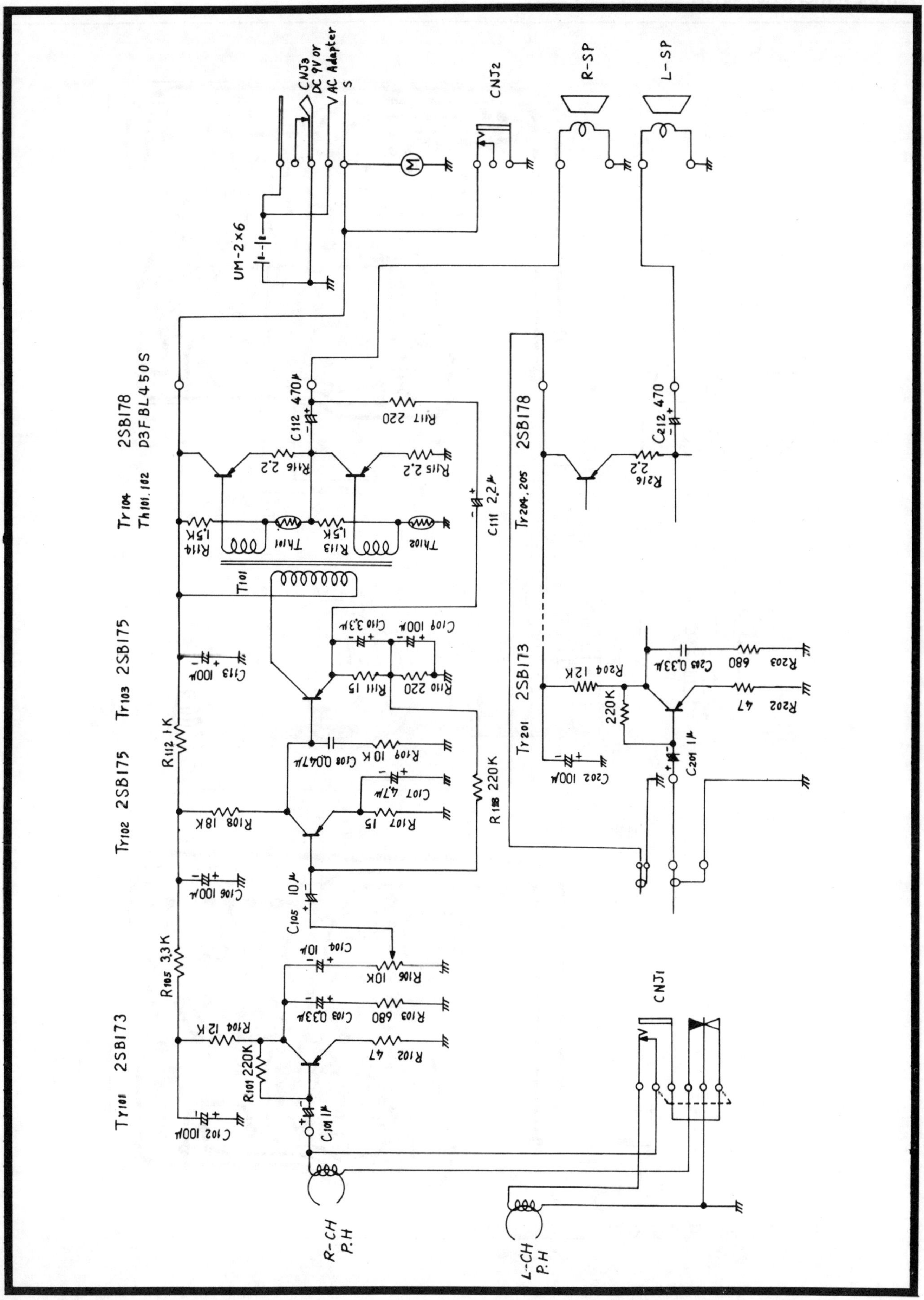
CNJ3
DC 9V or
VAC Adapter
CNJ2
R-SP
L-SP
UM-2×6
Tr104 2SB178
Th101.102 D3FBL450S
Tr103 2SB175
Tr102 2SB175
Tr101 2SB173
Tr204.205 2SB178
Tr201 2SB173
T101
CNJ1
R-CH
P.H
L-CH
P.H

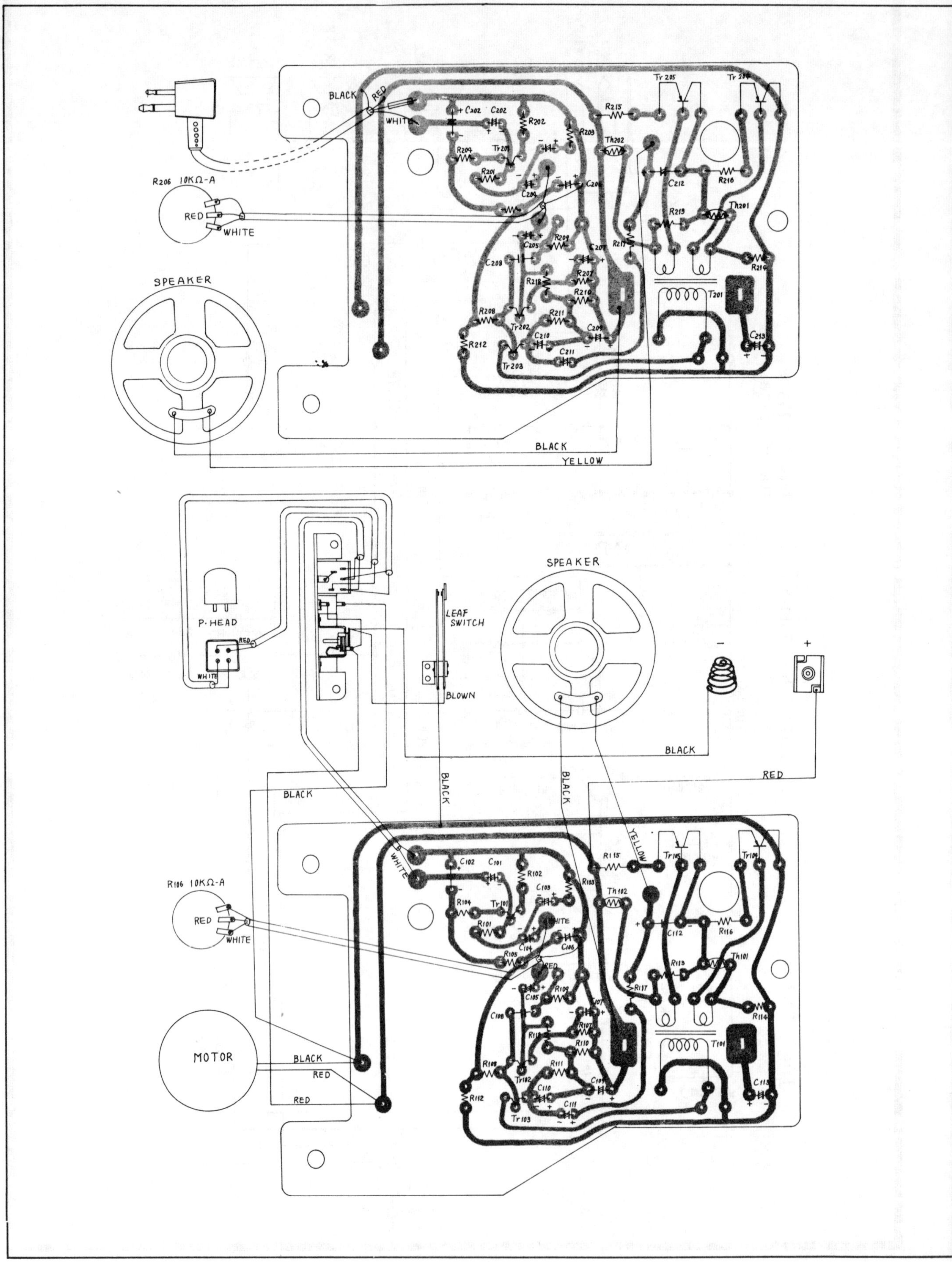
BLACK
RED
WHITE
R206 10KΩ-A
RED
WHITE
SPEAKER
BLACK
YELLOW
P. HEAD
RED
WHITE
LEAF SWITCH
BLOWN
SPEAKER
BLACK
RED
BLACK
BLACK
YELLOW
R106 10KΩ-A
RED
WHITE
MOTOR
BLACK
RED
RED

PARTS LIST

ELECTRICAL PARTS

TRANSFORMERS

Ref No.	Part Code.	Description.
T101	350360	Input transformer
T201	350360	Input transformer

TRANSISTORS & THERMISTORS

Ref No.	Part Code.	Description.
Tr101	340012	2SB173 (B.C.)
Tr102	340013	2SB175 (B.C.)
Tr103	340013	2SB175 (B.C.)
Tr104	340016	2SB178 (U.V.I.)
Tr105	340016	2SB178 (U.V.I.)
Tr201	340012	2SB173 (B.C.)
Tr202	340013	2SB175 (B.C.)
Tr203	340013	2SB175 (B.C.)
Tr204	340016	2SB178 (U.V.I.)
Tr205	340016	2SB178 (U.V.I.)
Th101	340064	D3FBL450S
Th102	340064	D3FBL450S
Th201	340064	D3FBL450S
Th202	340064	D3FBL450S

CONTROL

Ref No.	Part Code.	Description.
R106	350401	Variable resistor (10KΩ−A)
R206	350401	Variable resistor (10KΩ−A)

RESISTORS

Ref No.	Part Code.	Description.	
R101	312072	220KΩ	¼W
R102	312189	47Ω	〃
R103	312111	680Ω	〃
R104	312123	12KΩ	〃
R105	312082	3.3KΩ	〃
R107	312069	15Ω	〃
R108	312108	18KΩ	〃
R109	312090	10KΩ	〃
R110	312115	220Ω	〃
R111	312069	15Ω	〃
R112	312094	1KΩ	〃
R113	312089	1.5KΩ	〃
R114	312089	1.5KΩ	〃
R115	312092	2.2Ω	¼W
R116	312092	2.2Ω	〃
R117	312115	220Ω	〃
R118	312072	220KΩ	〃
R201	312072	220KΩ	〃
R202	312189	47Ω	〃
R203	312111	680Ω	〃
R204	312123	12KΩ	〃
R205	312082	3.3KΩ	〃
R207	312069	15Ω	〃
R208	312108	18KΩ	〃
R209	312090	10KΩ	〃
R210	312115	220Ω	〃
R211	312069	15Ω	〃
R212	312094	1KΩ	〃
R213	312089	1.5KΩ	〃
R214	312089	1.5KΩ	〃
R215	312092	2.2KΩ	〃
R216	312092	2.2KΩ	〃
R217	312115	220Ω	〃
R218	312072	220KΩ	〃

CONDENSERS

Ref No.	Part Code.	Description.		
C101	322037	1μf	50WV	Electrolytic
C102	322010	100μf	10WV	〃
C103	322055	0.33μf	25V	Aluminium
C104	322008	10μf	10WV	Electrolytic
C105	322008	10μf	10WV	〃
C106	322010	100μf	10WV	〃
C107	322030	4.7μf	10WV	〃
C108	322021	0.047μf	50V	Mylar
C109	322010	100μf	10WV	Electrolytic
C110	322057	3.3μf	25WV	〃
C111	322058	2.2μf	16V V	〃
C112	322011	470μf	10WV	〃
C113	322010	100μf	10WV	〃
C201	322037	1μf	50WV	〃
C202	322010	100μf	10WV	〃
C203	322055	0.33μf	25V	Aluminium
C204	322008	10μf	10WV	Electrolytic
C205	322008	10μf	10WV	〃
C206	322010	100μf	10WV	〃
C207	322030	4.7μf	10WV	〃
C208	322021	0.047μf	50V	Mylar
C209	322010	100μf	10WV	Electrolytic
C210	322057	3.3μf	25WV	〃
C211	322058	2.2μf	16WV	〃
C212	322011	470μf	10WV	〃
C213	322010	100μf	10WV	〃

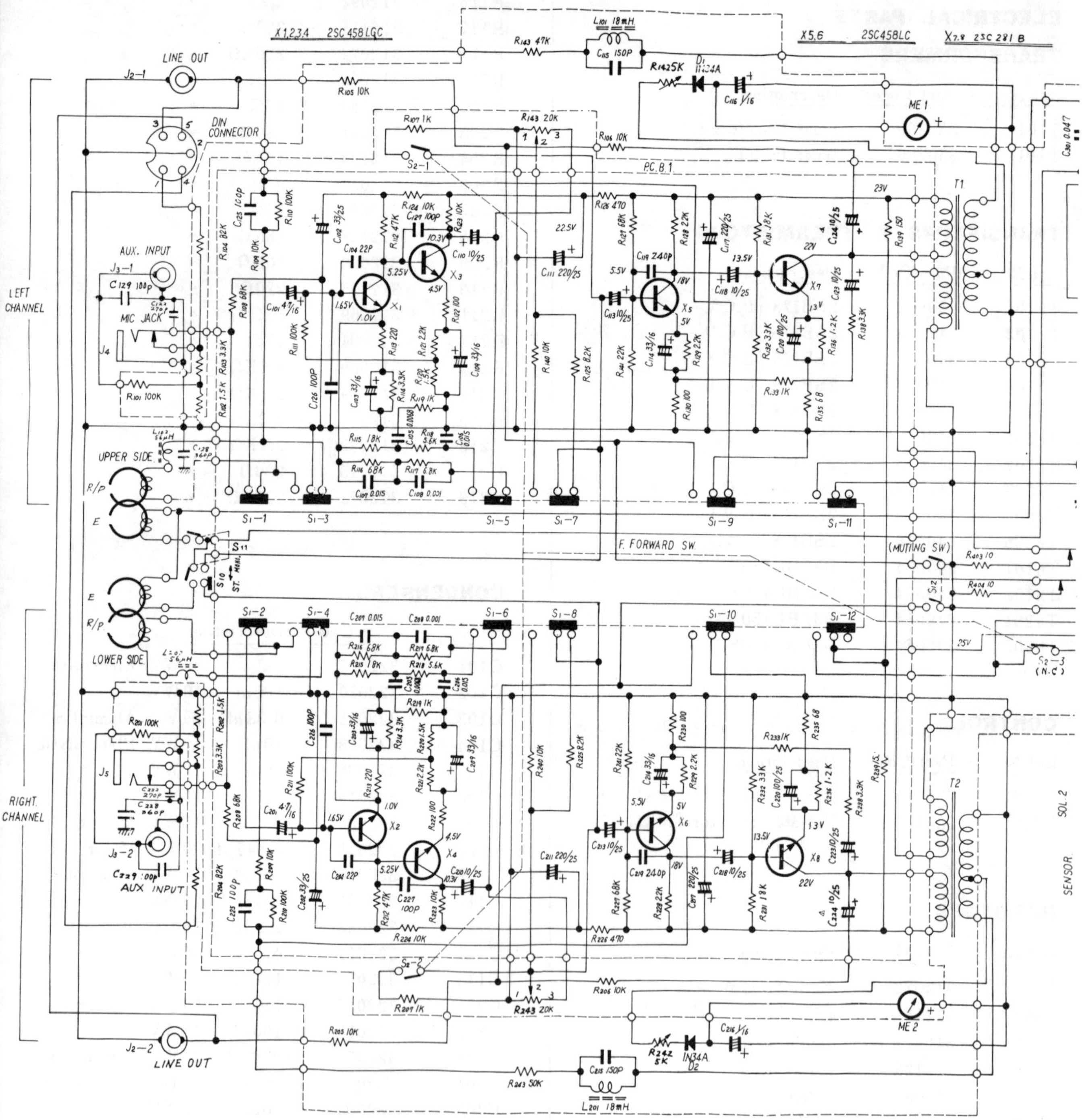

1. ( ) marked voltages are approximate D.C. values based on (E) point and on condition does not operate.
2. S-1 is set at play-back mode.
3. S-2 is not set at F.F. mode.
4. S-3 is set at OFF condition.
5. S-4 correlate the channel change-
6. S-9 is set at ON condition.
7. S-10 is set at stereo mode, S-11
8. (( )) marked voltages are vales on recording.
9. S-12 is set at OFF condition.

Schematic Diagram Model 1250

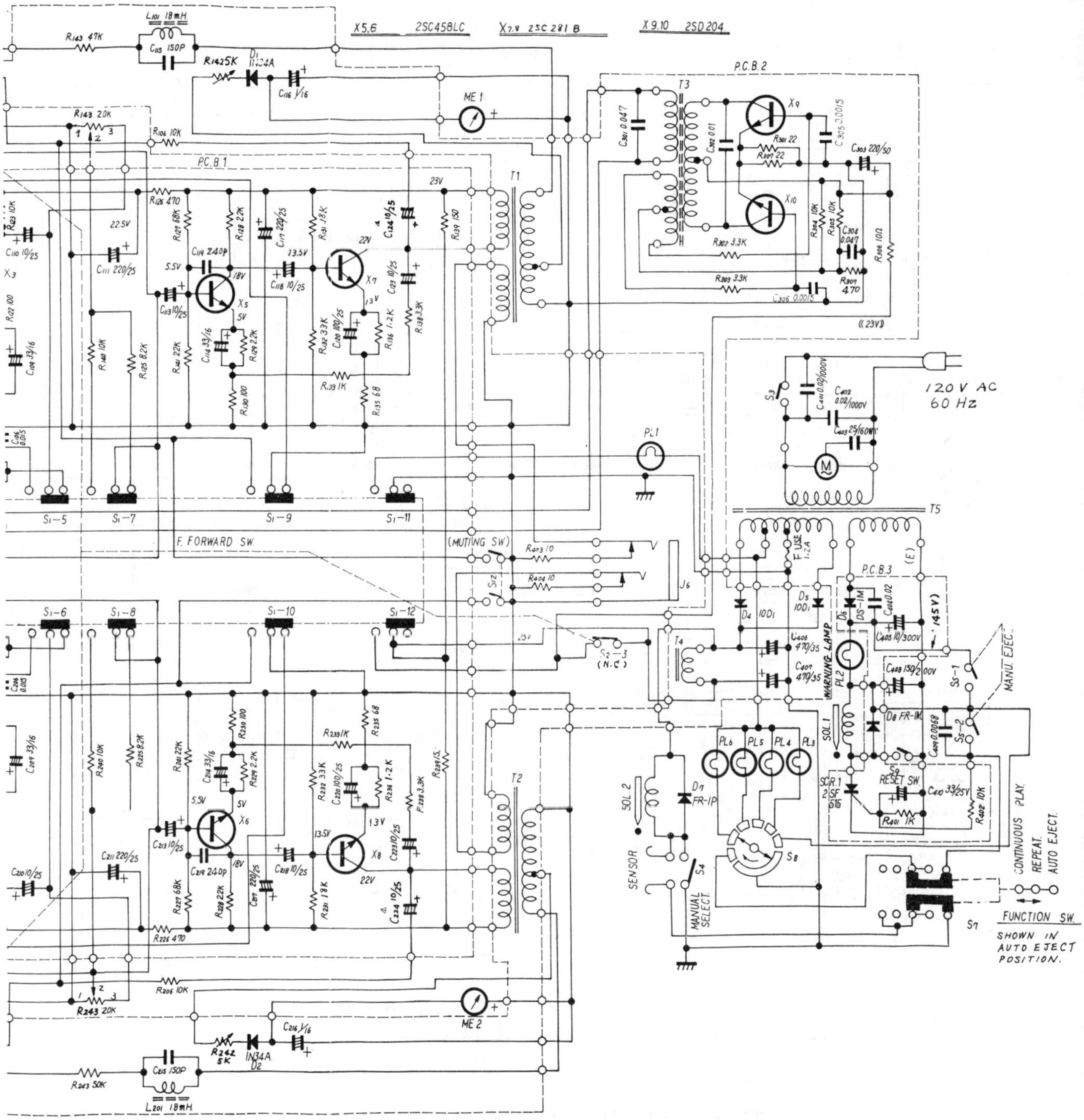

ltages are approximate D.C. values based on (E) point and on condition that EJECT SOLENOID

-back mode.
condition.
condition.
oltages are vales on recording.

3. S–2 is not set at F.F. mode.
5. S–4 correlate the channel change–over.
7. S–10 is set at stereo mode, S–11 is set at play–back mode.
9. S–12 is set at OFF condition.

Schematic Diagram Model 1250

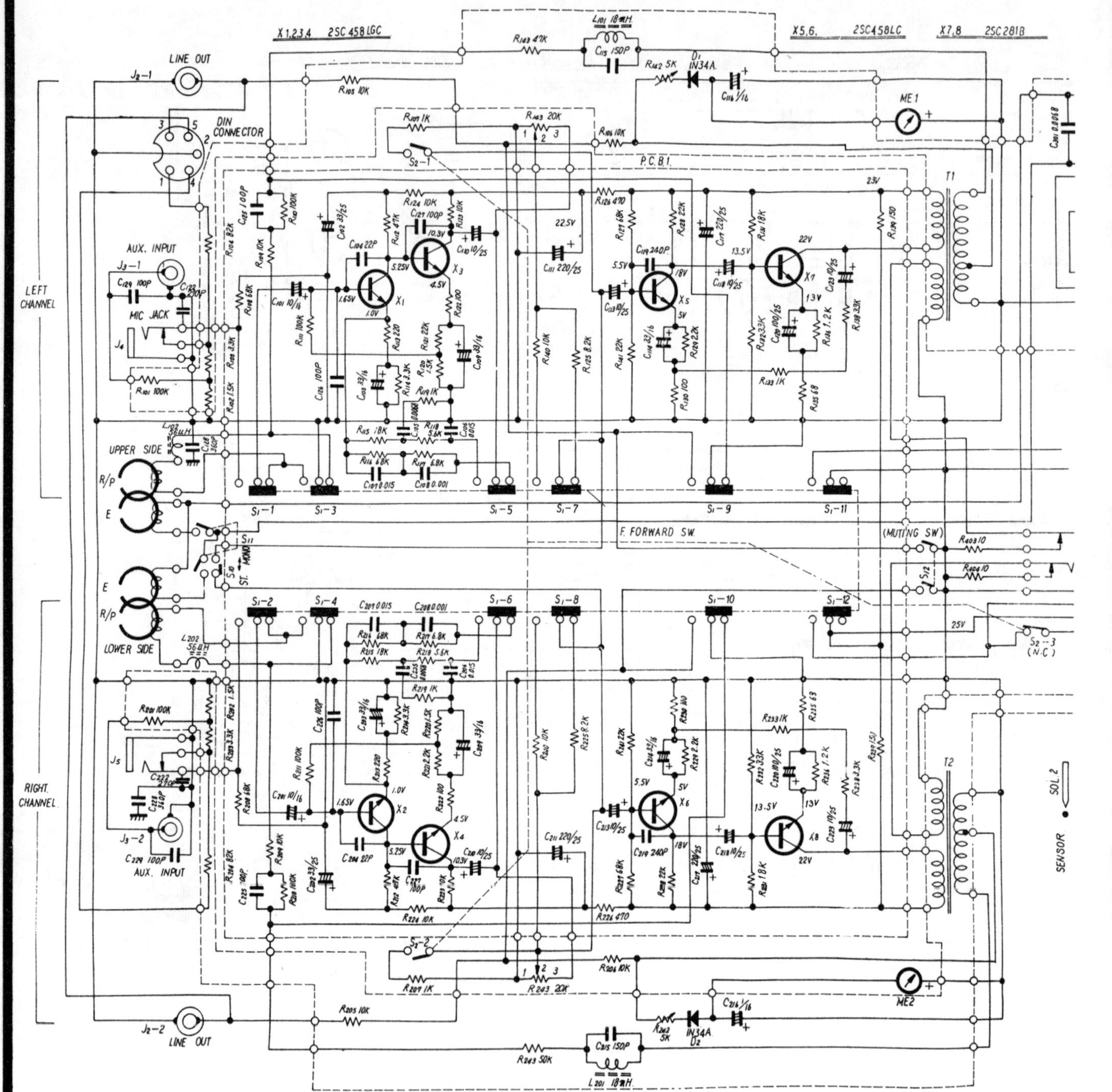

1. ( ) marked voltages are approximate D.C. values based on (E) point and on condition that EJECT SOLENOID does not operate.
2. S–1 is set at play–back mode.
3. S–2 is not set at F.F. mode.
4. S–3 is set at OFF condition.
5. S–4 correlate the the channel change–over.

Schematic Diagram CHR–250U

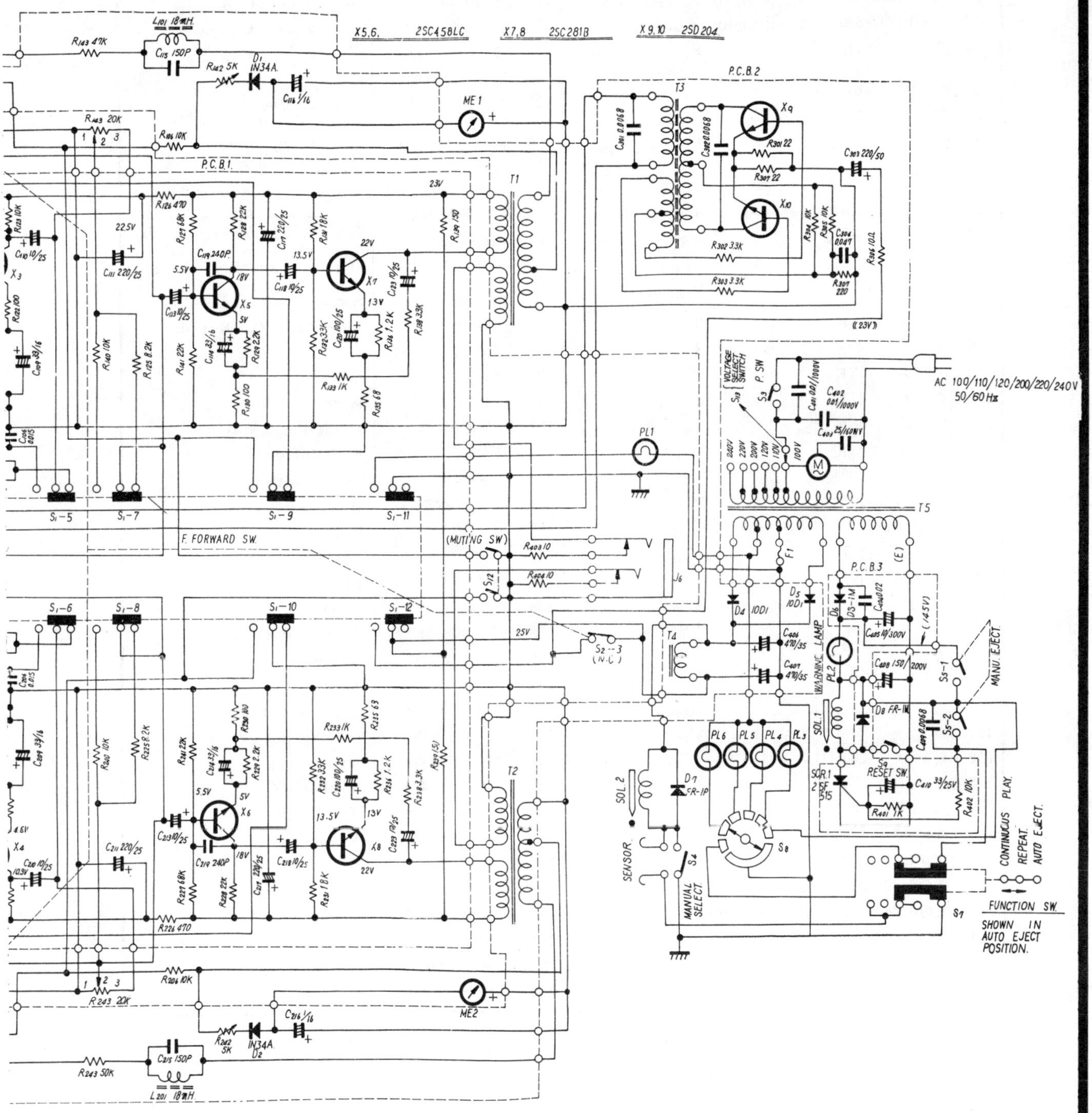

6. S–9 is set at ON condition.
7. S–10 is set at stereo mode, S–11 is set at play–back mode.
8. (( )) marked voltages are values on recording.
9. S–12 is set at OFF condition.

Dwg. No.	Parts Numper	Parts Name
86	※Q04802–470	C. Resistor $R_{123, 226, 102, 202}$
87	※Q04802–680	C. Resistor $R_{136, 236}$
88	※ 〃 –1K	〃 $R_{133, 233, 119, 219, 137, 237, 107, 207}$
89	※ 〃 –10K	〃 $R_{105, 205}$
90	※ 〃 –1.5K	〃 $R_{120, 220, 103, 203}$
91	※ 〃 –2.2K	〃 $R_{121, 221, 128, 228, 129, 229, 138, 238, 124, 234}$
92	※ 〃 –3.3K	〃 $R_{114, 214, 302, 303}$
93	※ 〃 –5.6K	〃 $R_{118, 218}$
94	※ 〃 –6.8K	〃 $R_{117, 217, 115, 215}$
95	※ 〃 –8.2K	〃 $R_{125, 225}$
96	※ 〃 10K	〃 $R_{124, 224, 123, 223, 109, 209, 140, 240, 304, 305}$
97	※ 〃 –18K	〃 $R_{132, 232}$
98	※ 〃 –22K	〃 $R_{141, 241}$
99	※ 〃 –33K	〃 $R_{131, 231}$
100	※ 〃 –47K	〃 $R_{112, 212}$
101	※ 〃 –68K	〃 $R_{127, 227, 108, 208}$
102	※ 〃 –82K	〃 $R_{104, 204}$
103	※ 〃 –100K	〃 $R_{111, 211, 116, 216, 110, 210, 101, 201}$
104	※Q04803–10	〃 R_{306}
105	※Q03108–1	E. Capacitor $C_{116, 216}$
106	※ 〃 –4.7	〃 $C_{121, 221}$
107	※ 〃 –10	〃 $C_{101, 201}$
108	※ 〃 –30	〃 $C_{109, 209, 103, 203, 114, 214}$
109	※ 〃 –100	〃 $C_{120, 220}$
110	※Q03110–10	〃 $C_{110, 210, 111, 211, 113, 213, 118, 218, 123, 223, 124, 224}$ 33/25V
111	※ 〃 –30	〃 $C_{102, 202, 410}$
112	※ 〃 –200	〃 $C_{111, 211, 117, 217}$
113	※Q03111–200	〃 C_{303}
114	※Q03152–10	〃 C_{405} 10/300V
115	※Q04112–1.2P	Fuse F_1 (CHR–250U)
	※F4974–1.2P	UL Fuse F_1 (1250)
116	※Q04305–22	Ceramic Capacitor $C_{104, 204}$
117	※Q03286–361	FM. Capacitor $C_{125, 225}$
118	※Q03287–151	〃 $C_{115, 215}$
119	※Q03244–102	M. Capacitor $C_{108, 208}$
120	※ 〃 –152	〃 $C_{106, 206}$

Dwg. No.	Parts Numper	Parts Nume
121	※Q03244–104	M. Capacitor $C_{112, 212}$
122	※ 〃 –103	〃 $C_{105, 205}$
123	※ 〃 –153	〃 $C_{107, 207}$
124	※ 〃 –223	〃 $C_{122, 222}$
125	※Q03205–682	O.F.T. Capacitor C_{409} 0.0068/400V
126	※ 〃 –223	O.F.T. Capacitor C_{404} 0.22/400V
127	※04091–K	Composition Resistor R_{401}
128	※04092–10K	〃 R_{402}
154	※Q03212–223	O.F.T. Capacitor 0.22/600V $C_{401, 402}$
155	※Q03210–103	O.F.T. Capacitor 0.01/1000 $C_{401, 402}$ (CHR–250U)
156	※Q30087–1	Socket Ass'y (CHR–250U)
157	※Q30052	Plug Ass'y (CHR–250U)
	Cabinet Parts	
129	※F3141	Lid
130	F4643	Wire
131	※F2101–B	Front Panel Ass'y
132	※F4953	F.F Button Ass'y
133	※F4859–1	Select Button
134	※F4859–2	Eject Button
135	※F4924	Record Button
136	※F4922	Knob (1)
137	※F4923	Knob (2)
138	F4910	Spacer
139	※F2100–D	Cabinet Ass'y
140	EYB7005	Eylet
141	LPSP3012Z	Screw
142	SBSB3006Z	Tapping Screw
143	REE3000	"E" Washer
144	LPSP3006Z	Screw
145	WNS3000N	Washer
146	REE5000	"E" Washer
147	WNS26000N	Washer
148	WNB2000	Nut
149	WNS2000	Washer
150	SPSP3008Z	Screw
151	SPSP4014N	〃
152	WNS4000N	Washer
153	SBSB3008Z	Tapping Screw

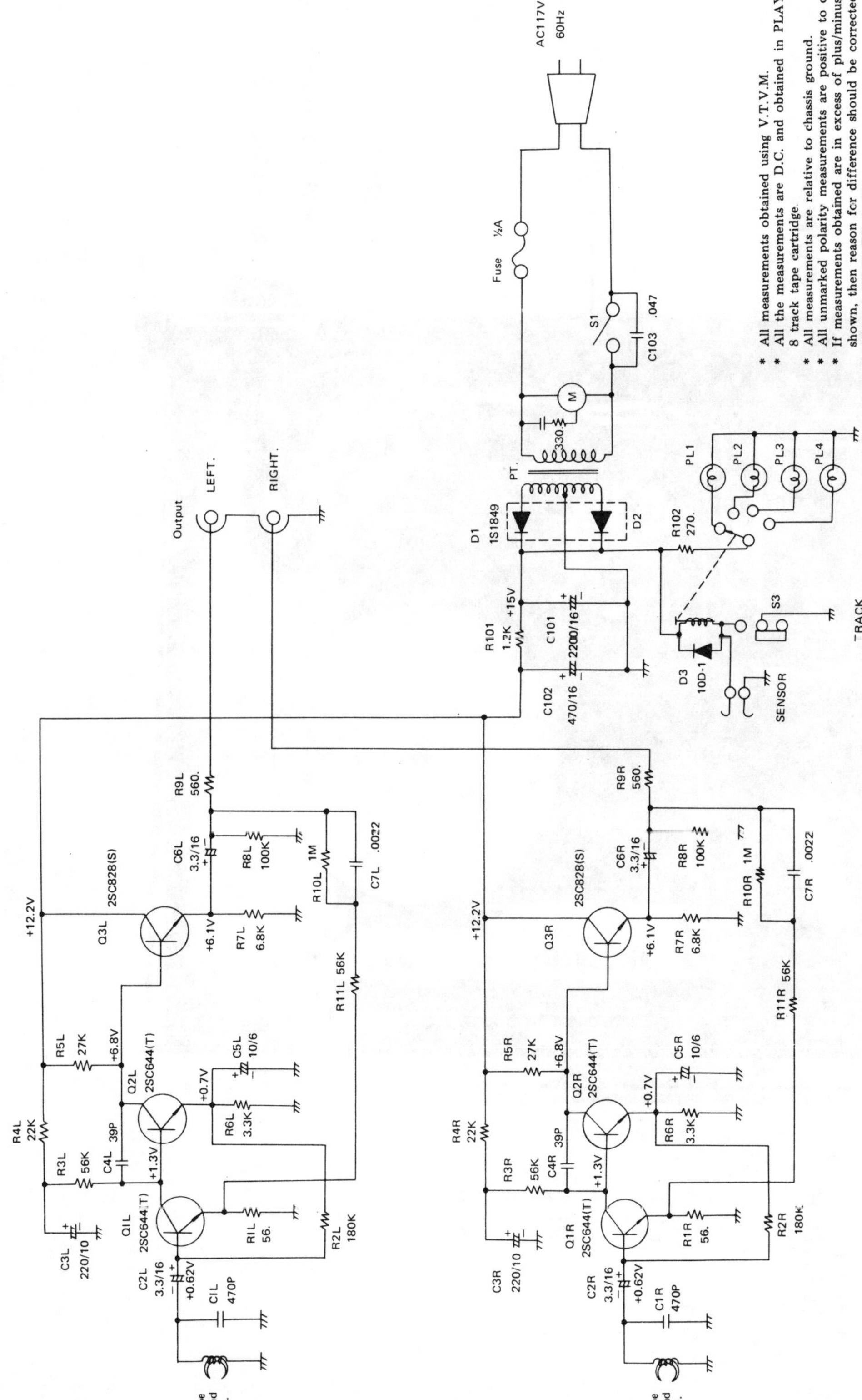

- * All measurements obtained using V.T.V.M.
- * All the measurements are D.C. and obtained in PLAY MODE inserting 8 track tape cartridge.
- * All measurements are relative to chassis ground.
- * All unmarked polarity measurements are positive to chassis ground.
- * If measurements obtained are in excess of plus/minus 20 percent of values shown, then reason for difference should be corrected.
- * SERIAL NUMBER 00001 —

LAFAYETTE 8-Track Stereo
Tape Cartridge Player

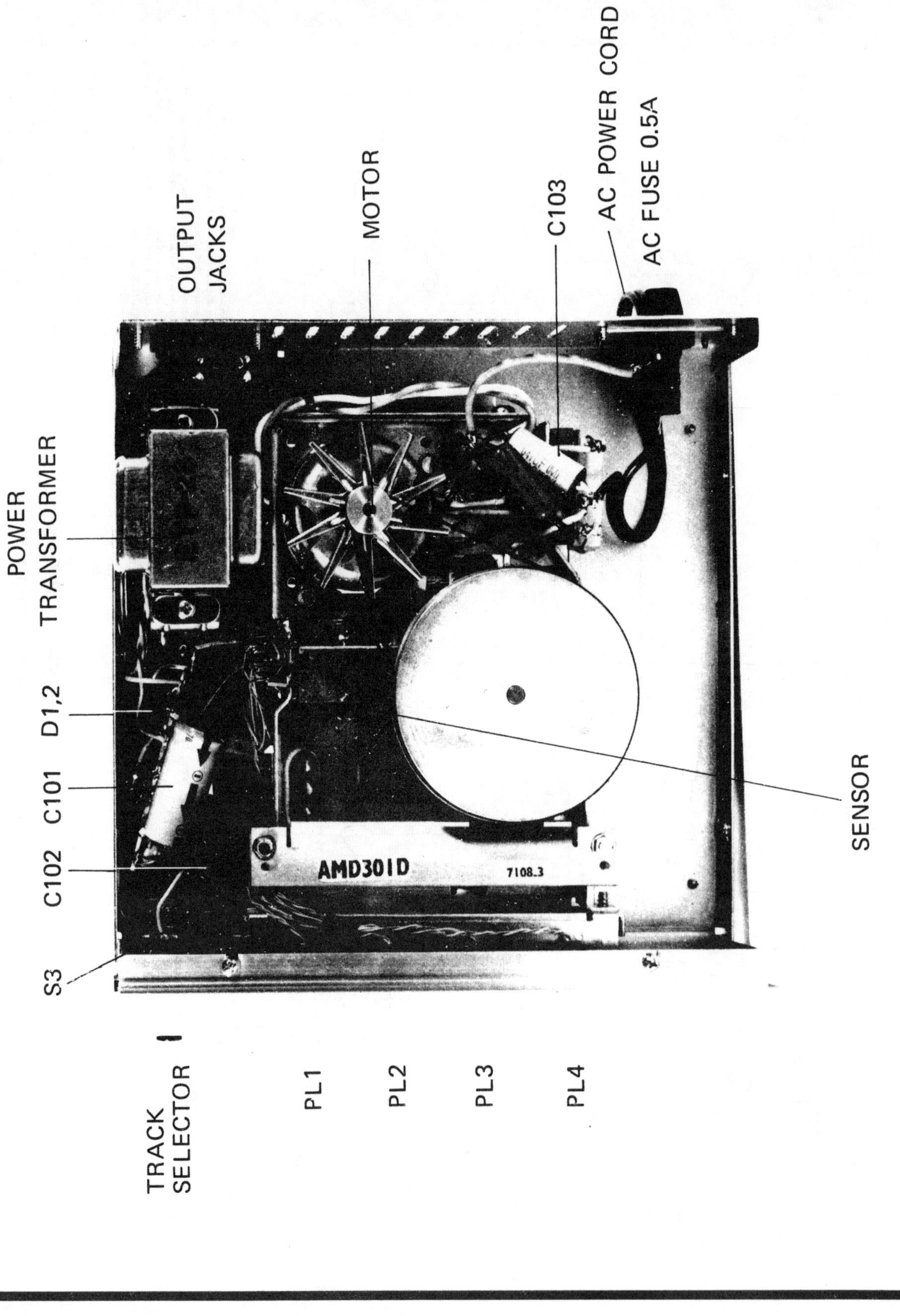

OUTPUT JACKS
MOTOR
C103
AC POWER CORD
AC FUSE 0.5A
POWER TRANSFORMER
D1,2
C101
C102
S3
SENSOR
AMD301D
7108.3
TRACK SELECTOR
PL1
PL2
PL3
PL4

IMPORTANT: USE THIS PARTS LIST ONLY WITH UNITS WITH SERIAL NUMBERS ABOVE H 121000

PARTS LIST

CABINET

Ref. Key	Part No.	Description
	302-1092-001	Cabinet with Bezel
	365-1410-001	Button Track Select
	368-1909-001	Spring
	366-1004-001	Clamp Button Retaining
	367-9110-002	Drive Screw #6 x 5/16
	369-1311-001	Decorative Label
5	373-1115-001	Switch Mounting Bracket
6	361-9006-001	Track Select Switch (S4)
7	360-9501-001	Lamp Holder
8	360-9401-002	Lamp No. 53-14 Volt (DSI)
	316-1004-001	Base
	369-9505-001	Bumper

CHASSIS ASSEMBLY

Ref. Key	Part No.	Description
25	309-1028-001	Tape & Sensing Switch Guide
26	342-1001-001	Solenoid (K1)
	373-1131-001	Solenoid Bracket
27	327-1002-001	Ratchet
28	367-1201-001	Head Adjustment Screw
	368-1001-001	Teflon Washer
	373-1113-001	Bracket Motor Mounting
30	302-1014-001	Head Support, Sub-Assembly
31	368-1907-001	Head Support Spring
32	342-9001-001	Playback Head
33	302-1096-001	Cart. Spring & Holder Assy.
	373-1120-001	Bracket, Motor Support
	354-9401-002	HC-30 Diode (CR6)
35	368-1502-001	Thrust Bearing Disc
36	374-9002-001	SH Jacketed Wire 2-Cond. Red/White
37	374-9002-002	SH Jacketed Wire 2-Cond. Blue/Yellow

MOTOR PLATE ASSEMBLY

Ref. Key	Part No.	Description
11	357-9002-001	AC Motor (M1)
	302-1067-001	Center Housing Assy.
12	373-1001-001	Pulley
13	369-9506-001	Motor Mounting Grommet
	368-9502-001	Tubular Spacer
15	356-9101-001	Power Transformer (T1)
16	366-1102-001	Grounding Strap
17	302-1095-002	Flywheel-Shaft Assembly

MOTOR PLATE ASSEMBLY (CONT.)

Ref. Key	Part No.	Description
	251-9101-229	Resistor 220Ω 2w 10%
	371-1207-001	Tape Barrier
	251-9102-219	Resistor 33Ω 1w 10%
18	373-9401-004	Drive Belt
19	302-1094-001	Flywheel Bracket and Bearing
22	374-9007-001	Power Cable
23	366-9203-001	Strain Relief Bushing
24	369-9102-001	Cable Grommet

PREAMPLIFIER ASSEMBLY 302-1064

Ref. Key	Part No.	Description
64	350-9504-001	Capacitor 10MF 2.5V. (C1, C8, C30, C23)
65	350-9504-002	Capacitor 25MF 6.4V. (C3, C29)
66	350-9504-008	Capacitor 100MF 2.0V. (C2, C26)
67	350-9504-009	Capacitor 125MF 16V. (C6)
68	350-9504-006	Capacitor 320MF 2.5V. (C10, C21)
69	350-9504-007	Capacitor 16MF 10V. (C11, C20)
70	350-9401-007	Capacitor .01MF 200V. 20% (C13, C27)
72	251-9104-225	Resistor, 100Ω 1/4W 10% R72, R73.
46	342-1005-001	RC Assembly P.E.C. 13 Leads
47	342-1006-001	RC Assembly P.E.C. 5 Leads
61	353-9306-001	Transistor Fairchild F 302-1 (Q2, Q11)
62	353-9306-002	Transistor Fairchild F302-2 (Q3, Q10)
63	353-9306-003	Transistor Fairchild SE 4010 (Q4, Q9)
71	351-9010-001	Potentiometer 2 Section (R13, R44)
52	251-9104-249	Resistor 10K 1/4W 10% (R29, R53)
53	369-1124-001	Insulator Preamp. P.C. Board
54	317-1019-001	Preamplifier Shield
55	374-1002-001	Audio Cable
	350-9107-001	Capacitor 180PF-1000V 20%
	361-9005-001	On-Off Switch (Cartridge) (S2)
58	350-9503-001	Capacitor 1000MF 15V. (C16) 500MF 15V. (C7) Dual
59	366-9006-001	Capacitor Mounting Clip
60	365-9204-001	Terminal Strip (10 Lug)
	354-9101-002	Diode HC-30 (CR3, CR4, CR6)
	251-9103-235	Resistor 680 OHM 1/2W 10% (R50)
	350-9507-001	Capacitor 500MF 25V. (C24)
	350-9509-001	Ceramic Disc Capacitor .05MF 500VDC–80-2070 (C4)

IMPORTANT: THIS SCHEMATIC APPLIES TO ALL UNITS WITH SERIAL NUMBERS ABOVE H-121000 (SEE SCHEMATIC #1 FOR UNITS WITH SERIAL NUMBERS FROM H-100000 THRU H-121000)

SCHEMATIC #2

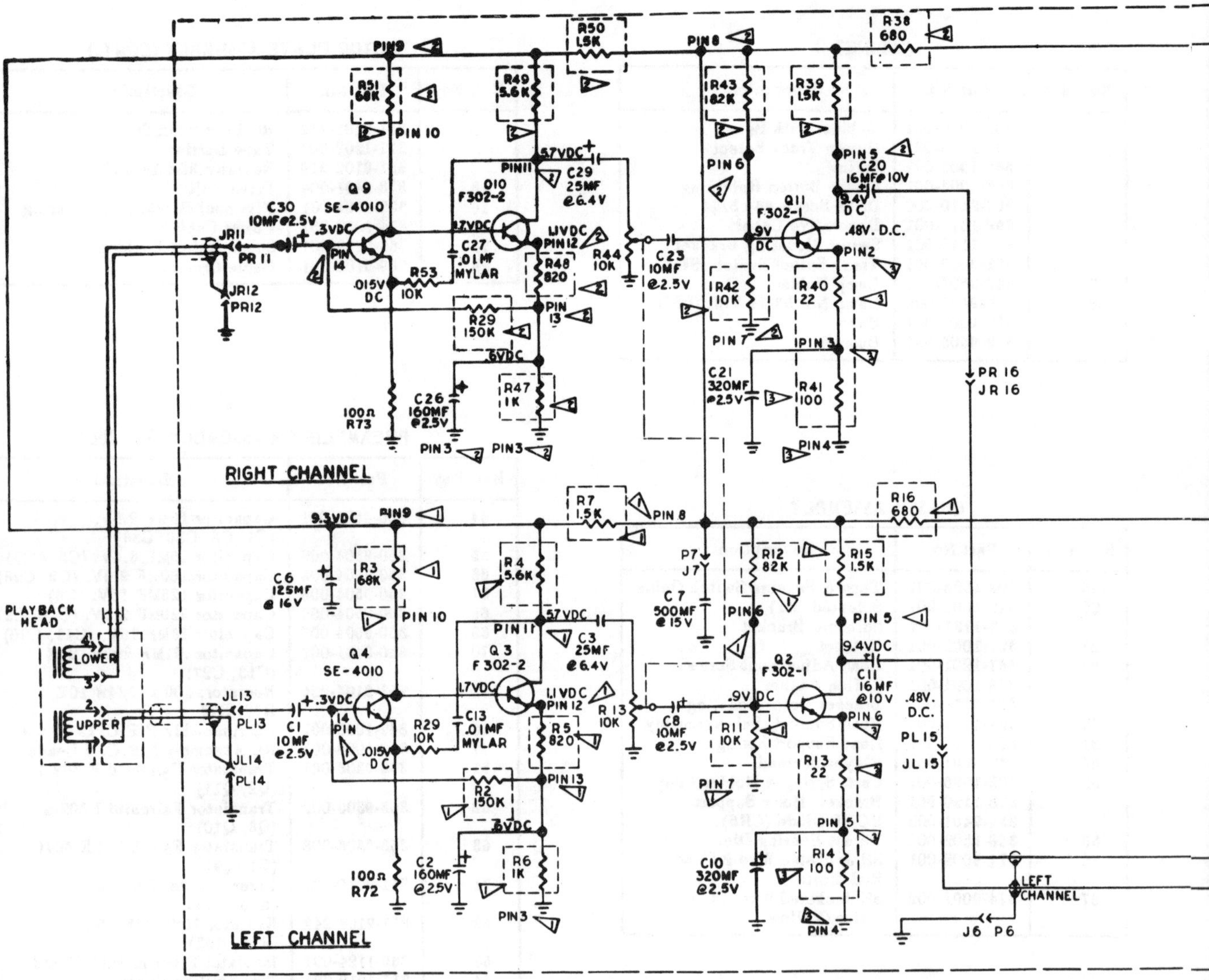

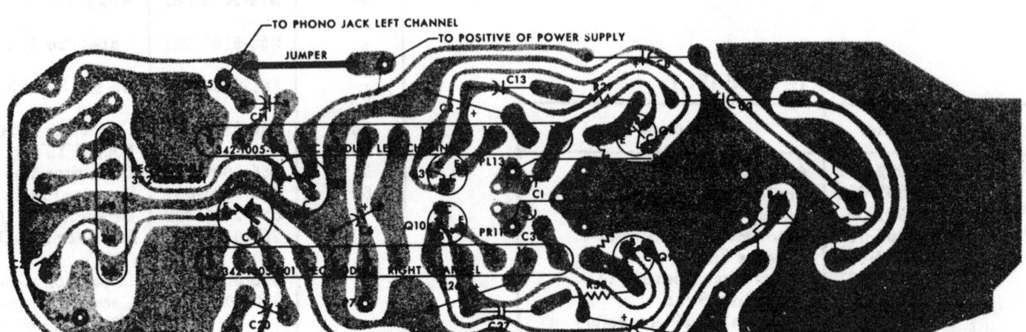

NOTE: WIRING DIAGRAM SHOWN FROM PRINTED SIDE OF BOARD

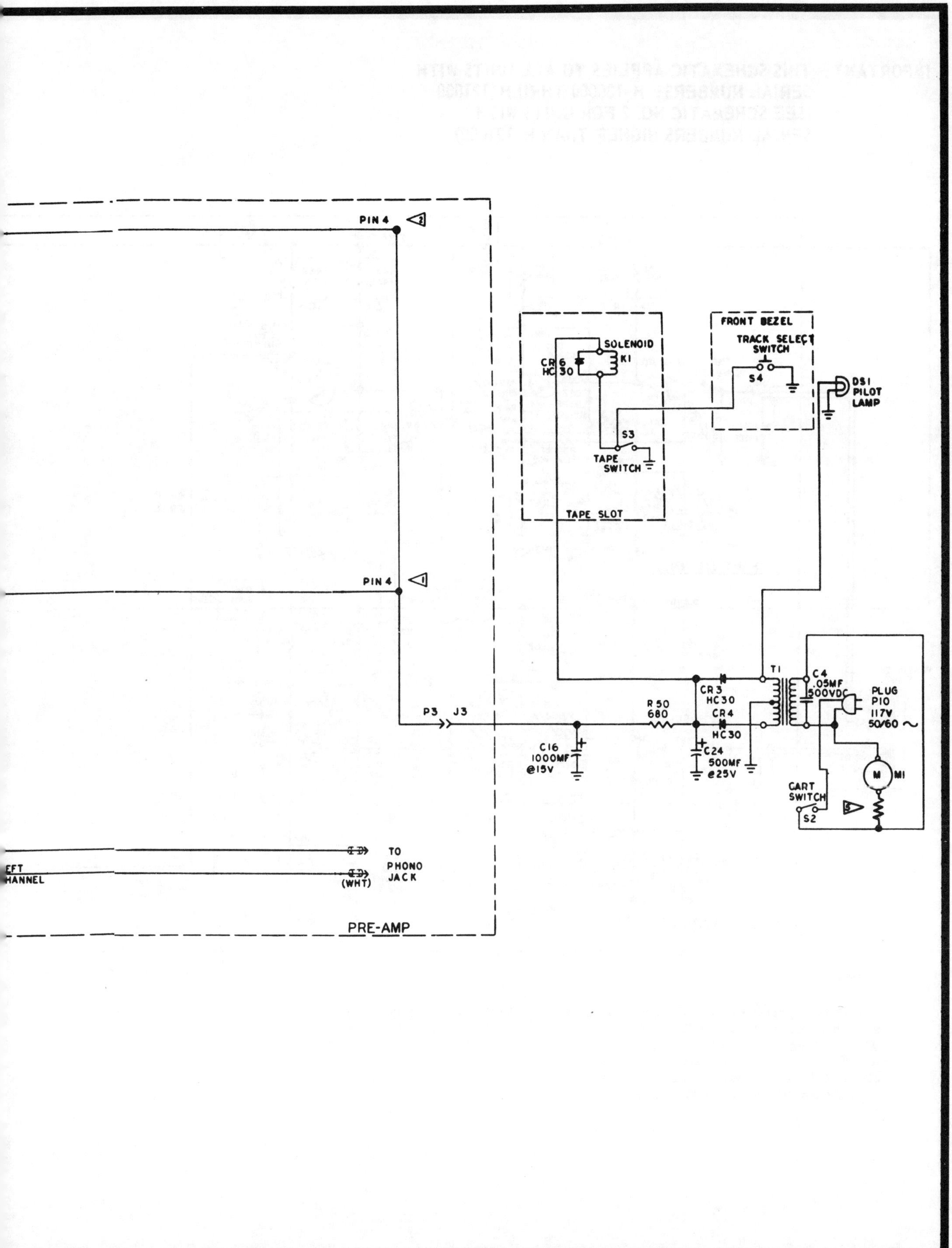
PIN 4
PIN 4
FRONT BEZEL
TRACK SELECT SWITCH
S4
SOLENOID
K1
CR 6 HC 30
S3
TAPE SWITCH
TAPE SLOT
DS1 PILOT LAMP
T1
C4 .05MF 500VDC
PLUG P10 117V 50/60 ~
CR3 HC30
CR4 HC30
R50 680
P3 J3
C16 1000MF @15V
C24 500MF @25V
M M1
CART SWITCH
S2
TO PHONO JACK
(WHT)
EFT
HANNEL
PRE-AMP

IMPORTANT: THIS SCHEMATIC APPLIES TO ALL UNITS WITH SERIAL NUMBERS: H-100000 THRU H-121000 (SEE SCHEMATIC NO. 2 FOR UNITS WITH SERIAL NUMBERS HIGHER THAN H-121000)

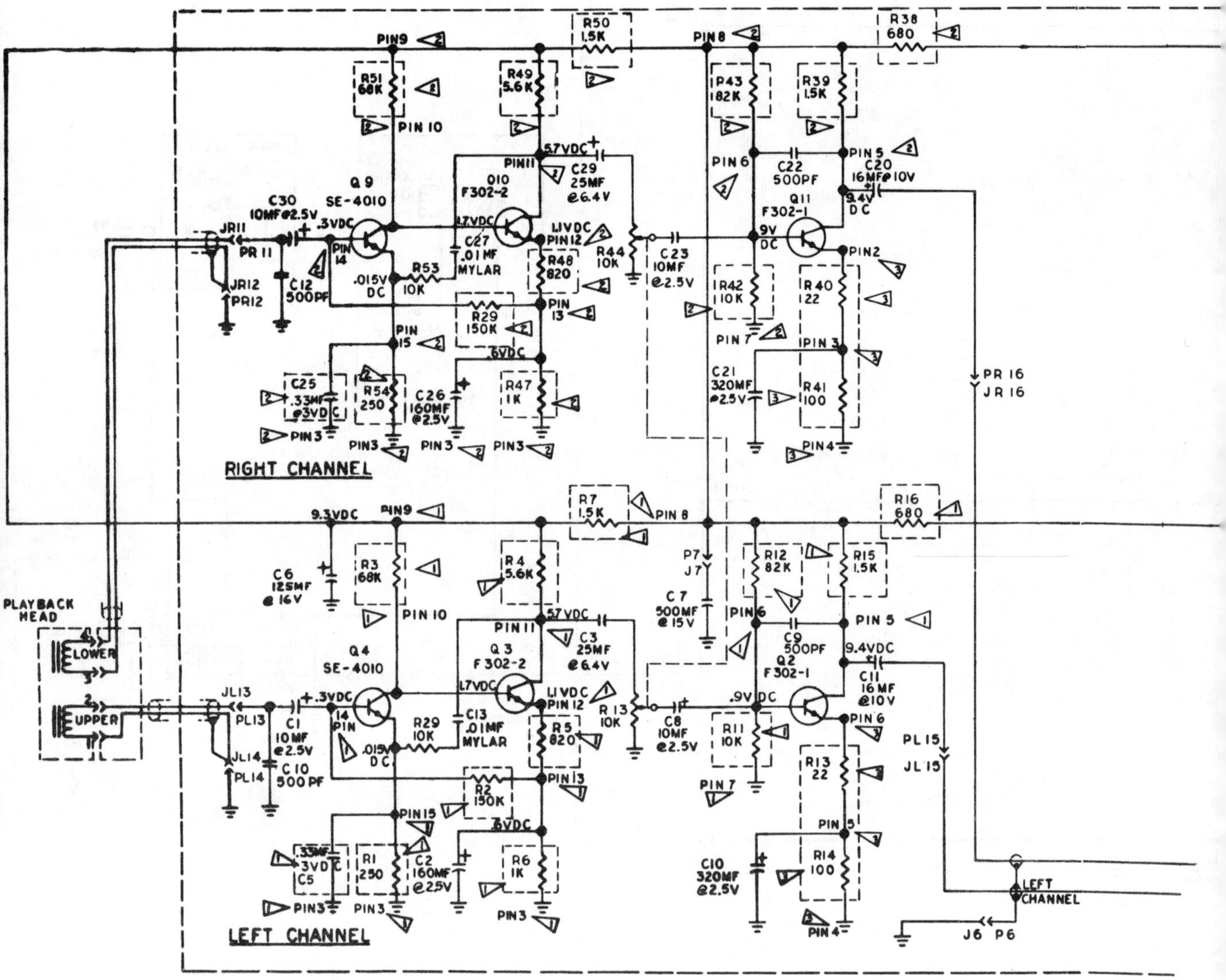

NOTES:

1. This component is part of P.E.C. module for left channel (No. 342-1005-001 Ref.)
2. This component is part of P.E.C. module for right channel (No. 342-1005-001 Ref.)
3. This component is part of P.E.C. module (No. 342-1006-001 Ref.)
4. Unless otherwise noted, resistors are 1/4W
5. 68 OHM 10 watt resistor used with 357-9001-001 motor only. Not used with 357-9002-006 motor.

NOTES:

1. This component is part of P.E.C. module for left channel (No. 342-1005-001 Ref.)
2. This component is part of P.E.C. module for right channel (No. 342-1005-001 Ref.)
3. This component is part of P.E.C. module (No. 342-1006-001 Ref.)
4. Unless otherwise noted, resistors are 1/4W

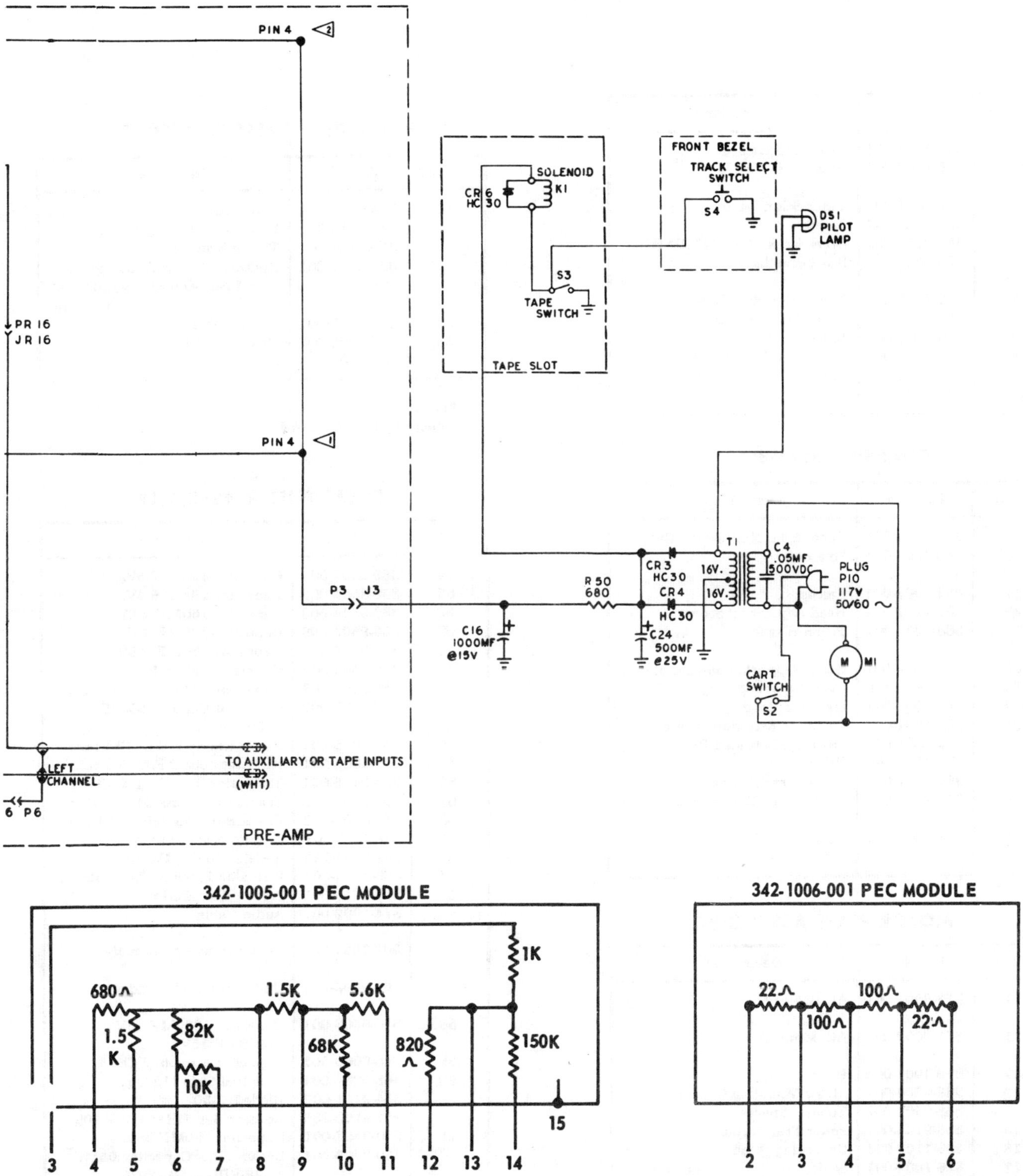

IMPORTANT: USE THIS PARTS LIST ONLY WITH UNITS WITH SERIAL NUMBERS H-100000 THRU H-121000

PARTS LIST

CABINET

Ref. Key	Part No.	Description
	316-1003-001	Cabinet (Fastened
	316-1005-001	Bezel Together)
	365-1401-001	Track Select Button
	368-1909-001	Spring
	367-9110-001	Drive Screw #6 x 1/4" RD-HD.
	369-1311-001	Decorative Label
5	373-1115-001	Switch Mounting Bracket
6	361-9006-001	Track Select Switch
7	360-9501-001	Lamp Holder
8	360-9401-002	Lamp No. 53 - 14 Volt
	316-1004-001	Base
	369-9505-001	Bumper

CHASSIS ASSEMBLY

Ref. Key	Part No.	Description
25	341-1001-001	Tape & Sensing Switch Guide
26	342-1001-001	Solenoid
27	327-1002-001	Ratchet
28	367-1201-001	Head Adjustment Screw
	368-1001-001	Teflon Washer
30	302-1014-001	Head Support, Sub-Assembly
31	868-1907-001	Head Support Spring
32	342-9001-001	Playback Head
33	368-1901-001	Cartridge Retention Spring
	373-1201-001	Cartridge Retention Roller
	354-9101-002	HC-30 Diode
35	368-1502-001	Thrust Bearing Disc
36	374-9002-001	SH Jacketed Wire 2-Cond. Red/White
37	374-9002-002	SH Jacketed Wire 2-Cond. Blu/Yellow

MOTOR PLATE ASSEMBLY

Ref. Key	Part No.	Description
11	357-9001-001	AC Motor*
11	357-9002-001	AC Motor
12	373-1001-001	Pulley
13	369-9506-001	Motor Mounting Grommet
	368-9502-001	Tubular Spacer
15	356-9101-001	Power Transformer
16	366-1102-001	Grounding Strap
17	326-1003-001	Shaft (Fastenea
	373-1002-001	Flywheel Together)

MOTOR PLATE ASSEMBLY (CONT.)

Ref. Key	Part No.	Description
18	373-9401-004	Drive Belt
19	373-1112-001	Flywheel Retaining Bracket
	369-1121-001	Thrust Bearing
	351-9202-002	Resistor (Use with 357-9001-001 AC Motor Only) 68 OHMS 10 Watt
22	374-9007-001	Power Cable
23	366-9203-001	Strain Relief Bushing
24	369-9102-001	Cable Grommet

* This Motor was used on First Production Units. Later Replaced by 357-9002-001.

PREAMPLIFIER ASSEMBLY

Ref. Key	Part No.	Description
64	350-9504-001	Capacitor 10MF 2.5V.
65	350-9504-002	Capacitor 25MF 6.4V.
66	350-9504-003	Capacitor 160MF 2.6V.
67	350-9504-009	Capacitor 125MF 16V.
68	350-9504-006	Capacitor 320MF 2.5V.
69	350-9504-007	Capacitor 16MF 10V.
70	350-9401-007	Capacitor .01MF 200V. 20%
	350-9509-002	Capacitor Ceramic 500PF 500VPC 20%
46	342-1005-001	RC Assembly P.E.C. 13 Leads
47	342-1006-001	RC Assembly P.E.C. 5 Leads
61	353-9306-001	Transistor Fairchild F 302-1
62	353-9306-002	Transistor Fairchild F 302-2
63	353-9306-003	Transistor Fairchild SE 4010
71	351-9010-001	Potentiometer 2 Section
52	251-9104-249	Resistor 10K 1/4W 10%
53	369-1124-001	Insulator Preamp. P.C. Board
54	317-1019-001	Preamplifier Shield
55	374-1002-001	Audio Cable
	302-1067-001	Center Housing Assembly
	361-9005-001	On-Off Switch (Cartridge)
58	350-9503-001	Capacitor 1000MF 15V - 500MF 15V.
59	366-9006-001	Capacitor Mounting Clip
60	365-9204-001	Terminal Strip (10 lug)
	354-9101-002	Diode HC-30
	251-9103-235	Resistor 680 OHM 1/2W 10%
	350-9507-001	Capacitor 500MF 25V.
	350-9509-001	Ceramic Disc Capacitor .05MF 500VDC + 80 - 2070

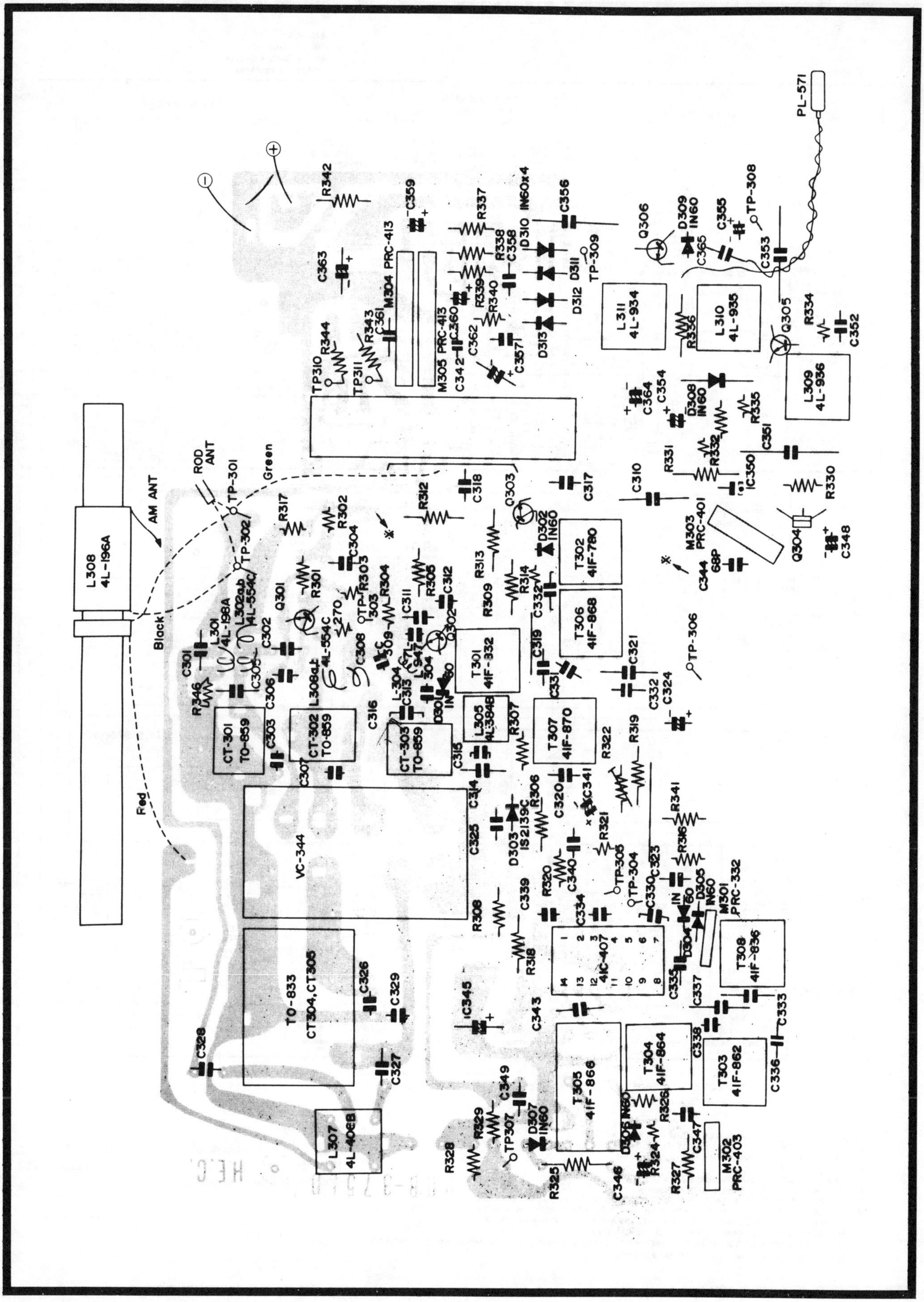
L308
4L-196A
AM ANT
ROD ANT
TP-301
TP-302
Green
Black
Red
VC-344
TO-833
CT304, CT305
CT-301
TO-859
CT-302
TO-859
CT-303
TO-859
L307
4L-406B
T305
4IF-866
T304
4IF-864
T303
4IF-862
T308
4IF-836
T307
4IF-870
T306
4IF-868
T301
4IF-332
T302
4IF-780
L311
4L-934
L310
4L-935
L309
4L-936
M301
PRC-332
M302
PRC-403
M303
PRC-401
M304 PRC-413
M305 PRC-413
IC-407
PL-571
Q301
Q303
Q304
Q305
Q306
TP-304
TP-305
TP-306
TP-307
TP-308
TP-309
TP310
TP311
D310 IN60x4
D303
IS2139C

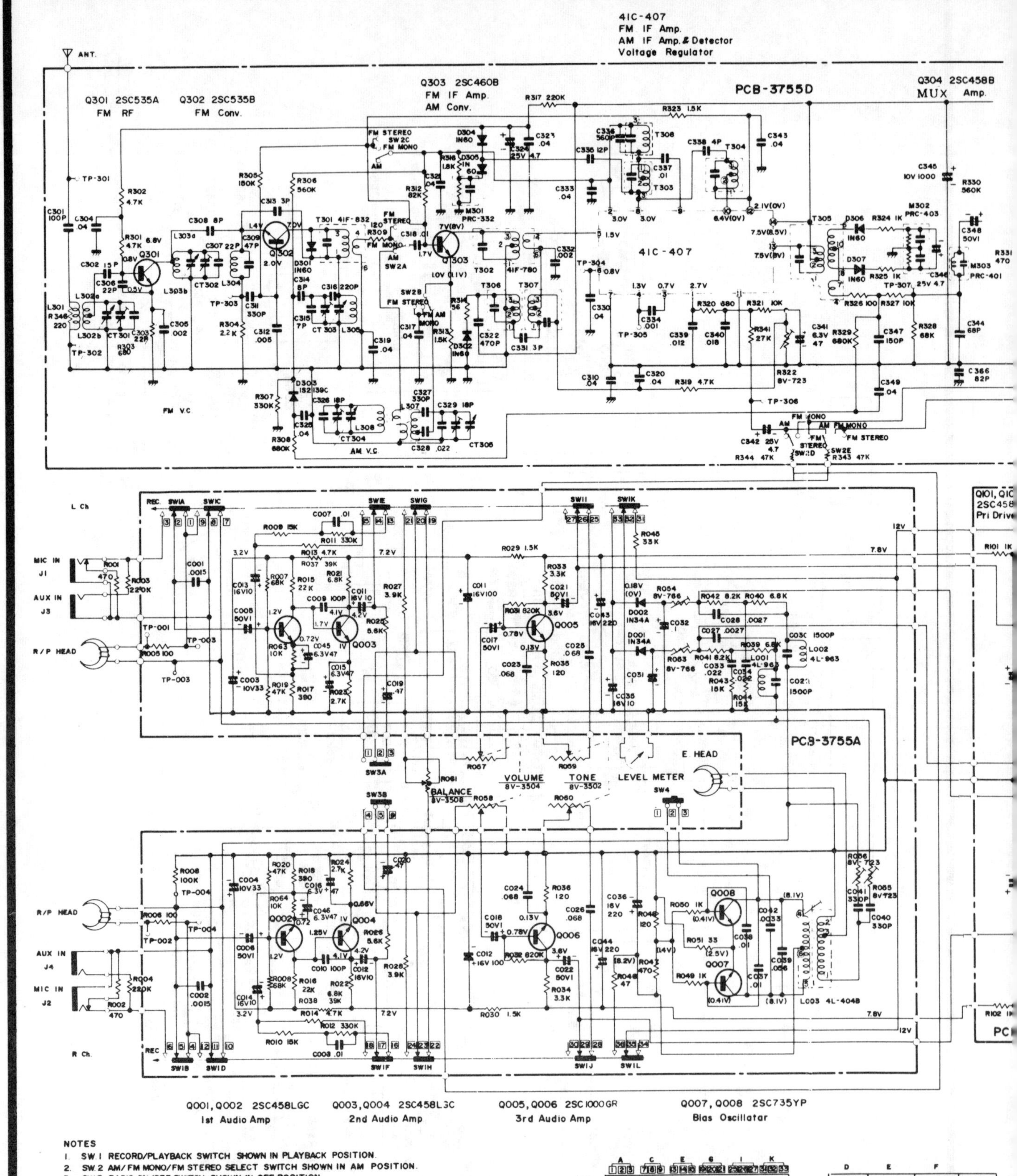

NOTES

1. SW.1 RECORD/PLAYBACK SWITCH SHOWN IN PLAYBACK POSITION.
2. SW.2 AM/FM MONO/FM STEREO SELECT SWITCH SHOWN IN AM POSITION.
3. SW.3 RADIO ON/OFF SWITCH SHOWN IN OFF POSITION.
4. SW.4 R.I.F. SWITCH
5. SW.5 SPEAKER ON/OFF SWITCH SHOWN IN ON POSITION.
6. SW.6 MOTOR SWITCH SHOWN IN OFF POSITION.
7. VOLTAGE IN PARENTHESES TAKEN WITH SW.I IN RECORD POSITION DURING DC OPERATION.
8. VOLTAGE WITHOUT PARENTHESES TAKEN WITH SW.I IN PLAYBACK POSITION DURING DC OPERATION.
9. VOLTAGE IN RECTANGLE TAKEN WITH SW.I IN PLAYBACK POSITION DURING AC OPERATION.
10. ALL RESISTORS ARE 1/4W UNLESS OTHERWISE NOTES
11. THE JUST SAME VOLTAGE WHEN THE UNIT RECEIVING FM STEREO AND FM MONAURAL BRORDCAST.
12. SHOW THE VOLTAGE IN ( ), WHEN AM RECEVING

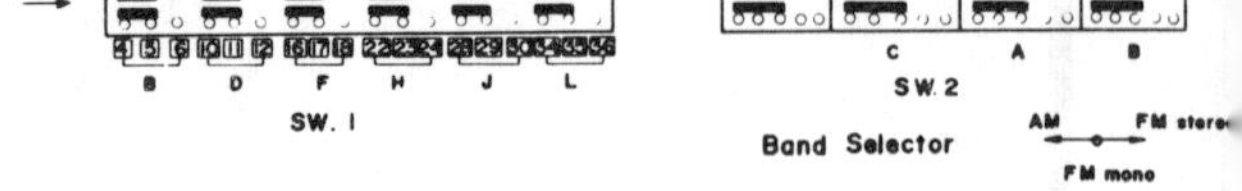

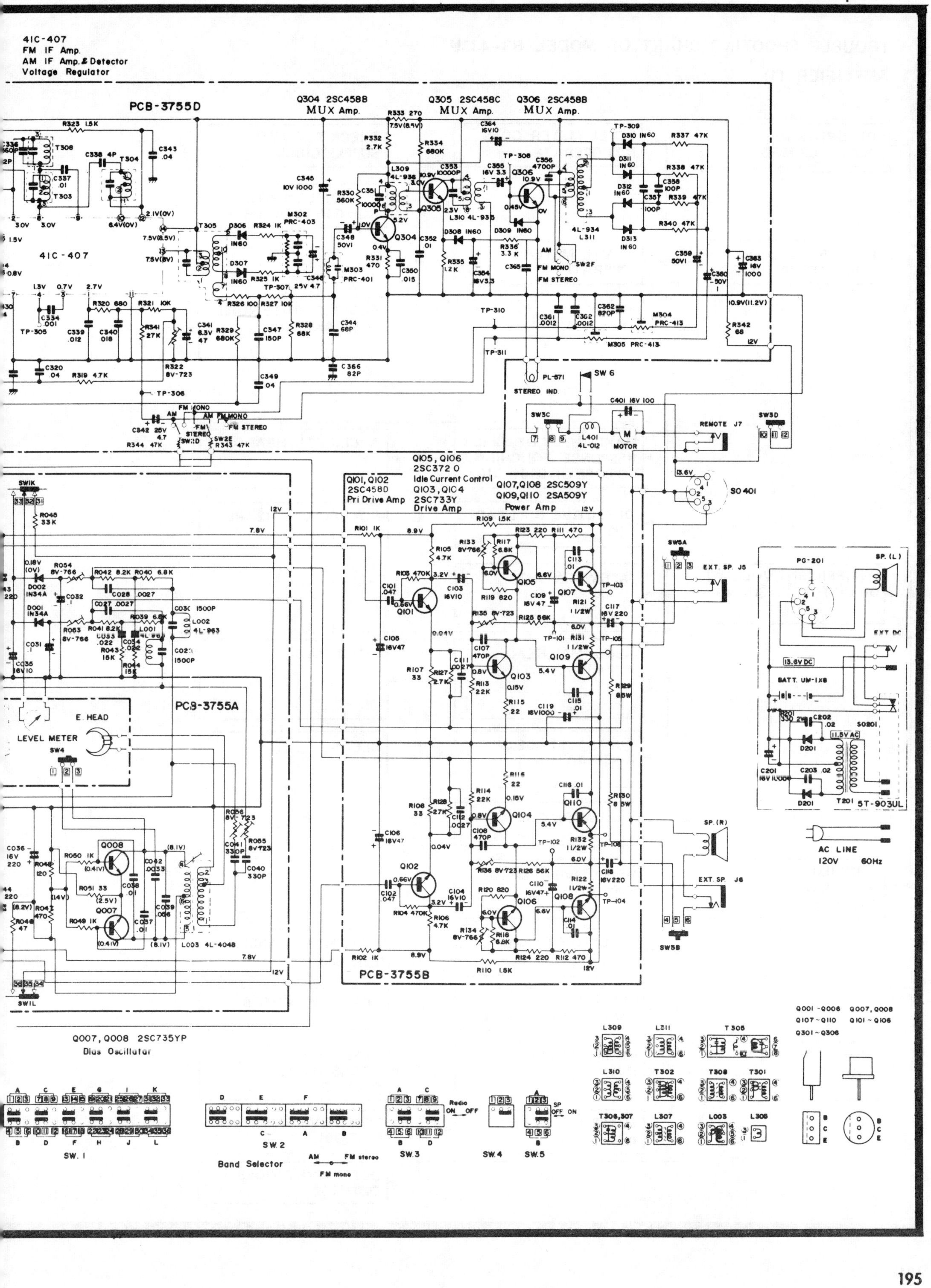

4IC-407
FM IF Amp.
AM IF Amp.& Detector
Voltage Regulator
PCB-3755D
Q304 2SC458B
MUX Amp.
Q305 2SC458C
MUX Amp.
Q306 2SC458B
MUX Amp.
STEREO IND.
SW 6
REMOTE J7
MOTOR
SO 401
Q105, Q106
2SC372 O
Idle Current Control
Q101, Q102
2SC458D
Pri Drive Amp
Q103, Q104
2SC733Y
Drive Amp
Q107, Q108 2SC509Y
Q109, Q110 2SA509Y
Power Amp
EXT. SP. J5
PG-201
SP. (L)
EXT. DC
BATT. UM-1X8
SO201
5T-903UL
E. HEAD
LEVEL METER
PCB-3755A
SP. (R)
EXT. SP. J6
AC LINE
120V 60Hz
PCB-3755B
Q007, Q008 2SC735YP
Bias Oscillator
SW. 1
SW. 2
Band Selector
SW. 3
SW. 4
SW. 5

TROUBLE SHOOTING CHART OF MODEL RS–433U

AMPLIFIER (1)

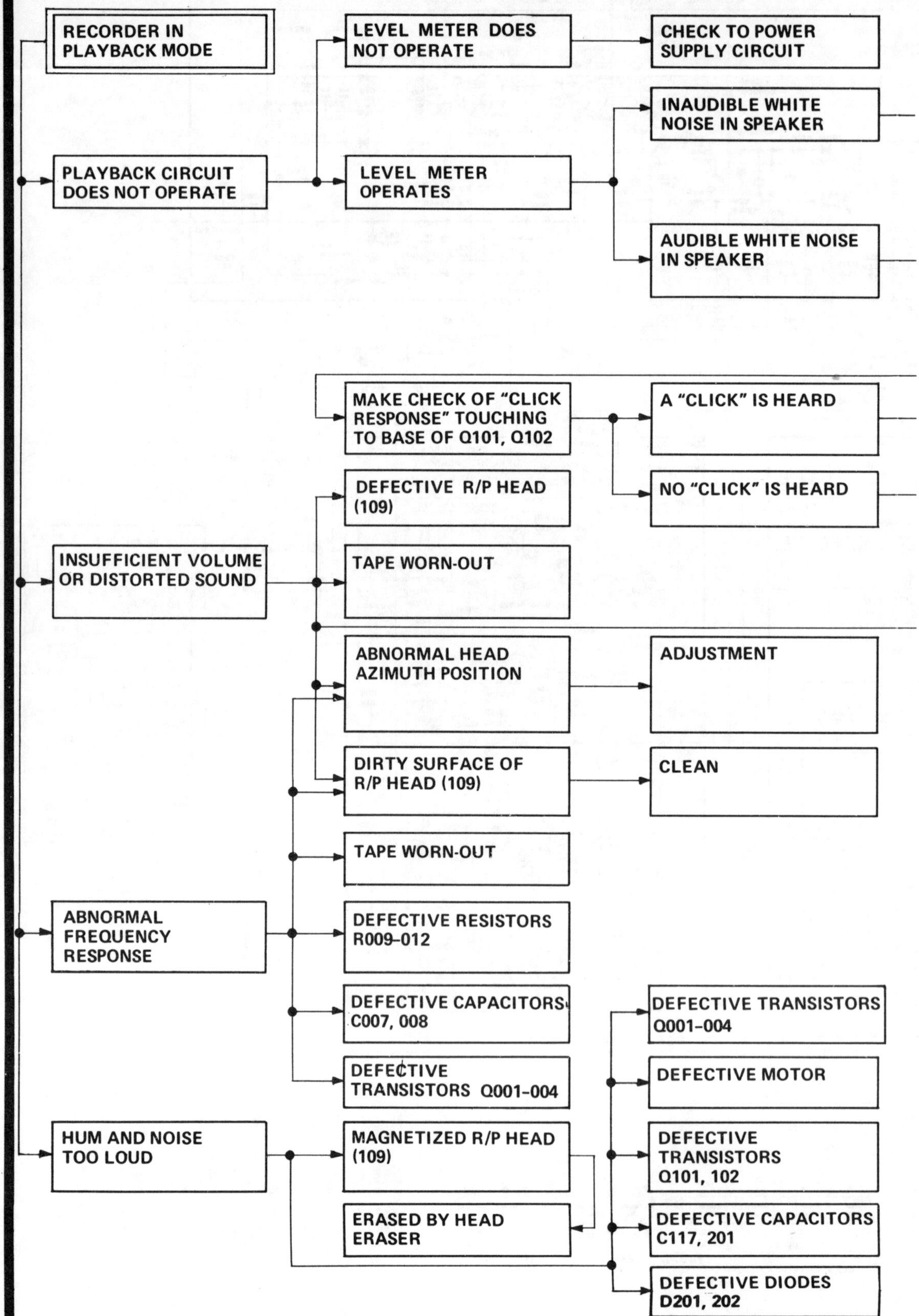

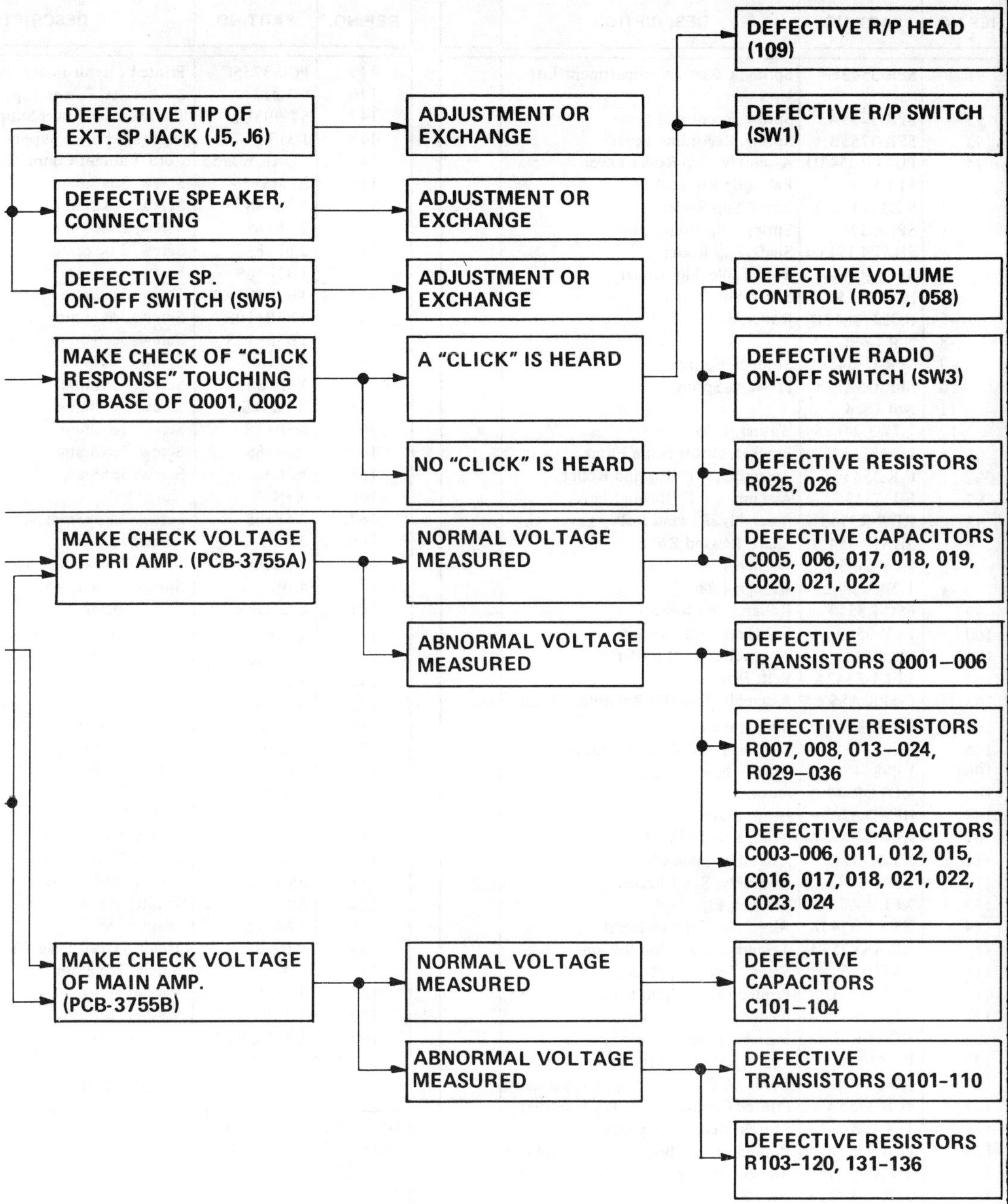
DEFECTIVE R/P HEAD (109)
DEFECTIVE TIP OF EXT. SP JACK (J5, J6)
ADJUSTMENT OR EXCHANGE
DEFECTIVE R/P SWITCH (SW1)
DEFECTIVE SPEAKER, CONNECTING
ADJUSTMENT OR EXCHANGE
DEFECTIVE SP. ON-OFF SWITCH (SW5)
ADJUSTMENT OR EXCHANGE
DEFECTIVE VOLUME CONTROL (R057, 058)
MAKE CHECK OF "CLICK RESPONSE" TOUCHING TO BASE OF Q001, Q002
A "CLICK" IS HEARD
DEFECTIVE RADIO ON-OFF SWITCH (SW3)
NO "CLICK" IS HEARD
DEFECTIVE RESISTORS R025, 026
MAKE CHECK VOLTAGE OF PRI AMP. (PCB-3755A)
NORMAL VOLTAGE MEASURED
DEFECTIVE CAPACITORS C005, 006, 017, 018, 019, C020, 021, 022
ABNORMAL VOLTAGE MEASURED
DEFECTIVE TRANSISTORS Q001–006
DEFECTIVE RESISTORS R007, 008, 013–024, R029–036
DEFECTIVE CAPACITORS C003-006, 011, 012, 015, C016, 017, 018, 021, 022, C023, 024
MAKE CHECK VOLTAGE OF MAIN AMP. (PCB-3755B)
NORMAL VOLTAGE MEASURED
DEFECTIVE CAPACITORS C101–104
ABNORMAL VOLTAGE MEASURED
DEFECTIVE TRANSISTORS Q101-110
DEFECTIVE RESISTORS R103-120, 131-136

REF.NO	PART NO.	DESCRIPTION	
91	SPR-3543P	Spring, Cassette Compartment Lift Up Lever	
92	SPR-3755A	Spring, Recording Lever	
93	SPR-3755B	Spring, Recording Lever	
94	PULLY3543D	Assembly, Slip Roller Lever	
-1	FELT-267	Felt, Slip Roller	
-2	S-L3543	Lever, Slip Roller	
-3	SPR-537F	Spring, Slip Roller Lever	
-4	SHAFT3543H	Shaft, Slip Roller	
-5	SLID-P3543	Turn Table, Slip Roller	
-6	BOS-3543C	Shaft, Slip Roller	
-7	ROLL-3543B	Roller, Slip	
-8	Not Used		
-9	SPR-U3543	Stopper, Spring	
-10	SPR-OSAE-B	Pressure, Spring	
-11	Not Used		
-12	1.7W3.2-0.25	Washer, 1.7ϕ	
95	L-ANG3543	Bracket, Motor Noise Filter	
96	F-R-3543	Assembly, F.F./Rewind Roller	
97	F-L-3543	Assembly, F.F./Rewind Lever	
98	REW-R3543	Assembly, Rewind Roller Lever	
-1	SHAFT3543L	Shaft, Rewind Roller Lever	
-2	1.0W4-0.5	Washer, 1.0ϕ	
-3	1.7W3.2-0.25	Washer, 1.7ϕ	
99	6SC-R3543	Roller, Sub Chassis	
100	FLY-3543	Flywheel with Capstan	
101	P-R-3543	Assembly, Pinch Roller	
102	BELT-3543A	Belt, Drive	
103	C-SPR-ASSY	Assembly, Cassette Retaining Spring	
104	R-DAI3543A	Reel Spindle	
105	BALL	Steel Ball, Sub Chassis Sliding	
106	13S-854	Switch, Power Supply	
107	MOTOR-3755	Motor	
108	HEAD-3265B	Head, Erase	
109	HEAD-3473A	Head, Record/Playback	
111	X6MC-3543	Assembly, Main Chassis	
112	X6SC-3543	Assembly, Sub Chassis	
113	XFLY-A3543	Bracket, Flywheel	
114	XMET-3543A	Assembly, Capstan Metal	
115	XB-ASY3543	Assembly, Function Button	
116	X5TOK-5	Screw, 5ϕx5mm, Special	
117	PCB-3755D	Printed Circuit Board, Tuner	
121	13S-128	Switch, Band Selector	
122	VC-344	Variable Capacitor	
123	PL-332	Lamp, FM Stereo Indicator Pilot	
126	ACC-285UL	Cord, AC Power Supply, UL Standard	
127	PCB-3755A	Printed Circuit Board, Pri-Amplifier	
128	13S-408	Switch, Record/Playback	
129	PCB-3755B	Printed Circuit Board, Main Amplifier	
130	TR-ANG3755	Bracket, Power Transistor	
131	XHON-P3755	Heat Sink	
132	X3P+6S	Screw, 3ϕx6mm	
133	SO-011	Socket, Connecting Cord, 5 pins	
134	T-227	Tip, Speaker Lead Wire	
135	13S-406	Switch, Radio on-off	
136	8V-3504	Control, Volume, 50K ohm/A type	
137	8V-3502	Control, Tone, 10K ohm/A type	
138	8V-3508	Control, Balance, 10K ohm/B type	
139	METER-3715	Meter, Battery/Record Level	
140	J-135	Jack, Microphone, External Speaker	
141	J-137	Jack, Line In	
142	J-3909	Jack, Remote Control with Jack Plate	
143	5S-51	Switch, R.I.F.	
144	5S-55	Switch, Speaker on-off	

REF.NO.	PART NO.	DESCRIPTION	
145	PCB-3755C	Printed Circuit Board, Power Supply	
146	SO-313	Socket, AC Power Supply	
147	5T-903UL	Transformer,Power Supply,UL Standard	
148	J-313	Jack, DC Power Supply	
149	CONE-W3755	Cord, Cabinet Connecting	
150	X3MS+8S	Screw, 3ϕx8mm	
151	X3B+8S	Screw, 3ϕx8mm	
152	2.6P+6S	Screw, 2.6ϕx6mm	
153	2.6P+8S	Screw, 2.6ϕx8mm	
154	TB3B+6S	Screw, 3ϕx6mm	
155	TB3B+8S	Screw, 3ϕx8mm	
156	TB3B+10S	Screw, 3ϕx10mm	
157	TB3B+12S	Screw, 3ϕx12mm	
159	X3P+6S	Screw, 3ϕx6mm	
160	X4P+6S	Screw, 4ϕx6mm	
161	X3MS+8S	Screw, 3ϕx8mm	
162	X3B+8S	Screw, 3ϕx8mm	
163	X3B+8S	Screw, 3ϕx8mm	
165	X3L-6S	Screw, 3ϕx6mm	
166	X3N	Nut, 3ϕ	
167	2.6P+8S	Screw, 2.6ϕx8mm	
168	2T+6S	Screw, 2ϕx6mm	
169	2.6TOK-6B	Screw, 2.6ϕx6mm, Special	
170	4SW	Spring Washer, 4ϕ	
171	E-STOPRING	"E" Stopring	
172	X3TOK-5	Screw, 3ϕx5mm, Special	
173	X2TOK-4	Screw, 2ϕx4mm, Special	
174	X8N	Nut, 8ϕ	
175	X3TOK-N	Nut, 3ϕ	
176	2K-6S	Screw, 2ϕx6mm	
177	2.6M+4S	Screw, 2.6ϕx4mm	
178	2.6M+6S	Screw, 2.6ϕx6mm	
179	2.6MS+4S	Screw, 2.6ϕx4mm	
180	2N	Nut, 2ϕ	
181	2SW	Spring Washer, 2ϕ	
182	2.6SW	Spring Washer, 2.6ϕ	
183	8SW	Spring Washer, 8ϕ	
184	8W	Washer, 8ϕ	
185	3W6-0.5	Washer, 3ϕ	
186	WH-12	Holder, Head Lead Wire	
187	2M+8S	Screw, 2ϕx8mm	
188	1.7W3.2-0.25	Washer, 1.7ϕ	
189	1.25W4-0.5	Washer, 1.25ϕ	
190	E-STOPRING	"E" Stopring	

RESISTORS

REF.NO.	PART NO.	DESCRIPTION	
R001, 002	1/4SA-470K	470 ohm	
R003, 004	1/4SA-220KK	220K ohm	
R005, 006	1/4SA-100K	100 ohm	
R007, 008	1/4SU-68KK	68K ohm	
R009, 010	1/4SA-15KK	15K ohm	
R011, 012	1/4SA 330KK	330K ohm	
R013, 014	1/4SA 4.7KK	4.7K ohm	
R015, 016	1/4SU-22KK	22K ohm	
R017, 018	1/4SU-390K	390 ohm	
R019, 020	1/4SA-47KK	47K ohm	
R021, 022	1/4SA 6.8KK	6.8K ohm	
R023, 024	1/4SA 2.7KK	2.7K ohm	
R025, 026	1/4SA 5.6KK	5.6K ohm	
R027, 028	1/4SA 3.9KK	3.9K ohm	
R029, 030	1/4SA 1.5KK	1.5K ohm	
R031, 032	1/4SU-820KK	820K ohm	
R033, 034	1/4SA 3.3KK	3.3K ohm	
R035, 036	1/4SA-120K	120 ohm	

REF.NO.	PART NO.	DESCRIPTION	
R037, 038	1/4SU-39KK	39K ohm	
R039, 040	1/4SU 6.8KK	6.8K ohm	
R041, 042	1/4SA 8.2KK	8.2K ohm	
R043, 044	1/4SU-15KK	15K ohm	
R045	1/4SA-33KJ	33K ohm, 1/4W, 5%, Carbon	
R046	1/4SA-47K	47 ohm	
R047	1/4SU-470K	470 ohm	
R048	1/4SU-120K	120 ohm	
R049, 050	1/4SU-1KK	1K ohm	
R051	1/4SA-33K	33 ohm	
R053, 054	8V-766	Control, Record Level Current	
R055, 056	8V-723	Control, Bias Current	
R057, 058	8V-3504	Control, Volume, 50K ohm/A type	
R059, 060	8V-3502	Control, Tone, 10K ohm/A type	
R061	8V-3508	Control, Balance, 10K ohm/B type	
R063, 064	1/4SU-10KK	10K ohm	
R101, 102	1/4SU-1KK	1K ohm	
R103, 104	1/4SU470KK	470K ohm	
R105, 106	1/4SU 4.7KK	4.7K ohm	
R107, 108	1/4SU-33K	33 ohm	
R109, 110	1/4SU 1.5KK	1.5K ohm	
R111, 112	1/4SU-470K	470 ohm	
R113, 114	1/4SU-22KK	22K ohm	
R115, 116	1/4SU-22K	22 ohm	
R117, 118	1/4SU 6.8KK	6.8K ohm	
R119, 120	1/4SA-820K	820 ohm	
R121, 122, R131, 132	1/2MR-1J	1 ohm, 1/2W, 5%, Carbon	
R123, 124	1/4SU-220K	220 ohm	
R125, 126	1/4SU-56KK	56K ohm	
R127, 128	1/4SU 2.7KK	2.7K ohm	
R129, 130	5LC-8K	8 ohm, 5W, 10%, Cement	
R133, 134	8V-766	Control, Bias Current, Balance	
R135, 136	8V-723	Control, Bias Current, Balance	
R301, 302	1/4SU 4.7KK	4.7K ohm	
R303, 320	1/4SU-680K	680 ohm	
R304	1/4SU 2.2KK	2.2K ohm	
R305	1/4SU 150KK	150K ohm	
R306, 330	1/4SU 560KK	560K ohm	
R307	1/4SU 330KK	330K ohm	
R308, 329, R334	1/4SU 680KK	680K ohm	
R309	1/4SU-120K	120 ohm	
R312	1/4SU-82KK	82K ohm	
R313	1/4SU 1.5KK	1.5K ohm	
R314	1/4SU-56K	56 ohm	
R316	1/4SU 1.8KK	1.8K ohm	
R317	1/4SU 220KK	220K ohm	
R318	1/4SU-82K	82 ohm	
R319	1/4M-4.7KK	4.7K ohm	
R321, 327	1/4SU-10KK	10K ohm	
R322	8V-723	Control, IC Current Adj.	
R323	1/4M-1.5KK	1.5K ohm	
R324, 325	1/4SU-1KK	1K ohm	
R326	1/4SU-100K	100 ohm	
R328	1/4SU-68KK	68K ohm	
R331	1/4SU-470K	470 ohm	
R332	1/4SU 2.7KK	2.7K ohm	
R333	1/4SU-270K	270 ohm	
R335	1/4SU 1.2KK	1.2K ohm	
R336	1/4SU 3.3KK	3.3K ohm	
R337, 338, R339, 340	1/4SU-47KK	47K ohm	
R341	1/4SU-27KK	27K ohm	

REF.NO.	PART NO.	DESCRIPTION
R342	1/4SU-68K	68 ohm
R343, 344	1/4M-47KK	47K ohm
R345	1/4SU-180K	180 ohm
R346	1/4SU-220K	220 ohm

Unless otherwise specified, resistors are 1/4W, 10%, carbon type.

CAPACITORS

REF.NO.	PART NO.	DESCRIPTION
C001, 002	ML-5-156M	1500PF, 50V, 20%, Mylar
C003, 004	CU-1-322Q	33MFD, 10V, $^{+100}_{-10}$%, Electrolytic
C005, 006, C017, 018, C021, 022	CU-5-103R	1MFD, 50V, $^{+150}_{-10}$%, Electrolytic
C007, 008, C037, 038	ML-5-105M	.01MFD, 50V, 20%, Mylar
C009, 010	D-5-107K	100PF, 50V, 10%, Discap
C011, 012	CU-1-101Q	100MFD, 10V, $^{+100}_{-10}$%, Electrolytic
C013, 014, C035	CU1.6-102Q	10MFD, 16V, $^{+100}_{-10}$%, Electrolytic
C015, 016, C045, 046	CU-06-472Q	47MFD, 6V, $^{+100}_{-10}$%, Electrolytic
C019, 020	AD-1-474M	.47MFD,10V,20%,Aluminum Electrolytic
C023, 024	ML-5-335M	.033MFD, 50V, 20%, Mylar
C025, 026	ML-5-685M	.068MFD, 50V, 20%, Mylar
C027, 028	ML-5-276K	2700PF, 50V, 10%, Mylar
C029, 030	S-5-156J	1500PF, 50V, 5%, Styrene
C031, 032	AD-1-104M	.1MFD,10V,20%,Aluminum Electrolytic
C033, 034	ML-5-225M	.022MFD, 50V, 20%, Mylar
C036, 043, C044	CU1.6-221Q	220MFD, 16V, $^{+100}_{-10}$%, Electrolytic
C039	ML-5-565K	.056MFD, 50V, 10%, Mylar
C040, 041	D-5-337K	330PF, 50V, 10%, Discap
C042	ML-5-336M	3300PF, 50V, 20%, Mylar
C101, 102	ML-5-475M	.047MFD, 50V, 20%, Mylar
C103, 104	CU1.6-102Q	10MFD, 16V, $^{+100}_{-10}$%, Electrolytic
C105, 106, C109, 110	CU1.6-472Q	47MFD, 16V, $^{+100}_{-10}$%, Electrolytic
C107, 108	D-5-477K	470PF, 50V, 10%, Discap
C111, 112	ML-5-276K	2700PF, 50V, 10%, Mylar
C113, 114, C115, 116	ML-5-105M	.01MFD, 50V, 20%, Mylar
C117, 118	CU1.6-221Q	220MFD, 16V, $^{+100}_{-10}$%, Electrolytic
C119	CU1.6-100Q	1000MFD, 16V, $^{+100}_{-10}$%, Electrolytic
C201	CU1.6-100Q	1000PF, 16V, $^{+100}_{-10}$%, Electrolytic
C202, 203	D-2.5-205Z	.02MFD, 25V, $^{+80}_{-20}$%, Discap
C301, 357, C358	D-5-107K	100PF, 50V, 10%, Discap
C302	D-5-158K	15PF, 50V, 10%, Discap
C303, 306, C307	D-5-228K	22PF, 50V, 10%, Discap
C304, 317, C319, 320, C323, 325, C330, 349 C365	D-2.5-405Z	.04MFD, 25V, $^{+80}_{-20}$%, Discap
C305, 332	D-5-206P	2000PF, 50V, $^{+100}_{-0}$%, Discap
C308, 314	D-5-809D	8PF, 50V, .5PF, Discap
C309	D-5-478K	47PF, 50V, 10%, Discap

REF. NO.	PART NO.	DESCRIPTION
C310, 321, C333, 343	D-5-405Z	.04MFD, 50V, $^{+80}_{-20}$%, Discap
C311, 327	D-5-337K	330PF, 50V, 10%, Discap
C312	D-5-506P	5000PF, 50V, $^{+100}_{-0}$%, Discap
C313	DTH-5-309C	3PF, 50V, .25PF, Ceramic
C315	DTH-5-709D	7PF, 50V, .5PF, Ceramic
C316, 347	D-5-157K	150PF, 50V, 10%, Discap
C318, 337	D-5-105P	.01MFD, 50V, $^{+100}_{-0}$%, Discap
C322	D-5-477K	470PF, 50V, 10%, Discap
C324, 342, C346	CU2.5-473R	4.7MFD, 25V, $^{+150}_{-10}$%, Electrolytic
C326, 329	D-5-188K	18PF, 50V, 10%, Discap
C328	ML-5-225M	.022MFD, 50V, 20%, Mylar
C331	D-5-309C	3PF, 50V, .25PF, Discap
C334	D-2.5-205Z	.02MFD, 25V, $^{+80}_{-20}$%, Discap
C335	D-5-128K	12PF, 50V, 10%, Discap
C336	D-5-567K	560PF, 50V, 10%, Discap
C338	D-5-409C	4PF, 50V, .25PF, Discap
C339	ML-5-125M	.012MFD, 50V, 20%, Mylar
C340	ML-5-185M	.018MFD, 50V, 20%, Mylar
C341	CU-06-472Q	470MFD, 6V, $^{+100}_{-10}$%, Electrolytic
C344	D-5-688K	68PF, 50V, 10%, Discap
C345	CU-1-100Q	1000MFD, 10V, $^{+100}_{-10}$%, Electrolytic
C348, 359, C360	CU-5-103R	1MFD, 50V, $^{+150}_{-10}$%, Electrolytic
C350	ML-5-155K	.015MFD, 50V, 10%, Mylar
C351, 353	S-5-105J	.01MFD, 50V, 5%, Styrene
C352	ML-5-105M	.01MFD, 50V, 20%, Mylar
C354, 355	CU1.6-333R	3.3MFD, 16V, $^{+150}_{-10}$%, Electrolytic
C356	S-5-476J	4700PF, 50V, 5%, Styrene
C361, 362	ML-5-126M	1200PF, 50V, 20%, Mylar
C363	CU1.6-100Q	1000MFD, 16V, $^{+100}_{-10}$%, Electrolytic
C364	CU1.6-102R	10MFD, 16V, $^{+150}_{-10}$%, Electrolytic
C366	D-5-828K	82PF, 50V, 10%, Discap

SWITCHES, SOCKETS AND JACKS

REF. NO.	PART NO.	DESCRIPTION
SW1	13S-408	Switch, Record/Playback
SW2	13S-128	Switch, Band Selector
SW3	13S-406	Switch, Radio on-off
SW4	5S-51	Switch, R.I.F.
SW5	5S-55	Switch, Speaker on-off
SW6	13S-854	Switch, Motor (Power Supply)
SO201	SO-313	Socket, AC Power Supply
SO401	SO-011	Socket, Connecting Cord, 5 Pins
PG201	CONE-W3755	Plug, Connecting Wire with Wire
J1, J2	J-135	Jack, Microphone
J3, J4	J-137	Jack, Line In
J5, J6	J-135	Jack, External Speaker
J7	J-3909	Jack, Remote Control with Jack Plate
	J-313	Jack, DC Power Supply

COILS AND TRANSFORMERS

REF. NO.	PART NO.	DESCRIPTION
L001, 002	4L-963	Coil, Bias Trap
L003	4L-404B	Coil, Bias Oscillator
L301	4L-198A	Coil, FM Antenna
L302a,b	4L-554C	Coil, FM Antenna
L303a,b	4L-554C	Coil, FM RF
L304	7L-947	Coil, FM Emitter
L305	4L-384B	Coil, FM Oscillator
L307	4L-406B	Coil, AM Oscillator
L308	4L-196A	Coil, AM Antenna
L309	4L-936	Coil, FM MPX.
L310	4L-935	Coil, FM MPX.
L311	4L-934	Coil, FM MPX.
L401	4L-012	Coil, Motor Noise Filter
T201	5T-903UL	Transformer, Power Supply
T301	4IF-832	Transformer, FM 1st IF
T302	4IF-780	Transformer, FM 2nd IF
T303	4IF-862	Transformer, FM 3rd IF
T304	4IF-864	Transformer, FM 4th IF
T305	4IF-866	Transformer, FM Discriminator
T306	4IF-868	Transformer, AM 1st IF
T307	4IF-870	Transformer, AM 2nd IF
T308	4IF-868	Transformer, AM Discriminator

SEMICONDUCTORS

REF. NO.	PART NO.	DESCRIPTION
Q001, 002	2SC458LGC	Transistor, 1st Audio Amplifier
Q003, 004	2SC458LGC	Transistor, 2nd Audio Amplifier
Q005, 006	2SC1000GR	Transistor, 3rd Audio Amplifier
Q007, 008	2SC-735Y	Transistor, Bias Oscillator
Q101, 102	2SC-458D	Transistor, Pri-Drive Amplifier
Q103, 104	2SC733Y	Transistor, Drive Amplifier
Q105, 106	2SC3720	Transistor, Idle Current Control
Q107, 108	2SC509PY	Transistor, Power Audio Amplifier
Q109, 110	2SA509PY	Transistor, Power Audio Amplifier
Q301	2SC535A	Transistor, FM RF
Q302	2SC535B	Transistor, FM Convertor
Q303	2SC460B	Transistor, FM IF/AM Convertor
Q304	2SC458B	Transistor, FM MPX Amp.
Q305	2SC458C	Transistor, FM MPX Amp.
Q306	2SC458B	Transistor, FM MPX Amp.
D001, 002	1N34A	Diode, Meter Signal Rectifier
D201	SI-RECT-74	Diode, Power Rectifier
D301	1N60	Diode, FM Over Load
D302	1N60	Diode, AM Over Load
D303	1S2139C	Diode, FM AFC
D304	1N60	Diode, FM AGC
D305	1N60	Diode, FM AGC
D306, 307	1N60P	Diode, FM Discriminator
D308	1N60	Diode, FM MPX.
D309	1N60	Diode, FM MPX.
D310, 311	1N60P	Diode, FM MPX.
D312, 313	1N60P	Diode, FM MPX.
IC	4IC-407	Integrated Circuit, FM IF/AM IF and Detector/Voltage Regulator

Chapter 7

Schematics, Adjustments, and Parts Lists for Representative Cassette Tape Players

CIRCUIT DIAGRAM

CODE 01

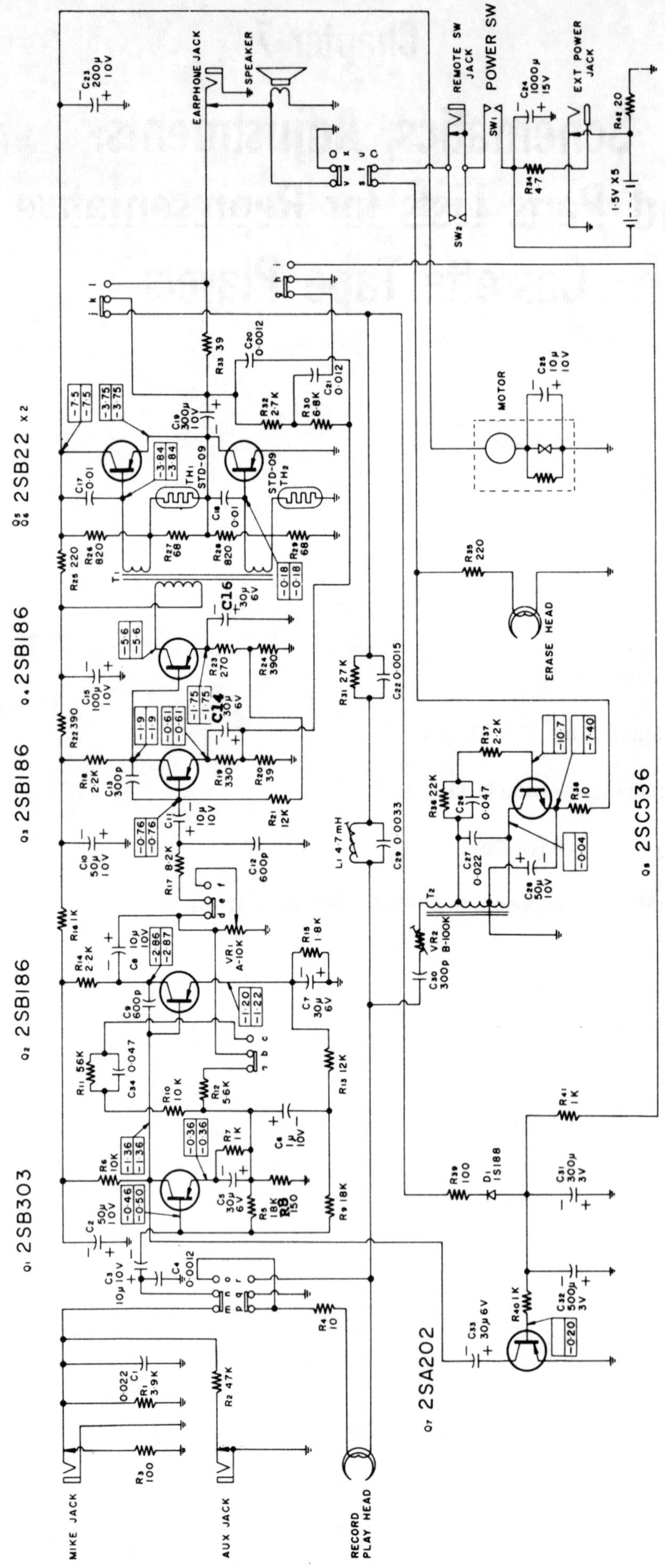

Notes

1. Values indicated in [box] are D.C. Voltage to the chassis.
   Upper in Playback and Lower in Recording Mode.
2. Record-Play SW (a~x) shown in Record Position.
3. SW_2 On at "Rewind", "Fast Forwrd" and "Stop".

Ref No. CODE 02	Ref No. CODE 01	Craig Key No.	Descrip.	Mfr's Sug. Rt. Price	Ref No. CODE 02	Ref No. CODE 01	Craig Key No.	Descrip.	Mfr's Sug. Rt. Price

AMPLIFIER

CAPACITORS, 50V OR NOTED

Ref No. CODE 02	Ref No. CODE 01	Craig Key No.	Descrip.	Mfr's Sug. Rt. Price	Ref No. CODE 02	Ref No. CODE 01	Craig Key No.	Descrip.	Mfr's Sug. Rt. Price
C30	C13,30	0026030196	Myl 330p	.20	C2,10, 13	C3,8, 11,25	0026030305	El 10/10V	.30
C4	C9,12	0026030197	Myl 560p	.20					
C21	C4,20	0026030198	Myl 0.0012	.20	C6,16, 11,17, 13	C5,7, 14,16, 33	0026030306	El 33/6V	.30
C28	C22	0026030199	Myl 0.0015	.20					
C29		0026030218	Myl 0.0022 ***	.20					
C5,23	C29	0026030300	Myl 0.0033	.20	C3,12, 27	C2,10, 28	0026030307	El 47/10V	.30
C18,19	C17,18	0026030301	Myl 0.01	.20					
C22	C21	0026030302	Myl 0.012	.20	C15	C15	0026030308	El 100/10V	.30
C34		0026030219	Myl 0.018 ***	.20	C24	C23	0026030309	El 220/10V	.40
C1	C1,27	0026030303	Myl 0.022	.20	C20	C19	0026030310	El 330/10V	.50
C26	C26,34	0026030304	Myl 0.047	.20	C31	C31	0026030311	El 330/3V	.50
C8		0026030330	Myl 0.033 ***	.20	C32	C32	0026030312	El 470/3V	.50
C25	C24	0026030313	El 1000/15V	.80	C7	C6	0026030314	El 1/10V	.30

RESISTORS, OHMS, ¼W OR NOTED. 0.20 each

Ref No. CODE 02	Ref No. CODE 01	Description	Ref No. CODE 02	Ref No. CODE 01	Description	Ref No. CODE 02	Ref No. CODE 01	Description
	R34	4.7	R23	R22,24	390	R17	R17	8.2K
R39,42, 4,25	R4,38	10	R24,34		680	R5,12,18	R6,10	10K
	R42	20	R27,29	R26,28	820	R14	R13,21	12K
R21,33	R20,33	39	R6,16,19 41	R7,16,40,41	1K	R7,9,20	R5,9	18K
R28,30	R27,29	68	R15	R15	1.8K	R38	R36	22K
R40,3	R3,39	100	R10,37	R14,18,37	2.2K	R22,35	R31	27K
R8	R8	150	R32	R32	2.7K	R2	R2	47K
R26,36	R25,35	220	R1,31	R1	3.9K	R11		68K
	R23	270	R13	R12	5.6K		R19	330
				R30	6.8K			

VARIABLE RESISTOR

VR1	VR1	0026030172	Control Vol. A10K	.90	VR2		0026030208	Ptntmtr.B1K ***	.30
VR3	VR2	0026030173	Potentiomtr. B100K	.30					

Ref No.	Craig Key No.	Description	Mfr's Sugg Ret. Price	Ref No.	Craig Key No.	Description	Mfr's Sugg Ret. Price
SWITCHES							
S (a-x)	0026030209	Ass'y, Slide Switch***	2.30	89(Sw2)	0026030120	Switch, Remote	.30
65(Sw1)	0026030097	Switch, Power	.30				
TRANSFORMERS AND COILS							
T1	0026030177	TRANSFORMER, DRIVER	.80	L1	0026030179	Bias Trap, 4.7mH	.70
T2	0026030178	OSC Coil	1.30				
L							
SEMICONDUCTORS							
Q1	00002SB303	Transistor	1.10	Q8	00002SC536	Transistor	2.00
Q2,3,4	00002SB186	Transistor	2.00	TH1,2	00000SDT09	Thermistor	.30
Q5,6	000002SB22	Transistor	1.60	D1	000001S188	Diode	.30
Q7	00002SA202	Transistor	1.30				
MISCELLANEOUS							
PCB-A	0026030210	Ass'y, PCB ***	19.50				

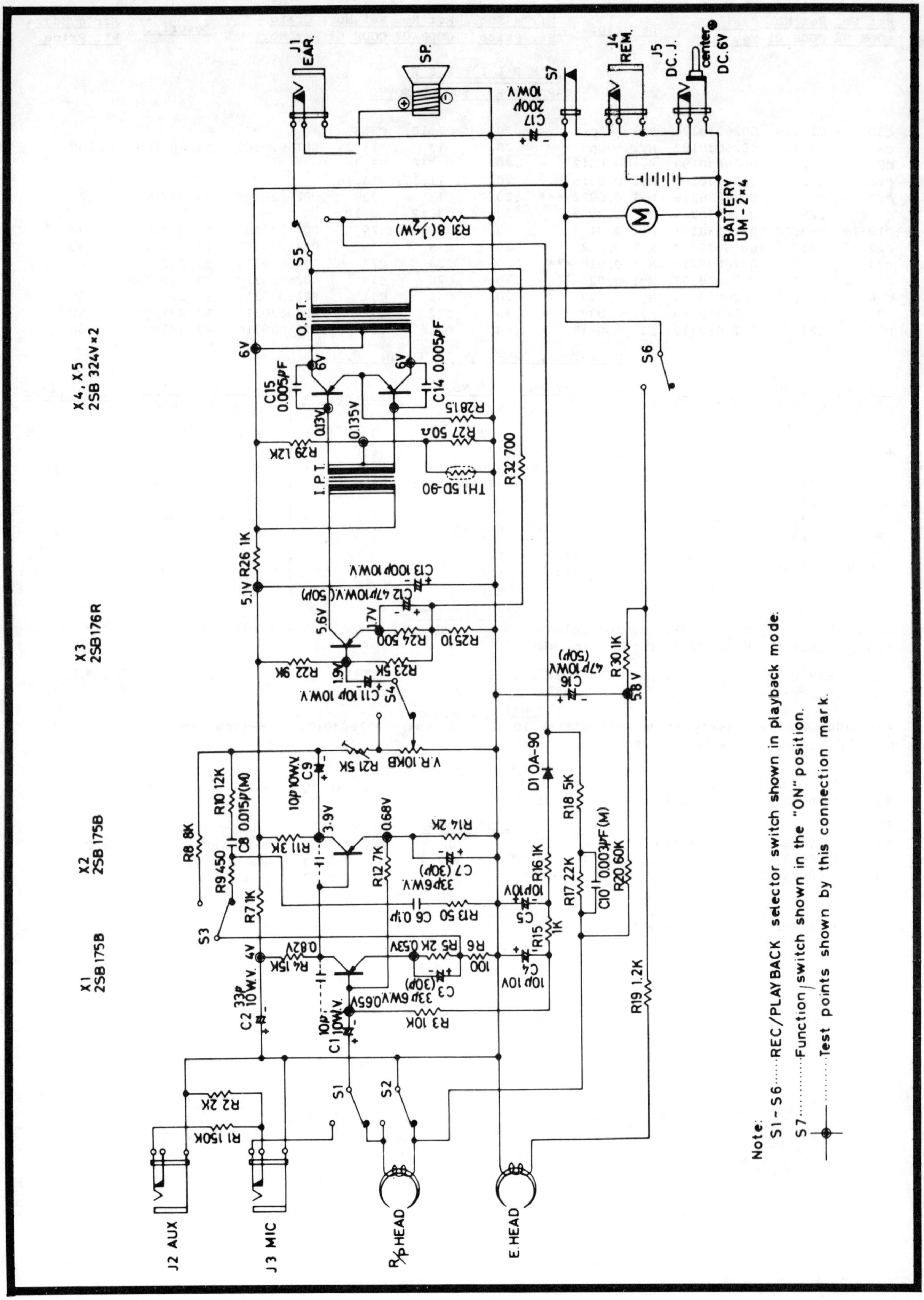
X1 2SB 175B
X2 2SB 175B
X3 2SB176R
X4, X5 2SB 324V×2
Note:
S1 - S6 ......REC/PLAYBACK selector switch shown in playback mode.
S7 ......Function switch shown in the "ON" position.
......Test points shown by this connection mark.

PARTS LIST

REF. NO:	DESCRIPTION:	PART NO:
	TRANSISTORS:	
X1,2	2SB-175	09-301020
X3	2SB-176	09-301026
X4,5	2SB-32A	09-301022
	THERMISTOR:	
TH1	5D-90	09-307015
	DIODE:	
D1	OA-90	09-306040
	RESISTORS: All resistors not shown on this parts list are carbon type, 1/8th watt. Refer to schematic for specific values.	
R31	8 Ohm, 1/2W, Wirewound	77-302809
	CAPACITORS: All capacitors not shown on this parts list are ceramic or mylar type, 50 W.V. Refer to schematic for specific values.	
C1,4, 5,9	10 MFD, 10 W.V., Electrolytic	77-336106
C2,3,7	33 MFD, 6.3 W.V., Electrolytic	77-332336
C12,16	47 MFD, 10 W.V., Electrolytic	77-336476
C13	100 MFD, 10 W.V., Electrolytic	77-336107
C17	220 MFD, 10 W.V., Electrolytic	77-336227
	COILS & TRANSFORMERS:	
OPT	Transformer, Audio Output	12-096061
IPT	Transformer, Audio Input	12-096062

SCHEMATIC DIAGRAM

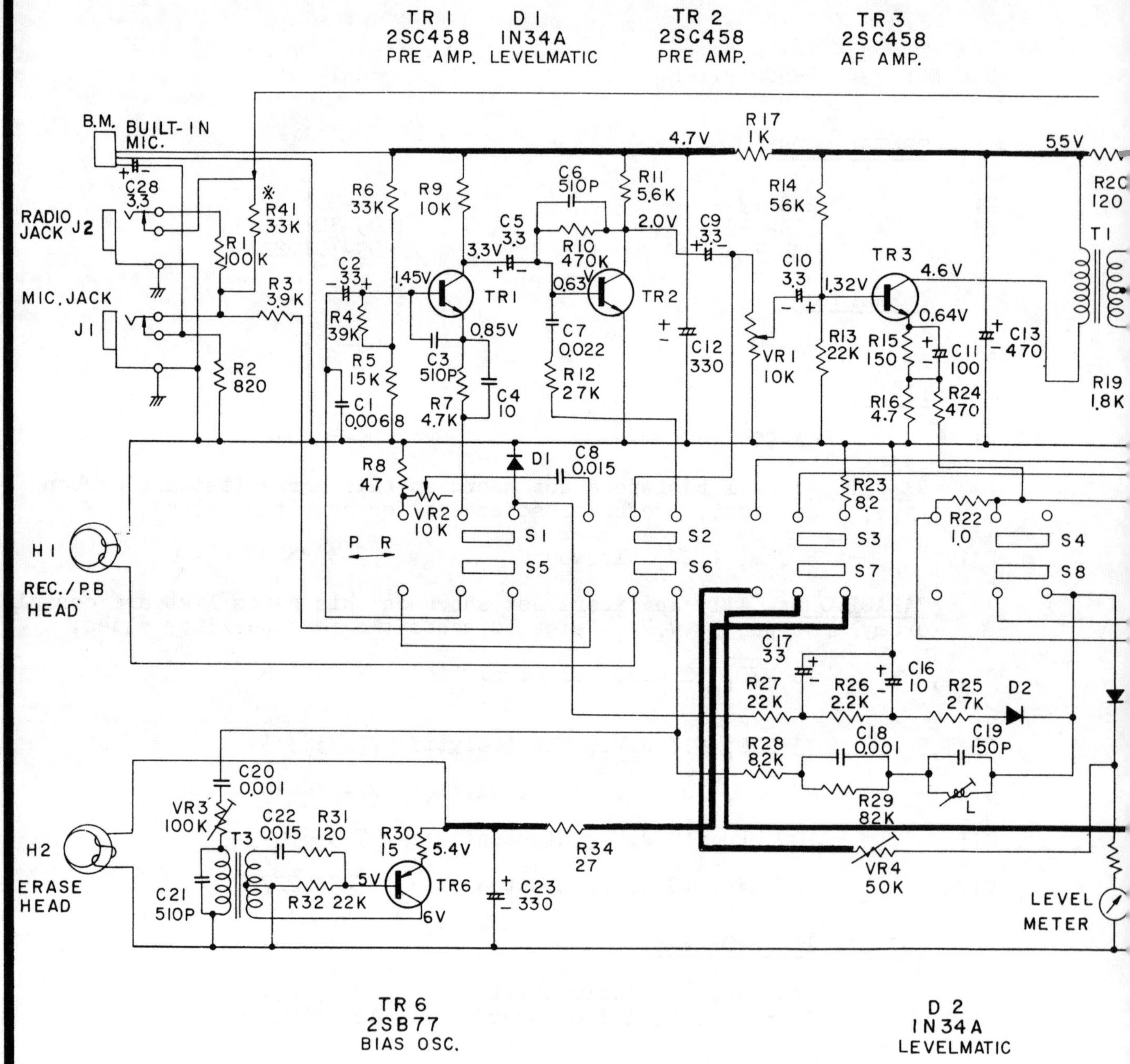

MODEL TRQ-286(A)(W) SERVICE MANUAL

1. VOLTAGE MEASURED WITH VOLTMETER AT 6V D.C. BASE OF ⊖ SIDE POWER SUPPly, VOLUME CONTROL AT MINIMUM AND NO SIGNAL
2. ALL RESISTANCE VALUES IN OHMS. K=1,000
3. ALL CAPACITANCE VALUES IN μF EXCEPT "P" NOTED. P=μμF=PF

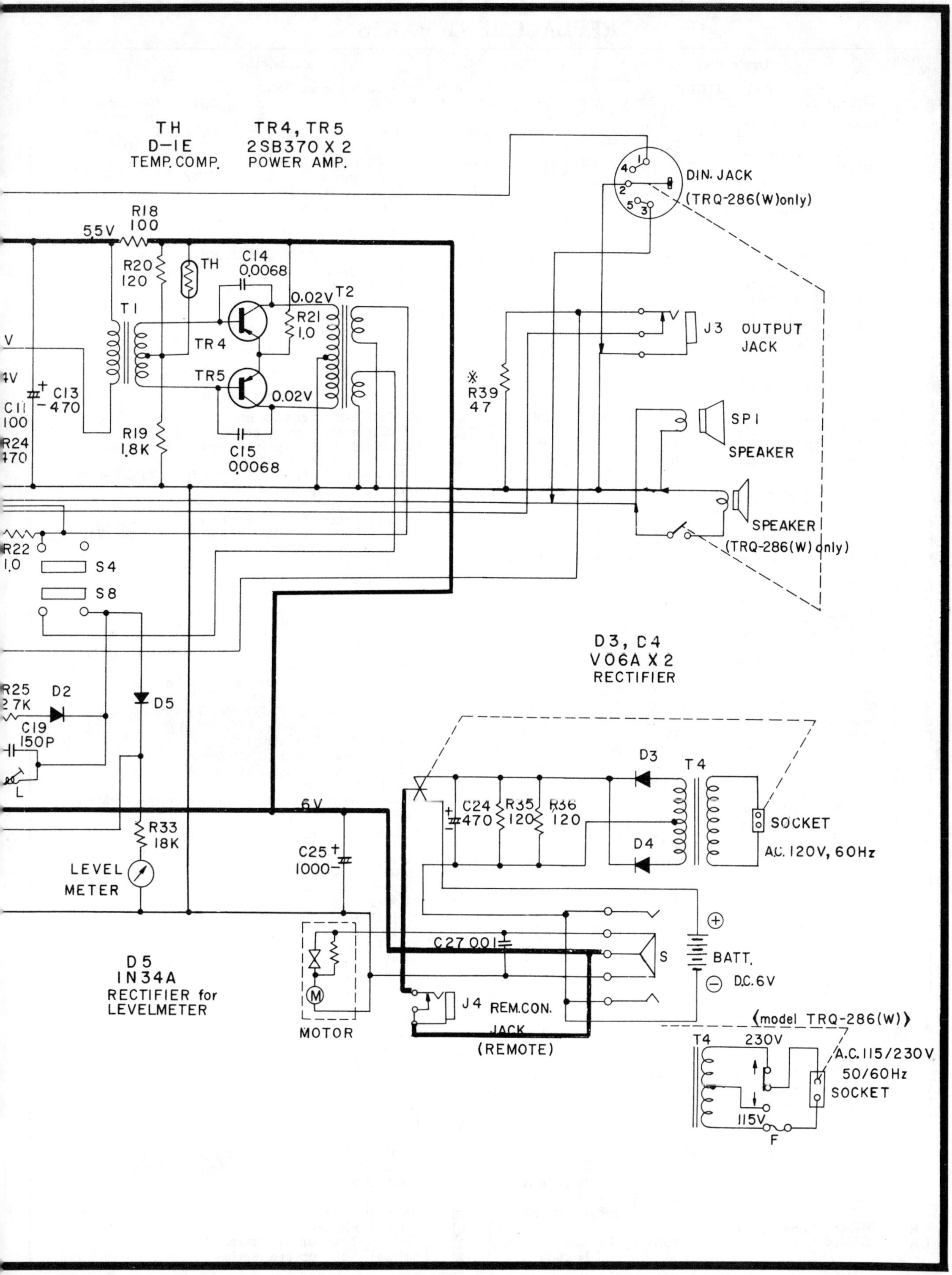
TH
D-1E
TEMP. COMP.
TR4, TR5
2SB370 X 2
POWER AMP.
DIN. JACK
(TRQ-286(W)only)
R18
100
5.5V
R20
120
TH
C14
0.0068
0.02V
T2
T1
R21
1.0
TR4
TR5
0.02V
C13
470
C11
100
R24
470
R19
1.8K
C15
0.0068
R39
47
J3
OUTPUT
JACK
SP1
SPEAKER
SPEAKER
(TRQ-286(W) only)
R22
1.0
S4
S8
D3, D4
V06A X 2
RECTIFIER
R25
2.7K
D2
D5
C19
150P
D3
T4
C24
470
R35
120
R36
120
SOCKET
A.C. 120V, 60Hz
D4
6V
R33
18K
C25
1000
LEVEL
METER
BATT.
D.C. 6V
C27 0.01
S
D5
IN34A
RECTIFIER for
LEVELMETER
J4
REM.CON.
JACK
(REMOTE)
MOTOR
(model TRQ-286(W))
T4
230V
A.C. 115/230V
50/60Hz
SOCKET
115V
F

MODEL TRQ-286(A)(W) SERVICE MANUAL

REPLACEMENT PARTS

Symbol No.	Stock No.	Description		
		CAPACITORS:		
C 1	0274016	Mylar	0.0068μF±10%	50V
C 2	0252613	Electrolytic	3.3μF	25V
C 3	0249515	Ceramic, discal	510pF±10%	
C 4	0252621	Electrolytic	10μF	25V
C 5	0252811	Electrolytic	1μF	50V
C 6	0249515	Same as C3		
C 7	0275013	Mylar	0.022μF±10%	50V
C 8	0275012	Mylar	0.015μF±10%	50V
C 9	0252811	Same as C5		
C 10	0252613	Same as C2		
C 11	0252231	Electrolytic	100μF	6.3V
C 12	0252233	Electrolytic	330μF	6.3V
C 13	0252335	Electrolytic	470μF	10V
C 14	0274016	Same as C1		
C 15	0274016	Same as C1		
C 16	0252621	Same as C4		
C 17	0252223	Electrolytic	33μF	6.3V
C 18	0274011	Mylar	0.001μF±10%	50V
C 19	0248688	Ceramic, discal	150pF±5%	50V
C 20	0274011	Same as C18		
C 21	0249515	Same as C3		
C 22	0275112	Mylar	0.015μF±20%	50V
C 23	0252233	Same as C12		
C 24	0252335	Electrolytic	470μF	10V
C 25	0256003	Electrolytic	1000μF	10V
C 27	0275111	Mylar	0.01μF±20%	50V
C 28	0252613	Same as C2	※mark: TRQ-286(W) only	
		RESISTORS:		
R 1	0131801	Composition	100kΩ±10%	¼GF
R 2	0131660	Composition	820Ω±10%	¼GF
R 3	0131698	Composition	3.9kΩ±10%	¼GF
R 4	0137908	Carbon film	39kΩ±10%	¼SD
R 5	0137903	Carbon film	15kΩ±10%	¼SD
R 6	0137907	Carbon film	33kΩ±10%	¼SD
R 7	0137859	Carbon film	4.7kΩ±10%	¼SD
R 8	0137767	Carbon film	47Ω±10%	¼SD
R 9	0137901	Carbon film	10kΩ±10%	¼SD
R 10	0137951	Carbon film	470kΩ±10%	¼SD
R 11	0137860	Carbon film	5.6kΩ±10%	¼SD
R 12	0137906	Carbon film	27kΩ±10%	¼SD
R 13	0137905	Carbon film	22kΩ±10%	¼SD
R 14	0137910	Carbon film	56kΩ±10%	¼SD
R 15	0137803	Carbon film	150Ω±10%	¼SD
R 16	0137755	Carbon film	4.7Ω±10%	¼SD
R 17	0137851	Carbon film	1kΩ±10%	¼SD
R 18	0137801	Carbon film	100Ω±10%	¼SD
R 19	0137854	Carbon film	1.8kΩ±10%	¼SD
R 20	0131650	Composition	120Ω±10%	¼GF
R 21	0149255	Wier wound	1.0Ω±10%	RU12L
R 22	0134289	Composition	10Ω±10%	½GF
R 23	0134288	Composition	8.2Ω±10%	½GF
R 24	0137809	Carbon film	470Ω±10%	¼SD
R 25	0137906	Same as R12		
R 26	0137855	Carbon film	2.2kΩ±10%	¼SD
R 27	0137905	Same as R13		
R 28	0137862	Carbon film	8.2kΩ±10%	¼SD
R 29	0137912	Carbon film	82kΩ±10%	¼SD
R 30	0137761	Carbon film	15Ω±10%	¼SD
R 31	0137802	Carbon film	120Ω±10%	¼SD
R 32	0137905	Same as R13		
R 33	0137904	Carbon film	18kΩ±10%	¼SD
R 34	0137764	Carbon film	27Ω±10%	¼SD
R 35	0134362	Composition	120Ω±10%	½GF
R 36	0134362	Same as R35		
※R 39	0134297	Composition	47Ω±10%	½GF
※ 41	0131739	Composition	33kΩ±10%	¼GF
VR 1	0151162	Variable resistor	10kΩ	
VR 2	0151162	Same as VR1		
VR 3	0151158	Semi variable	100kΩ(B)	
VR 4	0151825	Semi variable	50kΩ(B)	
		TRANSISTORS:		
TR 1	5320023		2SC458	Silicon 10KHz 200mW
TR 2	0573481		2SC458	Silicon 10KHz 200mW
TR 3	0573481	Same as TR2		
TR 4	0573024		2SB370	Germaniun 10KHz 200mW

Symbol No.	Stock No.	Description		
TR 5	0573024	Same as TR4		
TR 6	0573005		2SB77	Germanium 10KHz 150mW
D 1	0575001	Diode	1N34A	Germanium 455KHz
D 2	0575001	Same as D1		
D 3	5330104	Diode	VO6A	
D 4	5330104	Same as D3		
D 5	0575001	Same as D1		
TH	0576056	Thermistor	D-1E	
		TRANSFORMERS:		
T 1	0441084	Driver		
T 2	5250032	Audio output		
T 3	0316526	Oscillator		
T 4	5210491	Power(TRQ-286(A))		
	5210471	Power(TRQ-286(W))		
		COIL:		
L	5270001	Variable coil	33/22mH	
		for Final assembly		
①	6125362	Escutcheon assembly		
⑤	6125372	Bottom cover assembly (TRQ-286(A))		
	6125374	Bottom cover assembly (TRQ-286(W))		
⑥	6713931	Handle		
	6614134	Label (115/230V, 50/60Hz) (TRQ-286(W))		
③	6711033	Battery cover		
	7734561	Spacer		
	6614133	Label (120V, 60Hz) (TRQ-286(A))		
	6614134	Label(115/230V,50/60Hz)(TRQ-286(W))		
⑨	7511061	Case stud		
	8745410	Binding screw-3mm⌀×10mm		
	8815124	Locking washer-3mm⌀ (7 req'd) for case mounting		
	8711405	Pan head screw-3mm⌀×5mm for P.C.B mounting		
	0711305	Pan head screw-2.6mm⌀×5mm (2 req'd) for 2P socket mounting		
⑩	7196272	Socket holder (A)		
	8781436	Tapping screw-3mm⌀×6mm		
	5650092	2P socket (TRQ-286(A))		
	0541390	2P socket (TRQ-286(W))		
	7711671	Socket spacer(TRQ-286(A))		
	5660651	Plug (silencing)		
	5740621	Power cord (TRQ-286(A))		
	5743161	Power cord (TRQ-286(W))		
	0592101	Earphone		
	5420362	Dynamic microphone assembly (NDM-27S)		
	6150971	Lether case		
		for Chassis assembly		
⑪				
⑫	7221822	Head plate assembly		
⑬	6310511	Head plate spring		
⑭	0015595	Roller-5⌀		
⑮	0942042	Roller shaft		
	7501351	Head plate collar		
	0711308	Pan head screw-2.6mm⌀×8mm		
	8811113	Washer-2.6mm⌀		
	0948492	Ball-2⌀		
⑯	7193854	Function plate assembly		
⑰	6706743	Fast forward arm assembly		
	7512371	Function collar (B)		
⑲	7193875	Function plate guid		
⑳	7505342	Function collar		
	7771854	Screw with spring washer-2.6mm⌀		
	7778394	"E" ring		
	7778395	"E" ring		
	0940949	Staple		
㉑	6370791	Flywheel assembly		
㉒	7194161	Flywheel supporting plate assembly		
	0711305	Pan head screw-2.6mm⌀×5mm		
	7663971	Roller stopper		
㉓	6350621	Belt-1.2□		
㉔	7660854	FR belt		
㉕	6709473	Take up arm assembly		
㉖	6313581	Take up arm spring		

MAINTENANCE

CLEANING HEADS

For optimum performance, the tape heads, capstan and pinch roller surfaces should be cleaned after every 15-20 hours of use to remove oxide deposits.

CLEANING PROCEDURE

To clean the tape heads, you may use a cassette tape head cleaner available under Lafayette Stock No. 28-16015. This device is identical in construction to a cassette -- you simply insert it into the machine and run the cassette recorder in the normal manner for a few minutes and then remove.

Alternatively, you can use the following method. Remove the main top cover of the cassette recorder. This is achieved by removing 3 screws on the back edge of the cover and then carefully sliding the cover off toward the rear of the unit. Depress the UP button to raise the cassette compartment and remove any cassette that may be in the compartment. Using a soft, lint-free cloth moistened with denatured alcohol, carefully wipe the faces of the tape heads and the surfaces of the pinch roller and capstan which come into contact with the tape [see the diagram below].

DEMAGNETIZING HEADS

After a period of time, a residual magnetism will build up on the tape heads. It is, therefore, recommended that the heads be de-magnetized after every 20 hours of actual use. A head de-magnetizer available from Lafayette under Stock No. 28-72299 will perform this function quickly and easily.

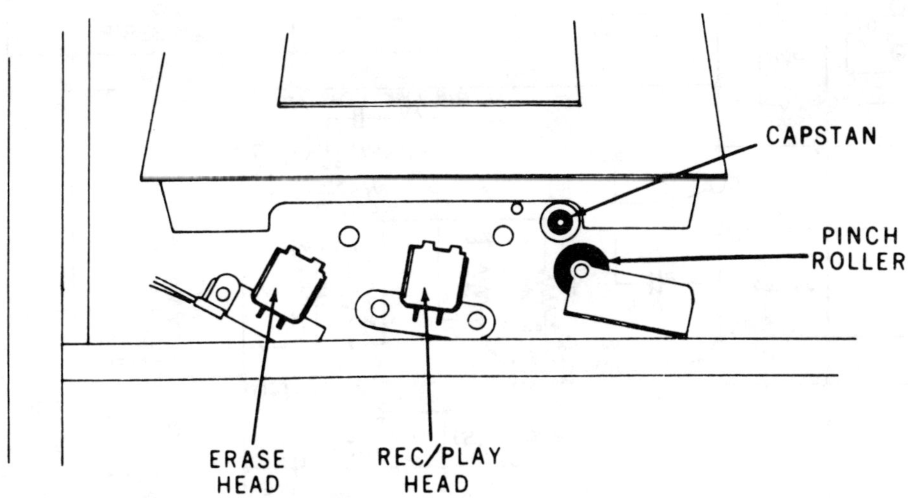

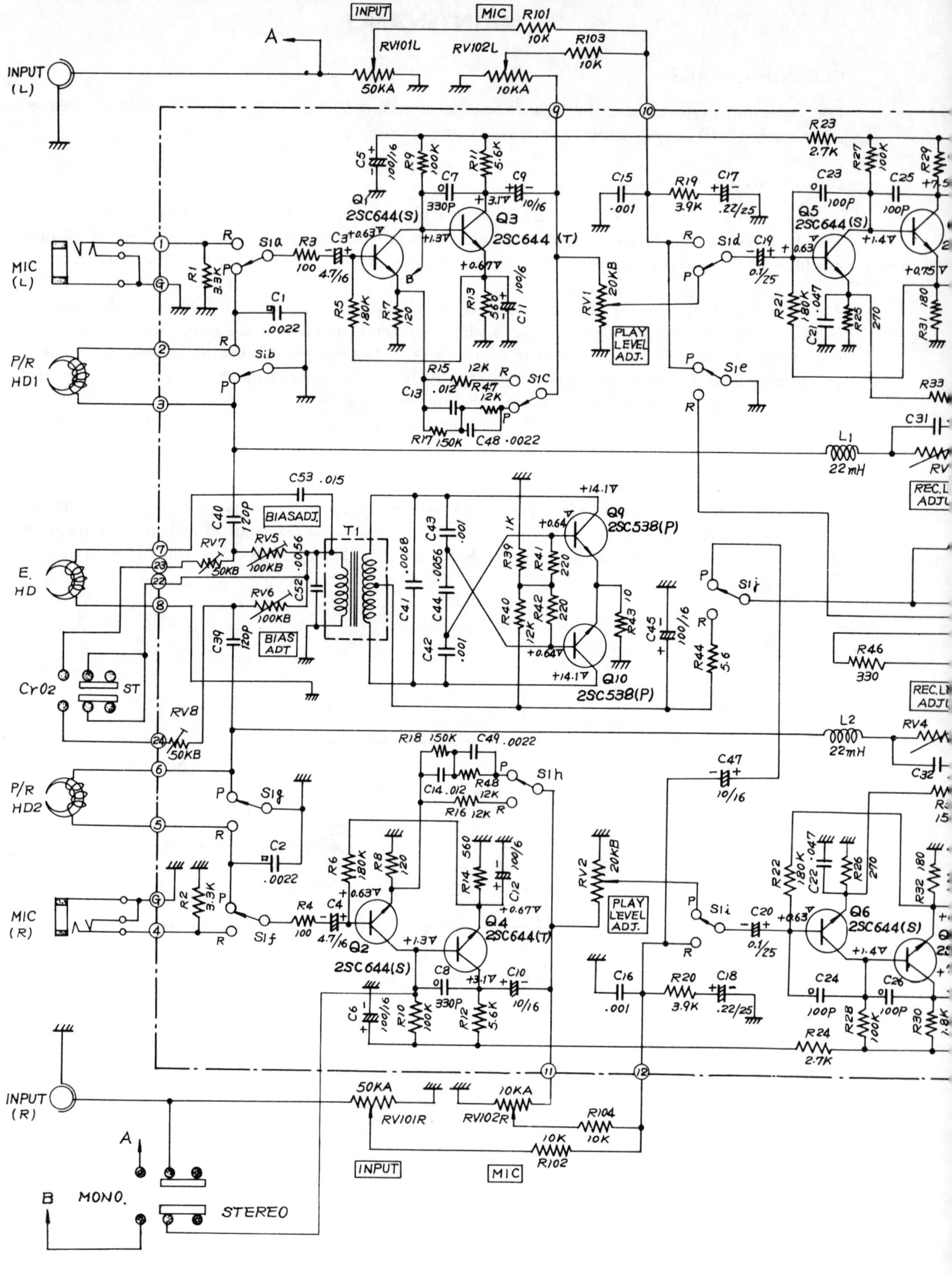
INPUT
MIC
R101
10K
R103
10K
RV101L
RV102L
50KA
10KA
INPUT (L)
MIC (L)
P/R HD1
E. HD
CrO2
ST
RV8
50KB
P/R HD2
MIC (R)
INPUT (R)
MONO.
STEREO
A
B
Q1 2SC644(S)
Q3 2SC644 (T)
Q5 2SC644(S)
Q9 2SC538(P)
Q10 2SC538(P)
Q2 2SC644(S)
Q4 2SC644(T)
Q6 2SC644(S)
T1
BIASADJ.
BIAS ADT
PLAY LEVEL ADJ.
RV1 20KB
RV2 20KB
L1 22mH
L2 22mH
RV4
REC.L ADJU
R23 2.7K
R24 2.7K
R46 330
C47 10/16
RV5 100KB
RV6 100KB
RV7 50KB
RV101R
RV102R
R104 10K
R102 10K

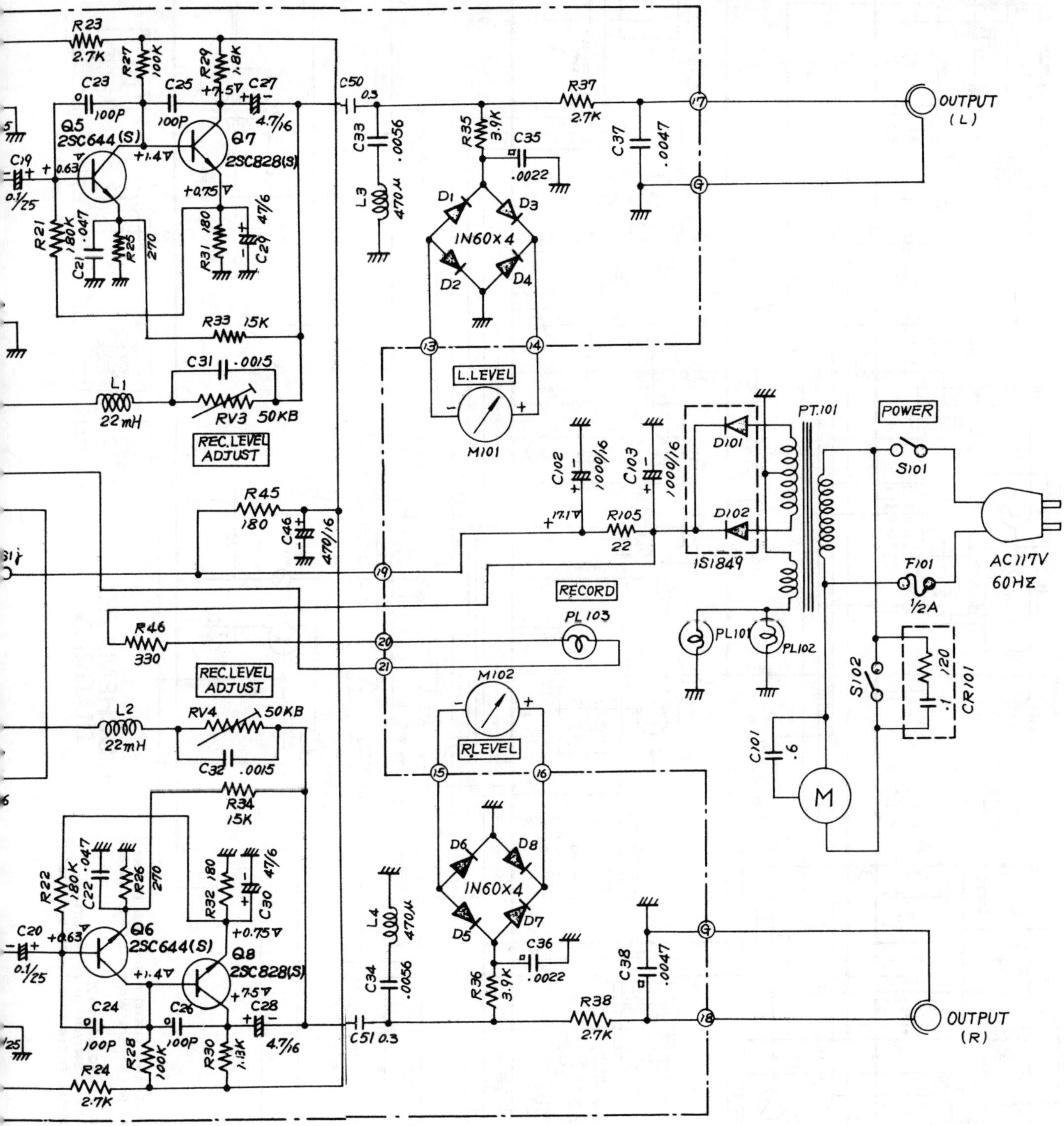

* All measurements obtained using V.T.V.M.
* All the measurements are D.C. and obtained in PLAY MODE with NO SIGNAL INPUT and volume controls at extreme minimum position.
And in this measurements, tape cartridge should not be inserted.
* All measurements are relative to chassis ground.
* All unmarked polarity measurements are positive to chassis ground.
* If measurements obtained are in excess of plus/minus 20 percent of values shown then reason for difference should be corrected.
* SERIAL NO. 00001 ~

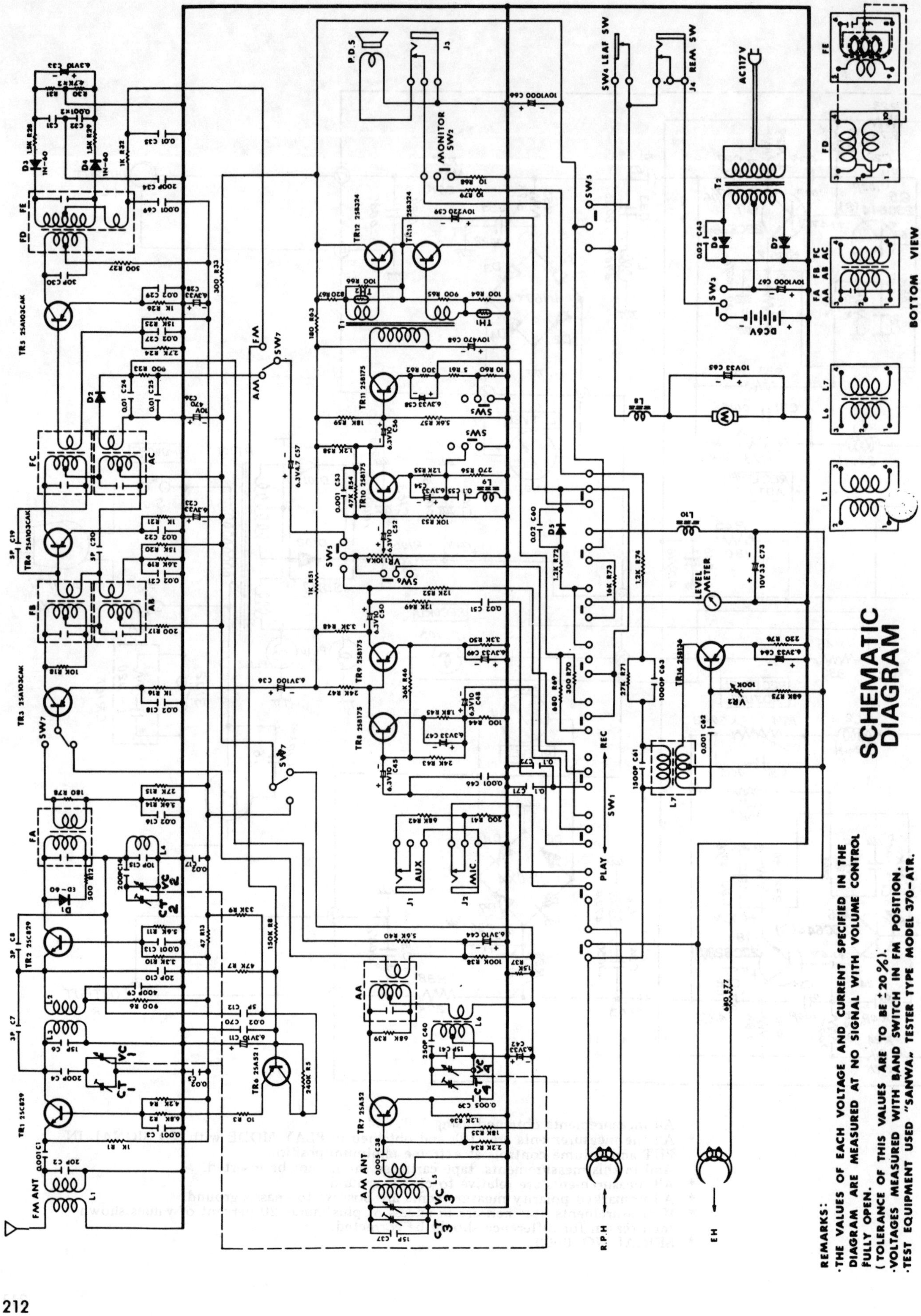

SCHEMATIC DIAGRAM

REMARKS:
- THE VALUES OF EACH VOLTAGE AND CURRENT SPECIFIED IN THE DIAGRAM ARE MEASURED AT NO SIGNAL WITH VOLUME CONTROL FULLY OPEN. (TOLERANCE OF THIS VALUES ARE TO BE ±20%).
- VOLTAGES MEASURED WITH BAND SWITCH IN FM POSITION.
- TEST EQUIPMENT USED "SANWA" TESTER TYPE MODEL 370-ATR.

REF.#	DESCRIPTION	PART NO.	LIST PRICE
M73	Mount, Flywheel Bushing	12-158203	$ 0.46
M74	Washer, Fiber	12-157184	0.46
M75	Clamp, Head Panel	12-158182	0.46
M76	Belt, Drive	12-198035	1.44
M77	Switch, Leaf	12-169055	0.74
M78	Screw	12-151129	0.20
M79	Washer-Lock	12-151272	0.20
M80	Screw	12-151234	0.20
M81	Nut	12-151725	0.46
M82	Screw	12-151624	0.46
M83	Screw	12-151136	0.20
M84	Screw	12-151234	0.20
M85	Screw	12-151171	0.20
M86	Screw	12-151127	0.20
M87	Screw	12-151131	0.20
M88	Nut	12-151726	0.46
M89	Nut	12-151727	0.46
M90	Lock Washer	12-151240	0.20
M91	Terminal	12-151728	0.46
M92	Washer	12-151729	0.46
M93	Spring, Record Lever	12-152371	0.46
SP	Speaker	12-06-056	1.44
RPH	Head, Record/Play	12-193078	2.54
EH	Head, Erase	12-193079	1.44
VR1	Control, Volume	12-160069	0.74
VR2	Control, Sensitivity	12-164023	0.46
SW1	Switch, Play/Record	12-183065	1.90
SW2	Switch, Monitor	12-183063	0.46
SW3	Switch, AC/DC	12-183063	0.46
SW5	Switch, Radio/Tape	12-183063	0.46
SW6	Switch, Tone	12-183064	0.46
SW7	Switch, Selector	12-180041	1.44
VC	Capacitor, Tuning	12-120008	3.28
J1,2,3	Jack, Aux/Mic/Earphone	12-153075	0.46
J4	Jack, Remote	12-153072	0.46
MO	Motor	12-190035	7.46
	Microphone	12-038053	3.54
	Earphone	12-038054	0.46
	Box, Battery	12-030085	1.44
	COILS & TRANSFORMERS		
L1	Coil, FM Antenna	12-176016	0.46
L2	Choke, FM	12-178025	0.46
L3	Coil, FM RF	12-176017	0.46
L4	Coil, FM Oscillator	12-170025	0.46
L5	Antenna, Loopstick	12-050009	0.96
L6	Coil, AM Oscillator	12-170026	0.46
L7	Coil, Bias Oscillator	12-170027	0.74
L8	Coil, Filter	12-178026	0.46
L9,10	Filter, Micro	12-178027	0.46
L11	Filter, Noise	12-178028	0.46
AA	IFT, AM 1st	12-090027	0.46
AB	IFT, AM 2nd	12-090028	$ 0.46
AC	IFT, AM 3rd	12-090029	0.46
FB	IFT, FM 2nd	12-090030	0.46
FC	IFT, FM 3rd	12-090030	0.46
FD	Discriminator, Primary	12-094013	0.46
FE	Discriminator, Secondary	12-094014	0.46
FA	IFT, FM 1st	12-090030	0.46
T1	Transformer, Input	12-096068	0.74
T2	Transformer, Power	12-098032	3.78
	TRANSISTORS		
TR1,2	2SC829	09-302033	2.40
TR3,4,5	2SA103	09-300028	2.10

REF.#	DESCRIPTION	PART NO.	LIST PRICE
TR6,7	2SA52	09-300011	1.40
TR8,9,10,11	2SB175	09-301020	1.22
TR12,13	2SB324	09-301022	1.70
TR14	2SB136	09-301007	3.00
	DIODES		
D1,2	ED-60	09-306047	0.60
D3,4	IN-60	09-306019	0.60
D5	ED-60	09-306047	0.60
D6,7	S-05-01	09-306162	0.46
	THERMISTOR		
TH1,2	TD-90	09-307036	0.46
	RESISTOR		
	ALL RESISTORS NOT SHOWN ON THIS PARTS LIST ARE EITHER SOLID TYPE, ¼ WATT OR CARBON TYPE, 1/8 WATT. FOR SPECIFIC VALUES OR TYPES CONSULT SCHEMATIC DIAGRAM OR PRINTED CIRCUIT BOARD LAYOUT DIAGRAM.		
	CAPACITORS		
	ONLY ELECTROLYTIC CAPACITORS ARE SHOWN ON THIS PARTS LIST. FOR SPECIFIC VALUES OR TYPES OF OTHER CAPACITORS CONSULT SCHEMATIC OR BOARD LAYOUT DIAGRAMS.		
C11,23,33,44, 45,48,49,50, 52	10MF/6.3V	77-332106	0.74
C26	47MF/10V	77-336476	0.74
C28,42,47,54, 56,57,58,65 66	33MF/6.3V	77-332336	0.74
C36	100MF/6.3V	77-332107	$ 0.74
C59	220MF/10V	77-336227	0.74
C66,67	100MF/10V	77-336107	0.74
C68	470MF/10V	77-336477	0.74
C73	33MF/10V	77-336336	0.74

Chapter 8
Schematics, Parts Layout, and Parts Lists for Two-Speed Reel-to-Reel Tape Recorder-Players

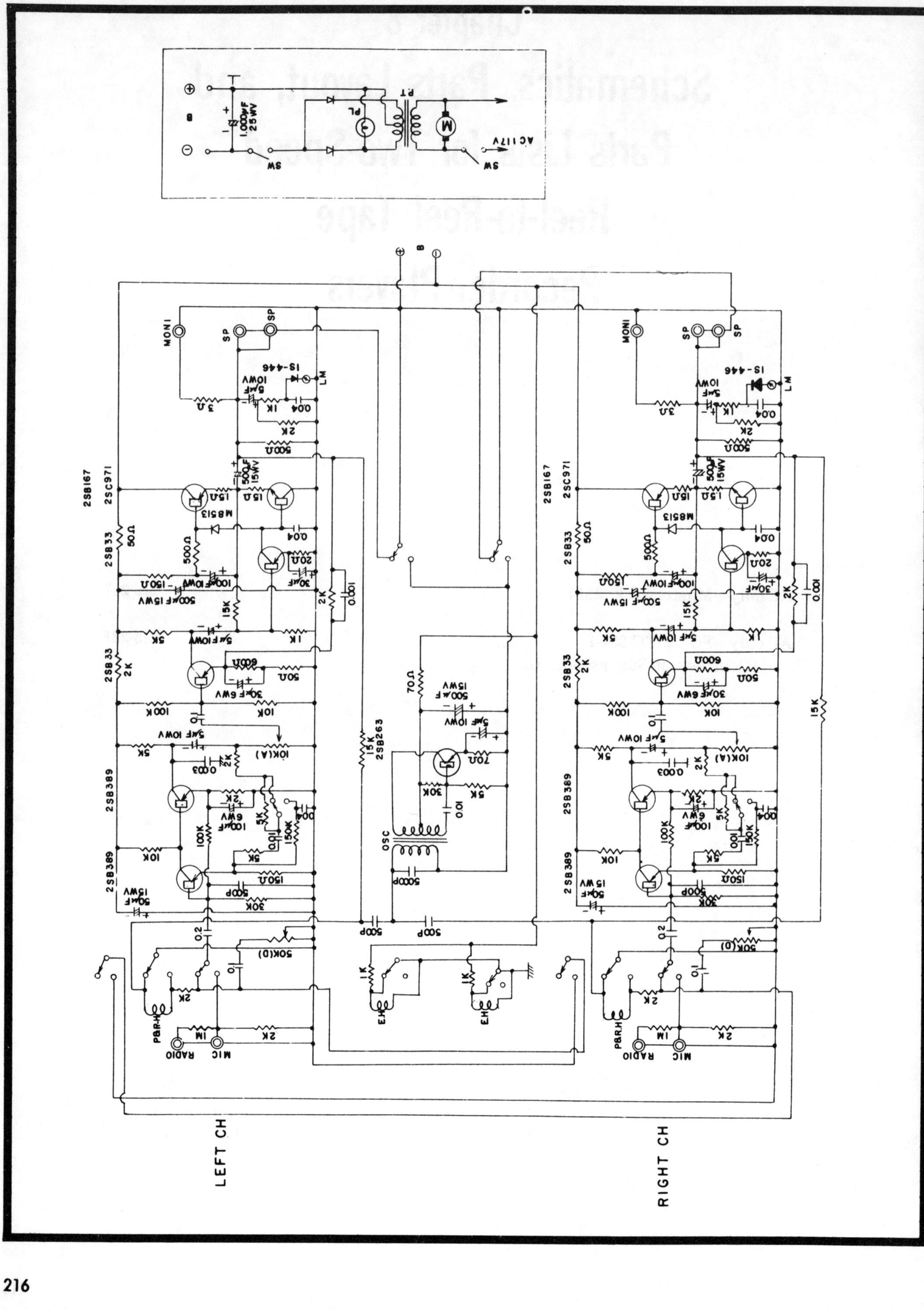
AC 117V
SW
PL
RT
M
B
1,000μF 25WV
MONI
SP
LM
1S-446
2SB167
2SC971
2SB33
2SB389
2SB263
M8513
OSC
EH
P&R-H
MIC
RADIO
LEFT CH
RIGHT CH

Ordering parts use 814 before item number

No.	Item
197	Capacitor - 0.04 MFD 50V
198	Capacitor - 0.1 MFD 50V
199	Capacitor - 0.2 MFD 50V
S11424	Capacitor - 5 MFD 10V
201	Capacitor - 30 MFD 6V
S11421	Capacitor - 50 MFD 15V
203	Capacitor - 100 MFD 6V
S11426	Capacitor - 100 MFD 10V
S11423	Capacitor - 500 MFD 15V
206	Resistor - 1.5 ohm 1/2W
207	Resistor - 3 ohm 1/2W
208	Resistor - 20 ohm 1/8W
209	Resistor - 50 ohm 1/8W
210	Resistor - 70 ohm 1/8W
211	Resistor - 150 ohm 1/8W
212	Resistor - 600 ohm 1/8W
213	Resistor - 500 ohm 1/8W
214	Resistor - 1K ohm 1/8W
215	Resistor - 2K ohm 1/8W
216	Resistor - 5K ohm 1/8W
217	Resistor - 10K ohm 1/8W
218	Resistor - 15K ohm 1/8W
219	Resistor - 30K ohm 1/8W
220	Resistor - 100K ohm 1/8W
221	Resistor - 150K ohm 1/8W
222	Resistor - 1 meg ohm 1/8W
223	Transistor - first pre-amp 2SB-389
224	Transistor - second pre-amp 2SB-389
225	Transistor - AF amp 2SB-33
226	Transistor - driven amp 2SB-33
227	Transistor - power amp 2SB-405
228	Transistor - power amp 2SB-72
229	Transistor - oscillator 2SB-263
S16186	Varistor - KV-1
N21193	Diode - 18-466
S9457	Variable resistor - 10K ohm
S9458	Variable resistor - 50K ohm
S21310	Coil - oscillator
235	Circuit board main (left channel)
236	Circuit board main (right channel)
237	Circuit board small (oscillator)
238	Bracket - main circuit board
239	Bracket - small circuit board

No.	Item
	Miscellaneous
S17459	Cabinet w/spkr. housing
S18291	Handle
160	Bracket - handle
S18292	Latch - lid and speaker housing
162	Jack assembly - speaker housing
163	Jack assembly - cabinet
164	Foot - rubber
S1121	Speaker - 4" X 6" 8ohm
S18293	Escutcheon
S18294	Escutcheon
S18295	Cloth - speaker housing front
S18296	Hindge - speaker housing left
S18297	Hindge - speaker housing right
171	Stopper - speaker housing
S18298	Hindge - lid
173	Bushing - cabinet motorboard
174	Screw - cabinet motorboard
175	Screw - cabinet amplifier
176	Washer - cabinet amplifier
177	Deck panel
178	Screw - deck panel mounting
179	Cover - head housing
180	Cover - pinchroller housing
181	Pad - top of capstan
S18299	Name plate - Masterwork
S18300	Name plate - model M-814
184	Decortion ring - turn table
185	Microphone
186	Stand - microphone
187	Cable - speaker housing cabinet
188	Retainer -mreel
189	Record tape 600FT
190	Cable - Auxiliary
191	Knob - AC power switch
192	Capacitor - 500PF 50V
193	Capacitor - 2500PF 50V
194	Capacitor - 0.001 MFD 50V
195	Capacitor - 0.003 MFD 50V
196	Capacitor - 0.01 MFD 50V

4-2 SCHEMATIC DIAGRAM (AMPLIFIER)

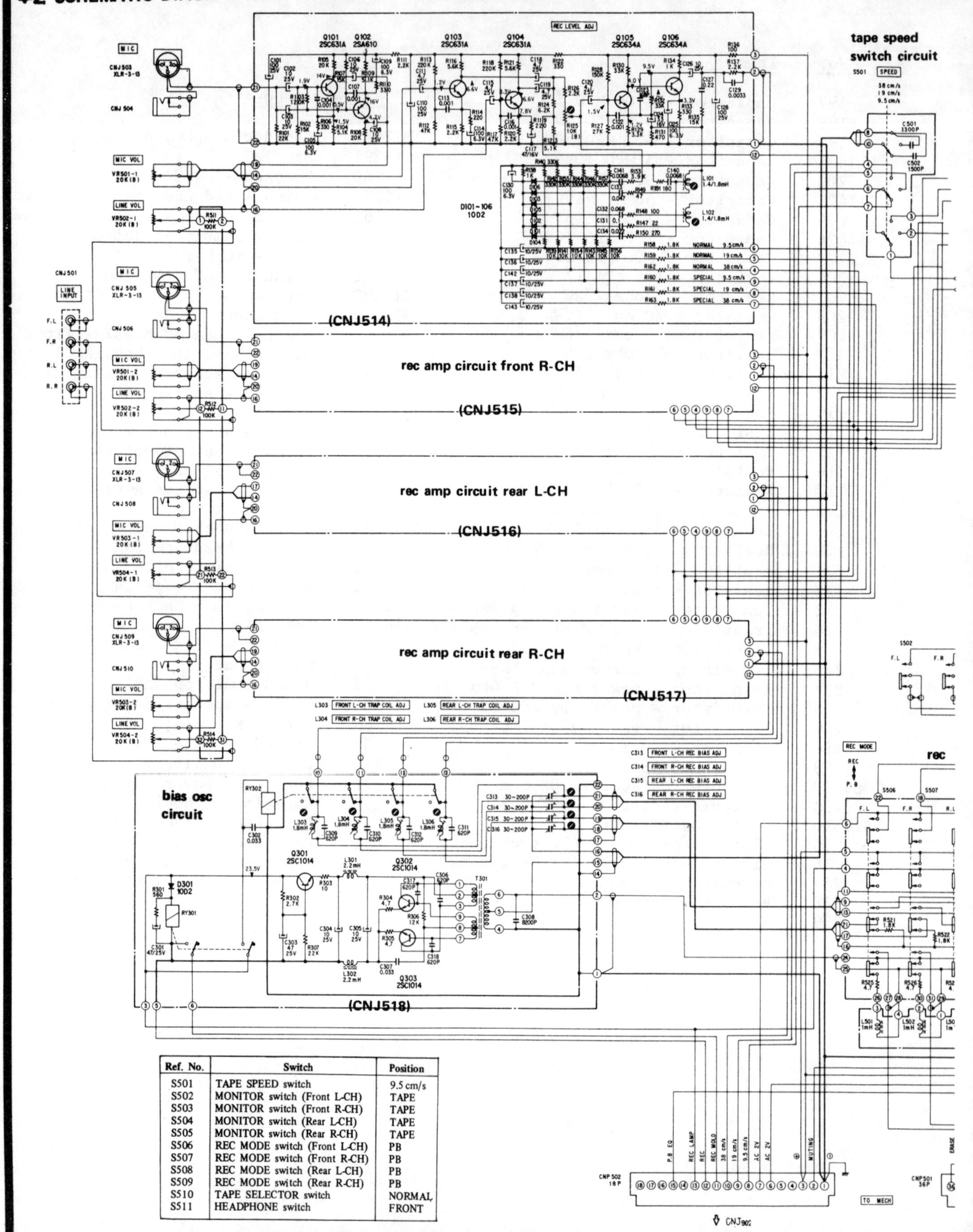

Ref. No.	Switch	Position
S501	TAPE SPEED switch	9.5 cm/s
S502	MONITOR switch (Front L-CH)	TAPE
S503	MONITOR switch (Front R-CH)	TAPE
S504	MONITOR switch (Rear L-CH)	TAPE
S505	MONITOR switch (Rear R-CH)	TAPE
S506	REC MODE switch (Front L-CH)	PB
S507	REC MODE switch (Front R-CH)	PB
S508	REC MODE switch (Rear L-CH)	PB
S509	REC MODE switch (Rear R-CH)	PB
S510	TAPE SELECTOR switch	NORMAL
S511	HEADPHONE switch	FRONT

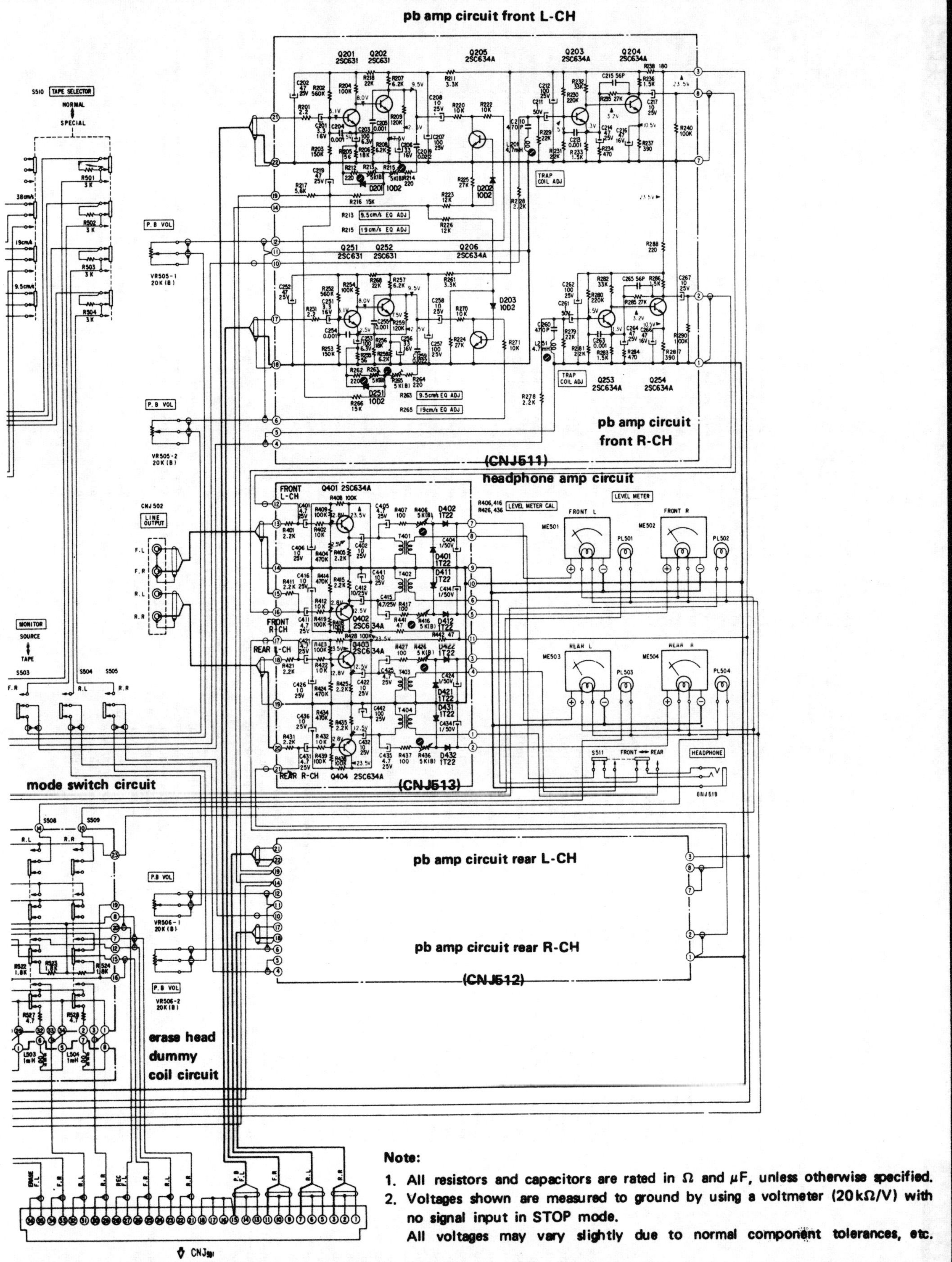

Note:

1. All resistors and capacitors are rated in Ω and μF, unless otherwise specified.
2. Voltages shown are measured to ground by using a voltmeter (20kΩ/V) with no signal input in STOP mode.
   All voltages may vary slightly due to normal component tolerances, etc.

SECTION 6
ELECTRICAL PARTS LIST

Ref. No.	Part No.	Description	
MOUNTED CIRCUIT BOARDS			
	X-31410-53	resistor terminal	
	X-31410-54	APS & ATS	
	X-31410-56	shut-off	
	X-31410-58	pb amp	
	X-31410-59	rec amp	
	X-31410-60	bias osc	
	X-31410-61	headphone amp	
	X-31410-62	system control	
	X-31410-63	capstan servo control	
	X-31410-64	erase head dummy coil	
	X-31410-65	REC MODE switch	
	X-31410-69	TAPE SPEED switch	
SEMICONDUCTORS			
Q101		transistor	2SC631A
Q102		transistor	2SA610
Q103		transistor	2SC631A
Q104		transistor	2SC631A
Q105		transistor	2SC634A
Q106		transistor	2SC634A
Q201, 251		transistor	2SC631A
Q202, 252		transistor	2SC631A
Q203, 253		transistor	2SC634A
Q204, 254		transistor	2SC634A
Q205		transistor	2SC634A
Q206		transistor	2SC634A
Q301		transistor	2SC1014
Q302		transistor	2SC1014
Q303		transistor	2SC1014
Q401~404		transistor	2SC634A
Q601		transistor	2SC634A
Q602		transistor	2SC634A
Q603		transistor	2SC634A
Q604		transistor	2SC634A
Q605		transistor	2SC634A
Q606		transistor	2SD28
Q607~611		transistor	2SC634A
Q701~712		transistor	2SC634A
Q801~808		transistor	2SC634A
Q809		transistor	2SA610
Q810		transistor	2SC634A
Q811		transistor	2SC1014
Q812~816		transistor	2SC634A
Q817		transistor	2SC895
Q818		transistor	2SC634A
Q901~907		transistor	2SD28
Q908		transistor	2SC807A
Q991		transistor	2SD28
Q992		transistor	2SC634
D101~106		diode	10D2
D201		diode	10D2

Ref. No.	Part No.	Description	
D202		diode	10D2
D203		diode	10D2
D251		diode	10D2
D301		diode	10D2
D401		diode	1T22
D402		diode	1T22
D411		diode	1T22
D412		diode	1T22
D421		diode	1T22
D422		diode	1T22
D431		diode	1T22
D432		diode	1T22
D601~619		diode	10D2
D620		diode	1T22
D621~632		diode	10D2
D633~637		diode	1T22
ZD601		diode, zener	1S334
D701		diode	10D2
D702		diode	1T22
D703		diode	10D2
D704		diode	10D2
D705		diode	1T22
D706		diode	10D2
D707		diode	10D2
D708		diode	1T22
D709		diode	10D2
D710		diode	10D2
D711		diode	10D2
D712		diode	10D2
ZD701		diode, zener	1S334
D801		diode	10D2
D802		diode	1T263
D803		diode	1S334
D804~811		diode	10D2
COILS & TRANSFORMERS			
L101	1-231-069	equalizer, 1.4/1.8 mH	
L102	1-231-069	equalizer, 1.4/1.8 mH	
L201	1-407-298	bias trap (TRAP COIL ADJ.)	
L251	1-407-298	bias trap (TRAP COIL ADJ.)	
L301	1-407-198-21	micro inductor, 2.2 mH	
L302	1-407-198-21	micro inductor, 2.2 mH	
L303	1-231-069	equalizer, 1.8 mH; FRONT L-CH TRAP COIL ADJ.	
L304	1-231-069	equalizer, 1.8 mH; FRONT R-CH TRAP COIL ADJ.	
L305	1-231-069	equalizer, 1.8 mH; REAR L-CH TRAP COIL ADJ.	
L306	1-231-069	equalizer, 1.8 mH; REAR R-CH TRAP COIL ADJ.	
L501~504	1-409-038	dummy, 1 mH	
L701	1-407-212	micro inductor, 33 mH	
T301	1-433-148	bias osc	

Ref. No.	Part No.	Description		
T401~404	1-427-284	headphone		
T901	1-441-628	power		
CAPACITORS				
C101	1-121-416	100 μF	25 V	electrolytic
C102	1-121-398	10 μF	25 V	electrolytic
C103	1-121-398	10 μF	25 V	electrolytic
C104	1-105-661-12	0.001 μF	50 V	mylar
C105	1-121-413	100 μF	6.3 V	electrolytic
C106	1-121-398	10 μF	25 V	electrolytic
C107	1-105-661-12	0.001 μF	50 V	mylar
C108	1-121-398	10 μF	25 V	electrolytic
C109	1-121-413	100 μF	6.3 V	electrolytic
C110	1-121-416	100 μF	25 V	electrolytic
C111	1-121-395	4.7 μF	25 V	electrolytic
C112				
C113	1-105-661-12	0.001 μF	50 V	mylar
C114	1-121-413	100 μF	6.3 V	electrolytic
C115	1-121-395	4.7 μF	25 V	electrolytic
C116	1-105-661-12	0.001 μF	50 V	mylar
C117	1-121-409	47 μF	16 V	electrolytic
C118	1-121-395	4.7 μF	25 V	electrolytic
C119	1-121-395	4.7 μF	25 V	electrolytic
C120	1-121-395	4.7 μF	25 V	electrolytic
C121				
C122	1-105-661-12	0.001 μF	50 V	mylar
C123	1-107-127	68 pF	50 V	silvered mica
C124	1-121-403	33 μF	16 V	electrolytic
C125	1-121-413	100 μF	6.3 V	electrolytic
C126	1-121-398	10 μF	25 V	electrolytic
C127	1-105-689-12	0.22 μF	50 V	mylar
C128	1-121-416	100 μF	25 V	electrolytic
C129	1-105-679-12	0.0033 μF	50 V	mylar
C130	1-121-413	100 μF	6.3 V	electrolytic
C131	1-105-685-12	0.1 μF	50 V	mylar
C132	1-105-683-12	0.068 μF	50 V	mylar
C133	1-105-681-12	0.047 μF	50 V	mylar
C134	1-105-677-12	0.022 μF	50 V	mylar
C135	1-121-398	10 μF	25 V	electrolytic
C136	1-121-398	10 μF	25 V	electrolytic
C137	1-121-398	10 μF	25 V	electrolytic
C138	1-121-398	10 μF	25 V	electrolytic
C139				
C140	1-105-671-12	0.0068 μF	50 V	mylar
C141	1-105-671-12	0.0068 μF	50 V	mylar
C142	1-121-398	10 μF	25 V	electrolytic
C143	1-121-398	10 μF	25 V	electrolytic
C201, 251	1-127-304	3.3 μF	16 V	electrolytic
C202, 252	1-121-410	47 μF	25 V	electrolytic
C203, 253	1-121-413	100 μF	6.3 V	electrolytic
C204, 254	1-105-661-12	0.001 μF	50 V	mylar
C205, 255	1-105-661-12	0.001 μF	50 V	mylar
C206, 256	1-121-403	33 μF	16 V	electrolytic

Ref. No.	Part No.	Description		
C207, 257	1-121-416	100 μF	25 V	electrolytic
C208, 258	1-121-398	10 μF	25 V	electrolytic
C209, 259	1-105-677-12	0.022 μF	50 V	mylar
C210, 260	1-107-244	470 pF	50 V	silvered mica
C211, 261	1-121-391	1 μF	50 V	electrolytic
C212, 262	1-121-416	100 μF	25 V	electrolytic
C213, 263	1-105-661-12	0.001 μF	50 V	mylar
C214, 264	1-121-410	47 μF	25 V	electrolytic
C215, 265	1-107-125	56 pF	50 V	silvered mica
C216, 266	1-121-409	47 μF	16 V	electrolytic
C217, 267	1-121-398	10 μF	25 V	electrolytic
C218				
C219	1-121-410	47 μF	25 V	electrolytic
C301	1-121-410	47 μF	25 V	electrolytic
C302	1-105-679-12	0.033 μF	50 V	mylar
C303	1-121-410	47 μF	25 V	electrolytic
C304	1-121-398	10 μF	25 V	electrolytic
C305	1-121-398	10 μF	25 V	electrolytic
C306	1-107-188	620 pF	500 V	silvered mica
C307	1-105-679-12	0.033 μF	50 V	mylar
C308	1-109-523	8,200 pF	500 V	silvered mica
C309	1-107-188	620 pF	500 V	silvered mica
C310	1-107-188	620 pF	500 V	silvered mica
C311	1-107-188	620 pF	500 V	silvered mica
C312	1-107-188	620 pF	500 V	silvered mica
C313~316	1-141-076	30~200 pF		trimmer
C317	1-107-188	620 pF	500 V	silvered mica
C318	1-107-188	620 pF	500 V	silvered mica
C401, 411 / C421, 431	1-121-395	4.7 μF	25 V	electrolytic
C402, 412 / C422, 432	1-121-398	10 μF	25 V	electrolytic
C403, 413 / C423, 433				
C404, 414 / C424, 434	1-121-391	1 μF	50 V	electrolytic
C405, 415 / C425, 435	1-121-395	4.7 μF	25 V	electrolytic
C406, 416 / C426, 436	1-121-398	10 μF	25 V	electrolytic
C441	1-121-416	100 μF	25 V	electrolytic
C442	1-121-416	100 μF	25 V	electrolytic
C501	1-109-506	3,300 pF	500 V	silvered mica
C502	1-109-503	1,500 pF	500 V	silvered mica
C601	1-121-810	470 μF	50 V	electrolytic
C602	1-121-810	470 μF	50 V	electrolytic
C603	1-121-417	100 μF	50 V	electrolytic
C604	1-121-388	1,000 μF	35 V	electrolytic
C605	1-121-388	1,000 μF	35 V	electrolytic
C606	1-121-422	220 μF	25 V	electrolytic

Ref. No.	Part No.	Description
C207, 257	1-121-416	100μF 25V electrolytic
C208, 258	1-121-398	10μF 25V electrolytic
C209, 259	1-105-677-12	0.022μF 50V mylar
C210, 260	1-107-244	470pF 50V silvered mica
C211, 261	1-121-391	1μF 50V electrolytic
C212, 262	1-121-416	100μF 25V electrolytic
C213, 263	1-105-661-12	0.001μF 50V mylar
C214, 264	1-121-410	47μF 25V electrolytic
C215, 265	1-107-125	56pF 50V silvered mica
C216, 266	1-121-409	47μF 16V electrolytic
C217, 267	1-121-398	10μF 25V electrolytic
C218		———
C219	1-121-410	47μF 25V electrolytic
C301	1-121-410	47μF 25V electrolytic
C302	1-105-679-12	0.033μF 50V mylar
C303	1-121-410	47μF 25V electrolytic
C304	1-121-398	10μF 25V electrolytic
C305	1-121-398	10μF 25V electrolytic
C306	1-107-188	620pF 500V silvered mica
C307	1-105-679-12	0.033μF 50V mylar
C308	1-109-523	8,200pF 500V silvered mica
C309	1-107-188	620pF 500V silvered mica
C310	1-107-188	620pF 500V silvered mica
C311	1-107-188	620pF 500V silvered mica
C312	1-107-188	620pF 500V silvered mica
C313~316	1-141-076	30~200pF trimmer
C317	1-107-188	620pF 500V silvered mica
C318	1-107-188	620pF 500V silvered mica
C401, 411, C421, 431	1-121-395	4.7μF 25V electrolytic
C402, 412, C422, 432	1-121-398	10μF 25V electrolytic
C403, 413, C423, 433		———
C404, 414, C424, 434	1-121-391	1μF 50V electrolytic
C405, 415, C425, 435	1-121-395	4.7μF 25V electrolytic
C406, 416, C426, 436	1-121-398	10μF 25V electrolytic
C441	1-121-416	100μF 25V electrolytic
C442	1-121-416	100μF 25V electrolytic
C501	1-109-506	3,300pF 500V silvered mica
C502	1-109-503	1,500pF 500V silvered mica
C601	1-121-810	470μF 50V electrolytic
C602	1-121-810	470μF 50V electrolytic
C603	1-121-417	100μF 50V electrolytic
C604	1-121-388	1,000μF 35V electrolytic
C605	1-121-388	1,000μF 35V electrolytic
C606	1-121-422	220μF 25V electrolytic

Ref. No.	Part No.	Description
VR501-1	1-222-314	20kΩ (B) variable MIC VOL (FRONT L-CH)
VR501-2	1-222-314	20kΩ (B) variable MIC VOL (FRONT R-CH)
VR502-1	1-222-314	20kΩ (B) variable LINE VOL (FRONT L-CH)
VR502-2	1-222-314	20kΩ (B) variable LINE VOL (FRONT R-CH)
VR503-1	1-222-314	20kΩ (B) variable MIC VOL (REAR L-CH)
VR503-2	1-222-314	20kΩ (B) variable MIC VOL (REAR R-CH)
VR504-1	1-222-314	20kΩ (B) variable LINE VOL (REAR L-CH)
VR504-2	1-222-314	20kΩ (B) variable LINE VOL (REAR R-CH)
VR505-1	1-222-314	20kΩ (B) variable PB VOL (FRONT L-CH)
VR505-2	1-222-314	20kΩ (B) variable PB VOL (FRONT R-CH)
VR506-1	1-222-314	20kΩ (B) variable PB VOL (REAR L-CH)
VR506-2	1-222-314	20kΩ (B) variable PB VOL (REAR R-CH)

Ref. No.	Part No.	Description
R812	1-242-685	3.3kΩ
R813	1-242-685	3.3kΩ
R814	1-242-701	15kΩ
R815	1-242-697	10kΩ
R816		———
R817	1-242-693	6.8kΩ
R818	1-242-673	1kΩ
R819	1-242-633	22Ω
R820	1-242-673	1kΩ
R821	1-242-669	680Ω
R822	1-242-680	2kΩ
R823	1-242-689	4.7kΩ
R824	1-242-713	47kΩ
R825	1-242-673	1kΩ
R826	1-242-697	10kΩ
R827	1-242-696	9.1kΩ
R828	1-242-685	3.3kΩ
R829	1-242-685	3.3kΩ
R830	1-242-666	510Ω
R831	1-242-701	15kΩ
R832	1-242-681	2.2kΩ
R833	1-242-701	15kΩ
R834	1-207-268	3.3Ω 1.5W metal oxide film
R835	1-242-721	100kΩ
R836	1-242-717	68kΩ
R837	1-242-697	10kΩ
R838	1-242-681	2.2kΩ
R839	1-206-093	1kΩ 1W metal oxide film
R840	1-242-649	100Ω
R841	1-242-653	150kΩ
R842	1-242-659	270Ω
R843	1-242-699	12kΩ
R844	1-242-681	2.2kΩ
R845	1-242-703	18kΩ
R846	1-242-689	4.7kΩ
R847		———
R848	1-242-710	36kΩ
RA801	1-242-693	6.8kΩ
RA802	1-242-693	6.8kΩ
RA803	1-242-699	12kΩ
VR801	1-221-175	20kΩ (B), semi-fixed
VR802	1-221-173	5kΩ (B), semi-fixed
VR803	1-221-172	2kΩ (B), semi-fixed
VR901	1-222-353	2kΩ, semi-fixed
R901	1-205-511	500Ω 40H wire wound
R902	1-205-502	100Ω 40H wire wound
R903	1-205-503	68Ω 40H wire wound
R904	1-205-503	68Ω 40H wire wound
R991	1-206-097	2.2kΩ 1W metal oxide film
R992	1-242-709	33kΩ ¼W

Ref. No.	Part No.	Description
R993	1-242-691	5.6kΩ ¼W
R994	1-242-691	5.6kΩ ¼W
R995	1-242-713	47kΩ ¼W
		SWITCHES
S501	1-514-693	lever, 3 position; TAPE SPEED
S502	1-514-692	lever, 2 position; MONITOR (FRONT L-CH)
S503	1-514-692	lever, 2 position; MONITOR (FRONT R-CH)
S504	1-514-692	lever, 2 position; MONITOR (REAR L-CH)
S505	1-514-692	lever, 2 position; MONITOR (REAR R-CH)
S506~509	1-514-769	lever, 2 position; REC MODE
S510	1-514-789	slide, TAPE SELECTOR
S511	1-514-324	slide, HEADPHONE
S901	1-514-734	seesaw; POWER
S902~906	1-514-057-10	micro; REW, STOP, PLAY, FF, REC
S907	1-514-057-10	micro; APS
S908	1-514-692	lever; 2 position; TAPE LIFT
S909	1-514-530-21	mirco; SHUT OFF
S910	1-514-735	reel selector; REEL SIZE
S911	1-513-149	push; AUTO STOP
		JACKS/CONNECTORS/PLUGS
CNJ501	1-507-328	phono; LINE INPUT
CNJ502	1-507-328	phono; LINE OUTPUT
CNJ503	1-509-303	cannon; MIC INPUT FRONT L-CH
CNJ504	1-507-281	phone; MICROPHONE FRONT L-CH
CNJ505	1-509-303	cannon; MIC INPUT FRONT R-CH
CNJ506	1-507-281	phone; MICROPHONE FRONT R-CH
CNJ507	1-509-303	cannon; MIC INPUT REAR L-CH
CNJ508	1-507-281	phone; MICROPHONE REAR L-CH
CNJ509	1-509-303	cannon; MIC INPUT REAR R-CH
CNJ510	1-507-281	phone; MICROPHONE REAR R-CH
CNJ511 518	1-507-300	connector, 22-P
CNJ5194	1-507-282	binaural, HEADPHONE
CNP501	1-539-674	connector plug, 36-P
CNP502	1-539-567	connector plug, 18-P
CNJ901	1-507-329	36-P
CNJ902	1-507-301	18-P
CNJ903	1-507-255	11-P
CNJ904	1-509-370	6-P
CNJ905	1-507-329	36-P
CNJ906	1-509-062	OUTLET
CNJ907	1-509-341	POWER SUPPLY
CNP901	1-508-399	6-P
CNP902	1-539-672	printed circuit board 36-P; head
CNP903	1-506-180	terminating; 11-P

Appendix 1
List of Major Importers with Addresses

Addresses for obtaining component parts from major importers. Also see Appendix 2 for alphabetical listing by brand name.

Automatic Radio
Melrose, Massachusetts 02176

Channel Master
Division of Avnet, Inc.
Ellenville, New York 12428

Broadmoor Industry, Ltd.
530 Santa Rosa Drive
Des Plaines, Illinois 60018

Craig Corporation
921 West Artesia Blvd.
Compton, California 90220

F.W. Woolworth Co.
Executive Office
Woolworth Bldg.
New York, N.Y. 10007

Gambles, Inc.
P.O. Box 458
Minneapolis, Minnesota 55440

Hitachi Sales Corp. of America
48-50 34th Street
Long Island City, New York 11101

Lafayette Radio Electronics Corp.
111 Jericho Turnpike
Syossett, L.I., N.Y. 11791

JVC America, Inc.
50-35 56th Road
New York, N.Y. 11378

Lloyds Electronics of California, Inc.
18601 So. Susana Road
Compton, California 90221

Micotron Parts Dept. (Midland)
P.O. Box 1903, 1909 Vernon Street
North Kansas City, Missouri 64141
(All 10, 11, 12 and 15 Series Models)

Medallion Parts Dept. (Midland)
1909 Vernon Street
North Kansas City, Missouri 64141

Lear Jet Stereo, Inc.
6868 South Plumer Avenue
Tucson, Arizona 85706

Matsushita Electric Corporation of America (Panasonic)
10-16 44th Drive
Long Island City, New York 11101

Masterwork Audio Service
1080 Goffle Road
Hawthorne, New Jersey 07506

Montgomery Ward (any store) or:
P.O. Box 8330
Chicago, Illinois 60680

Radio Shack
National Headquarters
2617 West Seventh Street
Fort Worth, Texas
76107

Regional Warehouses
ATTN: Parts Dept.

P.O. Box 276
Delair, New Jersey 08110

2160 Refugee Street
Columbus, Ohio 43207

12821 Knott Street
Garden Grove, California 92641

Campanelli Industrial Park
Granite Street
Braintree, Massachusetts 02184

Sharp Electronics Corporation
10 Keystone Place
Paramus, New Jersey 07652

Sony Corporation of America
47-47 Van Dam Street
Long Island, New York 11101

Superscope (Sony)
8150 Vineland Avenue
Sun Valley, California 91352

Appendix 2

List of Importers and Distributors of Japanese Consumer Electronic Products (Alphabetically by Brand Name)

Brand Name	Company	Address
ACME	Manhattan Novelty Co.	263 Canal St., N.Y., N.Y. 10013
ADELL	Adell International, Inc.	43700 Adell Blvd., Novi, Michigan 48050
ADONIS	A&S Trading Co.	124 W. 30th St., N.Y., N.Y. 10001
ADMIRAL	Admiral Corp.	3800 West Cortland St., Chicago, Ill. 60647
ADVANCO	Advance Transistor Corp.	1225 Broadway, N.Y., N.Y.
AFCO	AFCO Electronics, Inc.	P.O. Box 973, Oakland, Cal. 94604
AGS	Sterling Hi Fidelity, Inc.	22-20 40th Ave., L.I. City, N.Y. 11101
AIMOR	ABC Import & Export	47 West 39th St., N.Y., N.Y.
AIRCASTLE	Spiegel, Inc.	2511 West 23rd St., Chicago, Ill. 60608
AIRCORDEN	Ramson Trading Co.	1185 Broadway, N.Y., N.Y. 10001
AIRLINE	Montgomery Ward	P.O. Box 8339, Chicago, Ill. 60680
AIWA	AIWA Int'l Corp.	1 E. Wacker Dr., Chicago, Ill. 60611
AKA1	Roberts Div.—Rheem Mfg. Co.	5922 Bowcraft Ave., L.A., Cal. 90016
AKAL	AKAL America, Ltd	2139 E. Bell Amo Blvd. Compton, Cal. 90020
ALARON	B&B Import-Export Co.	15755 Wyoming Ave., Detroit, Mich. 48238
ALFRED	Capen Mercantile Corp.	109 Nelson Ave., Nogales, Ariz. 85621
ALLIANCE	Alliance Mfg. Co.	Lake Park Blvd., Alliance, Ohio
ALLIED	Allied Radio Corp. (Radio Shack)	2617 W. 7th St., Ft. Worth, Texas 76107
ALPHA	Nippon Alpha Elec. Co.	200 Park Ave., N.Y., N.Y. 10017
ALPS	Alps Electric Co. Ltd.	1 Whitehall St., N.Y., N.Y. 10004
AMBASSADOR	Allied Stores Corp.	401 5th Ave., N.Y., N.Y. 10016
AMC	Aimcee Wholesale Corp.	1440 Broadway, N.Y., N.Y. 10018
AMD	World Mark Electronics	663 Dowd Ave., Elizabeth, N.J. 07201
AMERTONE	Amerex Trading Co.	417 5th Ave., N.Y., N.Y.
AMICO	Exhibit Sales Co.	3633 I St., Philadelphia, Pa. 19134
AMPEX	Ampex Corp.	Landmeier Rd., Elk Grove Village, Ill. 60007
ANGEL	Arrow Trading Co.	7 West 26th St., N.Y., N.Y.
ANNABEL	B&B Import & Export Co.	15755 Wyoming Ave., Detroit, Mich. 48238
ANNABLE	J.M. Fields	450 West 33rd St., N.Y., N.Y. 10001
APOLEC	ABC Import & Export	47 W. 30th St., N.Y., N.Y.
ARISTO	Aristo Industries	240 5th Ave., N.Y., N.Y. 10001
ARISTOCRAT	Automative Associates	551 5th Ave., N.Y., N.Y. 10017
ARISTO-TONE	Aristo-Tone Electronics	240 5th Ave., N.Y., N.Y. 10001
ARROW	Arrow Trading Co.	220 5th Ave., N.Y., N.Y. 10001
ARVIN	Arvin Industries Inc.	1531 13th St., Columbus, Ind. 47201
ASTREX	Unelco Electronics Corp.	322 N. Eutaw St., Baltimore, Md. 21201
ASTROTONE	Lafayette Radio	111 Jericho Turnpike, Syosset, N.Y. 11101
AUDIO-LITE	Curtis-Electro Lighting Inc.	1536 S. Pawlene St., Chicago, Ill. 60608
AUDIO MAGNETICS	Audio Magnetics Corp.	14600 S. Broadway, Gardena, Cal. 90247
AUDIO-STEREO	Muntz Stereo Corp. of America	7715 Densmore Ave., Van Nuys, Cal. 91406
AUD-I-TONE	JJJ Merchandise	15 West 26th St., N.Y., N.Y. 10010
AUDION	Audion Corp.	200 5th Ave., N.Y., N.Y. 10001
AUDIOPAK	Audio Devices, Inc.	235 E. 42nd St., N.Y., N.Y.
AUDIO TAPE	Audio Devices, Inc.	235 E. 42nd St., N.Y., N.Y.
AUDIO VOX	Audiovox Corp.	300 Denton Ave., New Hyde Park, N.Y. 11040
AUDITRON	F.W. Woolworth	Woolworth Bldg., N.Y., N.Y. 10007
AUTOMATIC	Automatic Radio	Melrose, Massachusetts 02176
AUTO MUSIC	Original Industries Corp.	529 Jericho Turnpike, Mineola, N.Y. 11501
AUTO SONIC	Martel Electronics Sales	2356 S. Cotner Ave., Los Angeles, Cal. 90064
AZTEC	Aztec Sound Corp.	2140 S. Lipan St., Denver, Colorado
BASF	Basf Systems, Inc.	Crosby Drive, Bedford, Massachusetts 01730
BSR	ABC Import	47 West 30th St., N.Y., N.Y.
B&K	AFCO Electronics, Inc.	P.O. Box 973, Oakland, Cal. 94604
BARLEIGH	Barker Stores	815 Hutchinson River Parkway, Bronx, N.Y. 10465
BAYLOR	Vision Electronics, Inc.	1116 E. Huston St., San Antonio, Texas 78206
BELAIC	Fried Trading Co.	425 Bedford Ave., Brooklyn, N.Y. 11211
BELAIR	Hamway Import Co.	40 W. 29th St., N.Y., N.Y. 10001

Brand Name	Company	Address
	Belair Enterprises	5873 Rodeo Road, Los Angeles, Cal. 90016
	Mason Camera Corp.	1141 Broadway, N.Y., N.Y. 10001
	Toko New York, Inc.	350 5th Ave., N.Y., N.Y. 10001
	Toyo Radio Corp. of America	1842B S.W. 169th St., Gardena, Cal. 90247
BELAIRE	Fried Trading Co.	425 Bedford Ave., Brooklyn, N.Y. 11211
BELCOR	Belcor Corp.	457 Chancellor Ave., Newark, N.J.
BELCORDER	Hamway Import Co.	40 W. 29th St., N.Y., N.Y. 10001
BELL & HOWELL	Bell & Howell Co.	7235 N. Linder Ave., Skokee, Ill. 60076
BENJAMIN	Benjamin Sound Co.	40 Smith St., Farmingdale, Long Island, N.Y. 11736
BIG BEN	Tokyo Sansei, Inc.	1170 Broadway, N.Y., N.Y. 10001
BUCH-	BUCH-Tele-Tone Co., Inc.	444 S. 9th Ave., Mount Vernon, N.Y.
BAMAN	California Auto Radio, Inc.	9426 Stewart & Gary Rd., Downey, Cal. 90241
BOWMAN	California Auto Radio, Inc.	9426 Stewart & Gary Rd., Downey, Cal. 90241
BOZAK	Elpa Marketing Ind.	Thorens & Atlantic Ave., Garden City Park, N.Y. 11040
BRADFOR	W. T. Grant Co.	1441 Broadway, N.Y., N.Y. 10018
BRAVO	Recoton Corp.	46-23 Crane St., Long Island City, N.Y. 11101
BRENELL	Fen Tone International	106 5th Ave., N.Y., N.Y. 10011
BRIGHTONE	A&A Trading Co.	23 E. 26th St., N.Y., N.Y.
BROADMOR	Broadmoor Industries	530 Santa Rosa Dr., Des Plaines, Ill. 60218
BROWNE	Charles Brown & Co.	34 West 27th St., N.Y., N.Y. 10001
BULOVA	Bulova Watch Co., Inc. (no longer mfg.)	62-10 Woodside Ave., Woodside, N.Y. 11377
BUSH	Trans-America Electronics Co., Inc.	6479 N. Avondale, Chicago, Ill. 60631
CAISONIC	Radiant Corp.	8220 N. Austin Ave., Mortin Grove, Ill. 60053
CALIFCO	California Trading Co.	P.O. Box 3164, Torrance, Cal.
CANDLE	Candle America, Inc.	1457 Venice Blvd., Los Angeles, Cal. 90006
CAPEHART	Capehart Corp.	770 Lexington Ave., N.Y., N.Y. 10021
CAPTAIN	Captain Mercantile Corp.	109 Nelson Ave., Nogales, Ariz. 85621
CAR PAL	Audiovox Corp.	300 Denton Ave., New Hyde Park, N.Y. 11040
CAR TAPES	Car Tapes Inc.	320 W. Ohio St., Chicago, Ill. 60610
CHANNEL	Channel Marketing, Inc.	342 Madison Ave., N.Y., N.Y.
CHANNEL MASTER	Channel Master	Ellenville, N.J. 12428
CHEMICAN	Nippon Condenser Co.	86-16 60th Ave., Regis Park, N.Y. 11373
CIPHER	Metric TV Parts Repair Center	65 Lexington Ave., Passaic, N.J.
CITATION	Harman-Kardon Inc.	55 Ames Court, Plainview, N.Y. 11803
CLAIRTONE	Clairtone Electronic Corp.	100 Ranson Dr., Rexidale Ontario Canada
CLARICON	F.W. Woolworth Co.	Woolworth Bldg., N.Y., N.Y. 10007
	World Mark Electronics	663 Dowd Ave., Elizabeth, N.J. 07201
CLARION	World Mark Electronics	663 Dowd Ave., Elizabeth, N.J. 07201
CLARK	David Clark, Inc.	360 Franklin St., Worchester, Mass. 01604
COACH	Dictotape, Inc.	9090 Krier Place, Brooklyn, N.Y. 11236
COLUMBIA	Columbia Record Corp.	51 W. 52nd St., N.Y., N.Y. 10019
COMMODORE	Commodore Import Corp.	507 Flushing Ave., Brooklyn, N.Y. 11205
CONCERTINE	Concertine, Inc.	7035 Laurel Canyon Blvd., N. Hollywood, Cal. 91605
CONCORD	Concord Electronics Corp.	1935 Armacost Ave., W. Los Angeles, Cal. 90021
CONSTANT	Canton-Son, Inc.	12 W. 27th St., N.Y., N.Y.
CONSUL	GCL Mercantile Corp.	36 Utler Ave., Hawthorne, N.J. 07507
CANTELCEE	Continental Telephone	17 W. 46th St., N.Y., N.Y.
CONTINENTAL	California Auto Radio, Inc.	9426 Stewart & Gary Rd., Downey, Cal. 90241
CORONADO	Gambles, Inc.	P.O. Box 458, Minneapolis, Minn. 55440
CORONET	Arrow Trading Co.	7 W. 26th St., N.Y., N.Y.
CORTINA	Eice Electronic Instrument Co., Inc.	283 Malta St., Brooklyn, N.Y. 11207
CORVAIR	Transistor World Corp.	52 Broadway, N.Y., N.Y. 10004
CORVETTE	Micotron	P.O. Box 1903, Kansas City, Missouri 64141
COUNTRY	King Korn Marketing Div. of Fullview Industries	6001 N. Clark St., Chicago, Ill. 60626
CRAIG	Craig Corporation	921 W. Artesia Blvd., Compton, Cal. 90220
CREST	Alfred Toepfer	1 Broadway, N.Y., N.Y. 10004
CREST-LINE	Canton-Son, Inc.	12 W. 27th St., N.Y, N.Y.
CROWN	Crown Radio Corp.	228 E. Harris Ave., South San Francisco, Cal. 94080
CURTIS-MATHES	Curtis-Mathes Sales Co.	P.O. Box 151, Athens, Texas 75751
DALTONE	Dalamal & Sons, Inc.	107 Franklin St., N.Y., N.Y. 10013
DECCA	Decca Records Co.	441 Park Ave., N.Y., N.Y. 10022
DELMONICO	JVC America, Inc.	50-35 56th Road, Maspeth, N.Y. 11378
DEMPA	Delta International, Ltd.	P.O. Box 1946, Grand Junction, Colorado
DENON	Nippon Columbia Corp.	6 E. 43rd St., N.Y., N.Y. 10017
DIMENSION	Tenna Corp.	19201 Cranwood Pkwy., Warrensville Heights, Ohio 44128
DOKORDER	Pengo Traders	234 5th Ave., N.Y., N.Y.
DOMINO	Noam Electronics Corp.	60 W. 45th St., N.Y., N.Y. 10036
DORAL	Global Import	858 W. Flagler St., Miami, Florida 33134
DOUGLAS	Son Lee Electronics Corp.	1227 Broadway, N.Y., N.Y. 10001
DREXEL	Drexel Radio Corp.	P.O. Box 15156, New Orleans, La. 70115
DUAL	United Audio Products Inc.	1205 S. Columbus Ave., Mount Vernon, N.Y. 10553
DYN-SONEC	Dyn Association Imports	270 W. 22nd St., Hialeah, Fla. 33010
DYNAKETS	Dynaco, Inc.	3060 Jefferson St., Philadelphia, Pa. 11912
DYNATRONICS	Inland Dynatronics, Inc.	10 Horizon Blvd., S. Hackensack, N.J. 07606
EAGLE	Topp Electronics, Inc.	4201 N.W. 77th Ave., Miami, Fla. 33166
EDGE	Son-Lee Electronics Corp.	1227 Broadway, N.Y., N.Y. 10001
EICO	Eico Electronics Inst. Co.	283 Malta St., Brooklyn, N.Y. 11207
EICOCRAFT	Eico Electronics Inst. Co.	283 Malta St., Brooklyn, N.Y. 11207

Brand Name	Company	Address
ELDORADO	A. Cohen & Sons Inc.	27 W. 23rd St., N.Y., N.Y. 10010
ELECTRA	Electra Radio Corp.	30 W. 23rd St., N.Y., N.Y. 10010
ELECTRO	Electro-Brand Inc.	210-220 W. Chestnut, Chicago, Ill. 60610
ELECTROPHONIC	Morse-Electric Products Corp.	101-10 Foster Ave., Brooklyn, N.Y. 11236
ELGIN	Elgin National Watch Co.	847 West Joclan Blvd., Chicago, Ill. 60607
ELITE	Allied Radio Shack	2617 West 7th St., Ft. Worth, Texas 76107
EMBASSY	Embassy	5330 E. 38th St., Indianpolis, Ind. 46218
EMERALD	Irish Tape Div. Morhan National Sales Co.	458 Broadway, N.Y., N.Y. 10013
EMERSON	Emerson TV & Radio	51 W. 51st. St., N.Y., N.Y. 10019
EMPIRE	Trade Distributors	1199 Broadway, N.Y., N.Y. 10001
ENCORE	Coltrad	360 9th St., San Francisco, Cal. 94103
ESSEX	Lloyd Electronics of Calif.	18601 S. Susana Rd., Compton, Cal. 90221
ENCORE	Caltrad	360 9th St., San Francisco, Cal. 94103
ENGINEERS	Ross Electronics	2834 S. Lock St., Chicago, Ill. 60608
ESPO	Espo Eng. Co.	5334 W. 65th St., Chicago, Ill. 60638
EVER-PLAY	Gulton Industries	212 Durham Ave., Metuchen, N.J.
FABULLOYDS	Lloyds Electronics of Calif. Inc.	18601 S. Susana Rd., Compton, Cal. 90221
FAIRCREST	J.M. Fields	111 8th Ave., N.Y., N.Y.
FAIRFAX	Fairfax Industries, Inc.	165 Ward St., Paterson, N.J. 07505
FAIRMONT	Fairmont Electronics, Inc.	55 W. 39th St., Suite 506, N.Y., N.Y. 10018
FANON	Fanon Electronic Industry	100 Hoffman Rd., Hillside, N.J. 07205
FANTAVAX	Marshall Field Co.	111 North State St., Chicago, Ill. 60690
FACOM	Fujitsu, Ltd.	680 5th Ave., N.Y., N.Y. 10019
FEDERAL	Federal Transistor Co., Inc.	519 S. Broadway, Los Angeles, Cal. 90013
FEN-TONE	Fen-Tone International	106 5th Ave., N.Y., N.Y. 10011
FERROGRAPH	Elpa Marketing Ind., Inc.	Garden City Park, N.Y. 11040
FIDELITY	Ross Electronics	2834 S. Lock St., Chicago, Ill. 60608
FIX TUNED	Fix Tune Electronics	1 W. 30th St., N.Y., N.Y. 10001
FLEETWOOD	Transworld Industrial Corp.	5204 Hudson Ave., West New York, N.J. 07093
FOSTER	Foster Electric Co., Ltd.	230 N. Michigan Ave., Chicago, Ill. 60601
FOUR STAR	Fortune Star Products	1207 Broadway, N.Y., N.Y. 10001
FUJITSU	Fujitsu, Ltd.	680 5th Ave., N.Y., N.Y. 10019
GALAXY	Associated Importers, Inc.	270 W. 22nd St., Hialeah, Miami, Fla.
GAYTONE	Hamway Import Co.	40 W. 29th St., N.Y., N.Y. 10001
GC	GC Electronics	400 So. Wyman St., Rockford, Ill. 61101
GARRARD	Garrard, Div. of British Industries Co.	Westbury, N.Y. 11590
GELOSO	American Geloso Electronics, Inc.	251 Park Ave. So., N.Y., N.Y. 10010
GEMA	Associated Importers	34 Dore St., San Francisco, Cal.
GENERAL	Trans-American Electronics	6479 No. Avondale, Chicago, Ill. 60631
GENERAL ELECTRIC	General Electric	Electronics Park, Syracuse, N.Y. 13201
GIBBS	California Auto Radio, Inc.	9426 Stewart & Gary Rd., Downey, Cal. 90241
GIDGET	Automatic Radio Sales	Melrose, Mass. 02176
GLOBAL	Global Imports	858 W. Flagler St., Miami, Fla. 33130
GLOBE	G.C. Electronics	400 So. Wyman St., Rockford, Ill. 61101
GOLDEN SHIELD	Golden Shield Corp.	56 Hawester Ave., Batavia, N.Y.
GOTHAM	Gotham Electronics Inc.	170 Michael Dr., Syosset, N.Y. 11791
GRANADA	Fried Trading Co.	425 Bedford Ave., Brooklyn, N.Y. 11211
GRAND PIX	Charles Brown & Co.	34 W. 27th St., N.Y., N.Y. 10001
GRENDIER	Empire Scientific Corp.	1055 Stewart Ave., Garden City, N.Y. 11530
HALCO	Halen Associates	125 5th Ave., N.Y., N.Y. 10003
HAMMARLUND	Electronic Assistance Corp.	20 Bridge Ave., Red Bank, N.J. 07701
HARMAN-KARDON	Harman-Kardon, Inc.	55 Ames Court, Plainview, N.Y. 11803
HEMISPHERE	The Sampson Co.	2244 S. Western Ave., Chicago, Ill. 60608
HERITAGE	Heritage Int'l Trading Co.	1330 Stuyvesant Ave., Union, N.J. 07083
HIGHWAVE	Marvel International	30 East 42nd St., N.Y., N.Y. 10017
HIT	Heritage Int'l Trading Co.	1330 Stuyvesant Ave., Union, N.J. 07083
HITACHI	Hitachi Sales Corp.	48-50 34th St., Long Island City, N.Y. 11101
HOVEY TONE	Associated Importers Inc.	34 Dore St., San Francisco, Cal.
HTV	Hamasnatsu TV Co. c/o Kinsho-Matarchi Corp.	80 Pine St., N.Y., N.Y.
HUNTER	Allied Stores Purchasing	401 5th Ave., N.Y., N.Y. 10016
IDEAL	International Distributors, Inc.	1178 Pope St., Memphis, Tenn. 38102
IMPERIAL	Imperial Import Co.	1199 Broadway, N.Y., N.Y. 10001
IMPEX	Impex Electronics	213 S. Robertson Blvd., Beverly Hills, Cal.
INLAND	Inland Dynatronics Inc.	10 Horizon Blvd., S. Hackensack, N.J. 07060
INTERMART	Intermart Corp.	147 W. 42nd St., N.Y., N.Y.
INTERSONIC	Intersonic Corp.	P.O. Box 215, Rockaway, N.J. 07866
INVICO	Challenge Corp.	150 5th Ave., N.Y., N.Y. 10001
INVICTA	Toyomenka Inc.	200 5th Ave., N.Y., N.Y. 10010
IRISH	Irish Tape Div. Morhan National Sales	458 Broadway, N.Y., N.Y. 10013
JADE	Regency Electronics, Inc.	7900 Pendleton Pike, Indianapolis, Ind. 46206
J. B. HUNTER	Allied Stores Corp.	401 5th Ave., N.Y., N.Y. 10016
JJJ	JJJ Merchandise	15 W. 26th St., N.Y., N.Y. 10010
JULIETTE	Juliette-Tapp Electronics Co.	40 Whelan Rd., East Rutherford, N.J. 07073
KATONE	Kaytone Corp.	37 W. 28th St., N.Y., N.Y. 10001
KCK	KCK Co. Ltd.	528 W. Wellington Ave., Chicago, Ill. 60657
KENSINGTON	Etenel Import Co.	410 N. Milwaukee Ave., Chicago, Ill. 60610
KENT	Kent Overseas, Inc.	38 W. 33rd St., N.Y., N.Y.
KENWOOD	Kenwood Corp.	69-41 Calamus Ave., Woodside, N.Y.

Brand Name	Company	Address
KEYSTINE	Atlas-Rand Corp.	Keystone Place, Paramus, N.J. 07652
KITS	Kits Industries, Inc.	729 Ceres Ave., Los Angeles, Cal. 90021
KNIGHT	Allied Radio Shack	2617 W. 7th St., Ft. Worth, Texas 76107
KORONETTE	Fisher Sonic Co., Inc.	405 44th St., Brooklyn, N.Y.
KORTING	APE Electronics Inc.	515 Madison Ave., N.Y., N.Y. 10022
KOSS	Koss Electronics, Inc.	2227 N. 31th St., Milwaukee, Wis. 53208
KOWA	Kowa American Corp.	230 5th Ave., N.Y., N.Y. 10001
KOYO	Koyo International, Inc.	330 Madison Ave., N.Y., N.Y. 10017
KRON	Imperial Import Co.	1199 Broadway, N.Y., N.Y. 10001
KROY	York Radio Corp.	15 Empire Blvd., So. Hackensack, N.J. 07606
LAFAYETTE	Lafayette Radio	111 Jericho Turnpike, Syosset, Long Island, N.Y. 11791
LARK	Caltrade Mfg. and Trading Co.	360 9th St., San Francisco, Cal. 94103
LEADER	Leader Instrument Co.	24-20 Jackson Ave., Long Island City, N.Y. 11101
LEAK	Ercona Corp.	432 Park Ave. So., N.Y., N.Y. 10016
LEAR JET STEREO	Lear Jet Stereo, Inc.	6868 S. Plumer Ave., Tucson, Ariz. 85706
LIC	Lucky International	1155 Broadway, N.Y., N.Y. 10010
LEVTRON	Vison Electronics Inc.	1116 E. Huston St., San Antonio, Texas 78206
LEVINGTON	Ponder & Best Inc.	11201 W. Pico Blvd., Los Angeles, Cal. 90064
LLOYDS	Lloyds Electronics of Calif. Inc.	18601 S. Susana Rd., Compton, Cal. 90221
LONGINES SYMPHONETTE	Longines Symphonette	200 Myrtle Blvd., Larchmont, N.Y. 10538
LORD BARKLEIGH	Barker Stores	815 Hutchinson River Pky., Bronx, N.Y. 10465
LUCKY	Lucky International	1155 Broadway, N.Y., N.Y. 10010
LUXTONE	Luxor International Ind.	39 W. 29th St., N.Y., N.Y. 10010
LYNN	Lynn Stewart Co.	439 E. Ill. St., Chicago, Ill. 60611
MACROSONIC	Yewtic Corp.	1995 Palmer Ave., Larchmont, N.Y. 10538
MAGNAVOX	The Magnavox Co.	2131 Bueter Rd., Ft. Wayne, Ind. 46803
MAGATONE	Noam Electronics Corp.	60 W. 45th St., N.Y., N.Y. 10036
MAGNECORD	Telex-Waters Conly Co., Inc.	9600 Aldrich Ave. S., Minneapolis, Minn. 55420
MAGNUS ORGAN	Magnus Organ Corp.	1600 W. Edgar Rd., Linden, N.J. 07036
MAJOR	Major Electronics Corp.	649 39th St., Brooklyn, N.Y.
MAJORETTE	Major Electronics Corp.	649 39th St., Brooklyn, N.Y.
MALLORY	P.R. Mallory & Co.	3029 E. Washington St., Indianapolis, Ind. 46206
MANTONE	Manhattan Novelty Co.	263 Canal St., N.Y., N.Y. 10013
MAR-LIN	Mar-Lin Radio Corp.	45 W. 27th St., N.Y., N.Y. 10001
MARTEL	Martel Electronics Sales	2356 S. Cotner Ave., Los Angeles, Cal. 90064
MARVEL	Marvel International	30 E. 42nd St., N.Y., N.Y. 10017
MOSIOT	Ross Electronics	2834 S. Lock St., Chicago, Ill. 60608
MASTER CENTRAL	World Mart Inc.	201 N.W. 21st St., Miami, Fla. 33127
MASTERWORK	Columbia Records	1080 Goffle Rd., Hawthorne, N.J. 07506
MATSUSHITA	Matsushita Electronics Corp.	10-16 44th Dr., Long Island City, N.Y. 11101
MAYFAIR	Mayfair Electronics Corp.	666 W. Kinzie St., Chicago, Ill. 60610
MEDALLION	Medallion Parts Dept.	1909 Vernon St., North Kansas City, Mo. 64141
MEGATONE	A&S Trading Co.	124 W. 30th St., N.Y., N.Y. 10001
MERCURY	Pax Limited	5125 Church St., Skokie, Ill. 60076
METEX	Pan American Dynamic Corp.	118-21 Queens Blvd., Forest Hills, N.Y. 11375
METRO	Metro of California	4861 Santa Monica, Los Angeles, Cal. 90029
MEXHO	Puchi Properties Inc.	405 Grand Ave., Nogales, Ariz.
MGA	Mitsubishi International Corp.	7045 N. Ridgeway Ave., Lincolnwood, Ill. 60645
MIDLAND	Micotron Parts Dept.	P.O. Box 1903, 1909 Vernon St., N. Kansas City, Mo. 64141
MICROSONIC	Radiant Corp.	8220 N. Austin Ave., Morton Grove, Ill. 60053
MINI-SWINGEO	Tele-Tone Co. Inc.	444 S. 9th Ave., Mount Vernon, N.Y.
MINY	North American Foreign Trading Co.	1115 Broadway, N.Y., N.Y. 10010
MIRACON	Benjamin Electronic Sound Co.	40 Smith St., Farmingdale, Long Island, N.Y. 11736
MITSUBISHI	Mitsubishi International Corp.	7045 N. Ridgeway Ave., Lincolnwood, ILL. 60645
MONARCH	Monarch Electronics Corp.	7035 Laurel Canyon Blvd., N. Hollywood, Cal. 91605
MOORE	OTC International Inc.	205 W. Wacker Dr., Chicago, Ill. 60606
MORSE	Morse Electro Products Corp.	101 Foster Ave., Brooklyn, N.Y. 11236
MORTONE	Irish Tape Div. Morhan Sales Co.	458 Broadway, N.Y., N.Y. 10013
MOTOROLA	Motorola Inc.	9401 W. Grand Ave., Franklin Park, Ill. 60131
MUNTZ	Muntz Stereo Corp.	7715 Densmore Ave., Van Nuys, Cal. 91406
MUSHAT	Mushat Electronics Inc.	266 N.E. 70th St., Miami, Fla. 33138
NACO	North American Foreign Trading Co.	1115 Broadway, N.Y., N.Y. 10010
NASSAU	Telephone Dynamics Corp.	1333 Newbridge Rd., North Bellmore, N.J. 11710
NATIONAL	Panasonic Repair	43-20 24th St., Long Island City, N.Y.
NET	Net Electronics	8315 E. Firestone, Downey, Cal. 90241
NEWELL	ABC Import	47 W. 30th St., N.Y., N.Y.
NIVICO	JVC America Inc.	50-35 56th Rd., Maspeth, N.Y., N.Y. 11378
NOBILITY	New York Merchandise Co., Inc.	32 W. 23rd St., N.Y., N.Y. 10010
NOCTURNE	Harman-Kardon Inc.	55 Ames Court, Plainview, N.Y. 11803
NORDMENDE	Sterling Hi Fidelity, Inc.	22-20 40th Ave., Long Island City, N.Y. 11101
NORELCO	North American Philips, Inc.	100 E. 42nd St., N.Y., N.Y. 10017
NORTH AMERICAN	North American Foreign Trading Co.	1115 Broadway, N.Y., N.Y. 10010
NOVEL	Fuji Electrochemical Co.	437 5th Ave., N.Y., N.Y. 10016
NUVOX	Nuvox Electronics	150 5th Ave., N.Y., N.Y. 10011
OKI	OKI Electric Industry Co.	202 E. 44th St., N.Y., N.Y. 10017
OMGS	North American Foreign Trading Co.	1115 Broadway, N.Y., N.Y. 10010
ONKYO	World Mark, Inc.	201 N.W. 21st St., Miami, Fla. 33127
OLSON	Olson Electronic Inc.	260 S. Forge St., Akron, Ohio 44308

Brand Name	Company	Address
OLYMPIC	Olympic International	89-89 Union Turnpike, Ellendale, N.Y. 11227
ORION	Orion Electric Co.	1170 Broadway, N.Y., N.Y.
PACIFIC	Pacific Import Co.	37 W. 23rd St., N.Y., N.Y.
PANASONIC	Matsushita Electric Corp. of America	10-16 44th Dr., Long Island City, N.Y. 11101
PATRONIC	Patsen Trading Co.	1199 Broadway, N.Y., N.Y. 10001
PAX	Pax, Ltd.	5125 Church St., Skokie, Ill. 60076
PEERLESS	Peerless	37-15 61 St., Woodside, N.Y. 11377
PENCORDER	Ponder & Best Inc.	11201 W. Peco Blvd., Los Angeles, Cal. 90064
PENNCREST	J.C. Penney Co.	1301 Ave. of America, N.Y., N.Y. 10019
PETER PAN	Tele-Tone Co.	444 S. 9th Ave., Mount Vernon, N.Y.
PETLITE	Sterling Hi Fidelity, Inc.	22-20 40th Ave., Long Island City, N.Y. 11101
PHOENIX	Coltrade Mfg. & Trading Co.	360 9th St., San Francisco, Cal. 94103
PHILCO	Philco-Ford Corp.	Tioga & C Sts., Philadelphia, Pa. 19134
PHONOLA	Telex / Waters Conley Co.	9600 Aldrich Ave. S., Minneapolis, Minn. 55420
PIANOLA	California Auto Radio Inc.	9426 Stewart & Gary Rd., Downey, Cal. 90241
PICKERING	Pickering Co., Inc.	101 Sunnyside Blvd., Plainville, N.Y. 11803
PIONEER	U.S. Pioneer Electronics Corp.	178 Connesce Rd., Carlstadt, N.J. 07072
PLAYTAPE	Playtape Inc.	1115 Broadway, N.Y., N.Y. 10010
POLY-SONIC	The Magitron Co.	311 E. Part St., Moonachie, N.J. 07074
PRESIDENT	Peerless Telerad, Inc.	37-15 61 St., Woodside, N.Y. 11377
PRINCETON	Arlens Dept. Stores, Inc.	393 Seventh Ave., N.Y., N.Y. 10001
RADIANT	Radiant Corp.	8220 N. Austin Ave., Morton Grove, Ill. 60053
RADIOMASTER	Radiomaster, Inc.	1314 Blondell Ave., Bronx, N.Y. 10461
RALEIGH	Kaysons International	6500 Fiotilla St., Los Angeles, Cal. 90022
RANGER	Terna Corp.	19201 Cranwood Parkway, Warrensville Heights, Ohio 44128
REALISTIC	Allied Radio Shack	2617 W. 7th St., Ft. Worth, Texas 76107
REALTONE	Realtone Electronics	34 Exchange Place, Jersey City, N.J. 07302
RECORD-MATIC	Duotone Co., Inc.	6875 S.W. 81 St., Miami, Fla. 33143
RECOTON	Recoton Corp.	46-23 Crane St., Long Island City, N.Y. 11101
REGENCY	Regency Electronics Inc.	7900 Pendleton Pke., Indianapolis, Ind. 46206
RHAPSODY	B&B Import Co.	15755 Wyoming Ave., Detroit, Mich. 48238
RIVERIA	National Electronics	38-20 S.E. 8th, Miami, Fla.
ROBERTS	Roberts Division, Rheem Mfg. Co.	5922 Bowcraft, Ave., Los Angeles, Cal. 90016
ROBIN	Mason Camera Co.	1141 Broadway, N.Y., N.Y. 10001
ROMAR	D. Plough Dist. Co.	P.O. Box 2266, Memphis, Tenn. 38102
ROSS	Ross Electronics Corp.	2834 S. Lock St., Chicago, Ill. 60608
ROXY	Mason Camera Co.	1141 Broadway, N.Y., N.Y. 10001
ROYAL	Major Electronics Corp.	649 39th St., Brooklyn, N.Y.
SABA	Saba Corp.	724 N. 7th St., Allentown, Pa. 18102
SANDHURST	A&S Trading Co.	124 W. 30th St., N.Y., N.Y. 10001
SANSUI	Sansui Electronics Corp.	32-17 61 St., Woodside, N.Y. 11377
SANYO	Sanyo Electric, Inc.	1200 W. Walnut St., Compton, Cal. 90220
SATELLITE	Best of Tokyo	11 W. 42nd St., N.Y., N.Y.
SAXONY	Trans-Aire Electronics Corp.	393 Sagamore Ave., Mineola, N.Y.
SCOTT	H.H. Scott	111 Powder Mill Rd., Maynard, Mass. 01754
SEAVOX	Fried Trading Co.	425 Bedford Ave., Brooklyn, N.Y. 11211
SEEBURG	Seeburg Corp.	1500 N. Dayton St., Chicago, Ill. 60622
SELECTO	Duotone Co., Inc.	6875 S.W. 81 St., Miami, Fla. 33143
SEMINOLE	Same & Streiffe, Inc.	8400 Brookfield Ave., Brookfield, Ill. 60513
SENTRY	Audiovox Corp.	300 Denton Ave., New Hyde Park, N.Y. 11040
SETICO	Setico Importing Co.	2321 Liberty St., Jacksonville, Fla.
SHAMROCK	Irish Tape Div. Morhan National Sales	458 Broadway, N.Y., N.Y. 10013
SHANNON	Dictotape, Inc.	9090 Krier Place, Brooklyn, N.Y. 11236
SHARP	Sharp Electronics Corp.	10 Keystone Plaza, Paramus, N.J. 07652
SHERATON	Sheraton Electronics	310 5th Ave., N.Y., N.Y. 10001
SHERWOOD	Sherwood Electronics Inc.	4300 N. Calif. Ave., Chicago, Ill. 60618
SHURE	Shure Brothers, Inc.	222 Hartney Ave., Evanston, Ill. 60204
SILVER	Shin-Shirasuna Electronics Corp.	521 5th Ave., N.Y., N.Y. 10017
SINGER	Singer Consumer Products	30 Rockefeller Plaza, Room 6228, N.Y., N.Y. 10020
SKYMASTER	New York Merchandise Co.	32 W. 23rd St., N.Y., N.Y. 10010
SONY	Sony Corp. of America	47-47 Van Dam St., Long Island City, N.Y. 11101
SONY (Superscope)	Sony Superscope	8150 Vineland Ave., Sun Valley, Cal. 91352
SOUNDESIGN	Realtone Electronics	34 Exchange Place, Jersey City, N.J. 07032
SOVERIGN	Sheraton Electronics Inc.	401 Broadway, N.Y., N.Y.
SPECA	Penrosgar Inc.	862 6th Ave., N.Y., N.Y. 10001
SPINOLA	Spinola International Corp.	315 5th Ave., N.Y., N.Y. 10016
SPORTSMASTER	Sportsmaster Radio	2570 Devan Ave., Des Plaines, Ill.
STANDARD	Standard Radio Corp.	60-09 39th Ave., Woodside, N.Y. 11377
STANFORD	Stanford International	569 Laurel St., San Carlos, Cal. 94070
STANTON	Elpa Marketing Ind., Inc.	Thorens & Atlantic Ave., Garden City Park, N.Y. 11040
STARFIRE	Mar-Lin Enterprise	1472 Broadway, N.Y., N.Y.
STELLAR	Astro Trading Co.	175 5th Ave., N.Y., N.Y. 10010
STEREO-DYN	Dyn Associate Importer Inc.	270 W. 22nd St., Hialeah Fla. 33010
STEREOMATIC	A&S Trading Co.	124 W. 30th St., N.Y., N.Y. 10001
STERLING	Sterling Hi-Fidelity Inc.	22-20 40th Ave., Long Island City, N.Y. 11101
STEWART	Lynn Stewart Co.	439 E. Ill. St., Chicago, Ill. 60611
SUBLINE	Ramson Trading Co.	1185 Broadway, N.Y., N.Y. 10001

Brand Name	Company	Address
SUMMIT	Penrosgar Inc.	862 6th Ave., N.Y., N.Y. 10001
SUN GLASS	American Geloso Electronics, Inc.	251 Park Ave., S., N.Y., N.Y. 10010
SUPER-DYNAMIC	TDK Electronics Corp.	23-73 48th St., Long Island City, N.Y. 11101
SUPEREX	Superex Electronics	4 Radford Place, Yonkers, N.Y.
SUPERSONIC	Spiegel, Inc.	1061 W. 35th St., Chicago, Ill. 60609
SUPERSCOPE	Sony Superscope	8150 Vineland Ave., Sun Valley, Cal. 91352
SUPRE-MACY	Macy's Dept. Store	Herald Sq., N.Y., N.Y. 10001
SYLVANIA	Sylvania Entertainment Product	700 Elliott St., Batavia, N.Y. 14021
SYMPHONIC	Symphonic Electronic Corp.	470 Park Ave. S., N.Y., N.Y. 10018
TACT	Hanway Import Co.	40 W. 29th St., N.Y., N.Y. 10001
TAMRADIO	Tamradio Co.	241 Northern Blvd., Douglaston, N.Y. 11368
TANDBERG	Tandberg of America	8 Third Ave., Pelham, N.Y. 10803
TAYA	Topp Imports	40 Wheeland Rd., E. Rutherford, N.J.
TDK	TDK Electronics Corp.	23-73 48th St., Long Island City, N.Y. 11101
TDK CASSETTE	TDK Electronics Corp.	23-73 48th St., Long Island City, N.Y. 11101
TEAC	Teac Corp. of America	1547 18th St., Santa Monica, Cal. 90404
TECHICORDER	OKI Electric Industry	202 E. 44th St., N.Y., N.Y. 10017
TELE-TONE	Tele-Tone Co., Inc.	444 S. 9th Ave., Mount Vernon, N.Y.
TELEX	Telex / Waters Conley Co., Inc.	9600 Aldrich Ave. So., Minneapolis, Minn. 55420
TELEX-PHONOLA	Telex / Waters Conley Co., Inc.	9600 Aldrich Ave. So., Minneapolis, Minn. 55420
TELMAR	Martel Electronics Sales, Inc.	1199 Broadway, N.Y., N.Y. 10001
TEMPEST	Azad International	22 W. 27th St., N.Y., N.Y. 10001
TEN	Fryitsu California Co.	16410 S. Figueroa, Gardena, Cal. 90247
TENNA	Tenna Corporation	19201 Cranwood Parkway, Cleveland, Ohio 44128
TENPHONIC	ABC Imports	47 W. 30th St., N.Y., N.Y. 10001
THORENA	Elpa Marketing Ind.	Thorens & Atlantic Ave., Garden City Park, N.Y. 11040
THREE-STAR	Nasen Trading Co.	230 5th Ave., N.Y., N.Y.
TOHO	Tokyo Sanei, Inc. (N.Y.)	1170 Broadway, N.Y., N.Y. 10001
TOKAI	Tokai Corp. of America	500 5th Ave., N.Y., N.Y. 10036
TOKO	Toko New York, Inc.	350 5th Ave., N.Y., N.Y. 10001
TONECREST	Metro Wholesale Corp.	53 W. 43rd St., N.Y., N.Y. 10036
TONEMASTER	Broadmoor Ind.	530 E. Santa Rosa Dr., Des Plaines, Ill. 60018
TONEX	W.P. Hemenway Co.	P.O. Box 4032, Portland, Oregon 97208
TOSHIBA	Toshiba America, Inc.	41-06 Delong St., Flushing, N.Y. 11355
TOSHRI	Fedco, Inc.	P.O. Box 2605, Terminal Annex, Los Angeles, Cal. 90054
TOWNLEY	The Englishtown Corp.	42 Broadway, N.Y., N.Y. 10004
TRANSCRIBER	Transcriber Co., Inc.	P.O. Box 478, Attleboro, Mass. 02703
TRANSETTE	Microtron Parts Dept.	P.O. Box 1903, 1909 Vernon St., N. Kansas City, Mo. 64141
TRANSONIC	Transistor World Corp.	52 Broadway, N.Y., N.Y. 10004
TRANSWORLD	Electronic Ind. Inc.	901 Daniel St., Kenner, Louisiana 70062
TRIO	Trio Corp.	212 5th Ave., N.Y., N.Y.
TRIUMPH	Fannon Electronics, Ltd.	439 Frelinghuysen Ave., Newark, N.J. 07114
TROPICANA	Bell Electronics	1180 Broadway, N.Y., N.Y. 10009
TROUBADOR	Empire Scientific Corp.	1055 Stewart Ave., Garden City, N.Y. 11530
TRUETONE	Western Auto Supply	2107 Grand Ave., Kansas City, Mo. 64108
TUSSAH	Trade Unlimited	75-03 Main St., Flushing, N.Y.
UHER	Martel Electronic Sales	2356 S. Cotner Ave., Los Angeles, Cal. 90064
UNELCO	Unelco Electronics Corp.	322 N. Eutaw St., Baltimore, Md. 21201
UNICORD	American Geloso Electronics Inc.	251 Park Ave., S., N.Y., N.Y. 10010
UNIDYNE	Shure Brothers Inc.	222 Hartrey Ave., Evanston, Ill. 60204
UNIMETRUS	Lafayette Radio	111 Jericho Turnpike, Syosset, L.I., N.Y. 11791
UNISPHERE	Shure Brothers, Inc.	222 Hartrey Ave., Evanston, Ill. 60204
UNITED AUTO	United Auto Products	1205 S. Columbus Ave., Mount Vernon, N.Y. 10553
USAC	United States Antenna Corp.	P.O. Box 46462, Cleveland, Ohio 44146
VALIANT	Valiant Radio Corp.	380 Second Ave., N.Y., N.Y. 10010
VANTAGE	Fairfax Dist. Co.	1328 N.Y. Ave. N.W., Washington, D.C.
VESPER	Tokai Corp.	500 5th Ave., N.Y., N.Y. 10036
VIKING	Telex / Waters Conley Co.	9600 Aldrich Ave. S., Minneapolis, Minn. 55420
VIKOA	Vikoa, Inc.	400 9th Ave., Hoboken, N.J. 07030
VICOUNT	Consolidated Merchandise Co.	5950 Queens Midtown Expressway, Maspeth, N.Y. 11378
VISION	Vision Electronics, Inc.	1116 East Huston St., San Antonio, Texas 78206
VISTA	Craig Corp.	921 W. Artesia Blvd., Compton, Cal. 90220
VOCA	Voca Dictating Machine	274 Madison Ave., N.Y., N.Y.
VORNADO	Automatic Radio	Two Main St., Melrose, Mass. 02176
WALTHAM	Internation Importers	2242 S. Western Ave., Chicago, Ill. 60608
WATTS	Elpa Marketing Industries	Thorens & Atlantic Ave., Garden City Park, N.Y. 11040
WEALTH	Funai Electric Co.	3004 W. Logan Blvd., Chicago, Ill. 60647
WEBCOR	Consolidated Merchandise Corp.	5950 Queens Midtown Expressway, Maspeth, N.Y. 11378
WELTRON	Weltron Co., Inc.	514 E. Peabody, P.O. Box 217, Durham, N.C. 27702
WESTINGHOUSE	Westinghouse Electric	3 Gateway Centre, P.O. Box 2278, Pittsburgh, Pa. 15230
WILCO	OTC International Inc.	205 W. Wacker Dr., Chicago, Ill. 60606
WILSON	Wilson Import Co.	1199 Broadway, N.Y., N.Y. 10001
WINDSOR	Windsor Pen Corp.	88 Third Ave., Brooklyn, N.Y. 11217
WINSTON	Western Auto Supply	2107 Grand Ave., Kansas City, Mo. 64108
XAM	Selectron International	55 Veteran Blvd., Carlstadt, N.J. 07072
YAMAHA	Yamaha International Corp.	7733 Telegraph Rd., Montebello, Cal. 90054
YEW	Yewtec Corp.	1995 Palmer Ave., Larchmont, N.Y. 10538
YORK	York Radio Corp.	15 Empire St., S. Hackensack, N.J. 07606